The Cutting and Polishing of Electro-optic Materials

The Cutting and Polishing of Electro-optic Materials

G W Fynn and W J A Powell

Adam Hilger Ltd, Bristol

British Library Cataloguing in Publication Data
Fynn, G W
The cutting and polishing of electro-optic materials.
1. Electrooptical devices 2. Cutting
3. Grinding and polishing
I. Title II. Powell, W J A
621.36 TA1750

ISBN 0-85274-302-5

Published by Adam Hilger Ltd,
Techno House, Redcliffe Way,
Bristol BS1 6NX, England.
The Adam Hilger book-publishing imprint is owned by The Institute of Physics

Printed in Great Britain by
Page Bros (Norwich) Ltd, Norwich NR6 6SA

Contents

Foreword

Behind every new solid state device there stands a single crystal and the explosion in solid state device development which followed the invention of the transistor in 1948 meant that many new crystals had to be grown and fabricated in order to assess their device properties. Many of the crystalline solids which subsequently proved to be of interest had quite different physical and chemical properties from the classical metals and glasses of established mechanical and optical workshop technology. Some were hard, others soft, some brittle, others ductile, some reacted with standard processing fluids others generated poisonous by-products.

This book embodies the knowledge and experience gained by the authors over many years in devising ways and means of overcoming the many different problems arising from the fabrication of these important new materials and their work was often an essential precursor to the actual assessment of the new materials being generated by the crystal growers of the Electronic Materials Division at Malvern.

The authors have devised both instruments and machines that can readily be made and used in laboratory workshops as well as new processes. Their book is likely to be a standard text for years to come, not only for the invaluable information on the different processes so clearly described in great technical detail, but also for the spirit of creativity that emanates from its pages.

William Bardsley BSc, PhD, CEng., AInstP, MIEE
Senior Principal Scientific Officer,
Superintendent,
Electronic Materials Division, RSRE.
Malvern, 1979.

Preface

The purpose of this book is to give some technical guidance to those engaged in the working of electronic materials. For many centuries, optical surfaces have been produced on components for use in the visible portion of the electro-magnetic spectrum, and, as interest increased in new electro-optic and semiconducting materials, many of which are being developed by crystal growers at the Royal Signals and Radar Establishment, Malvern, there came an expanding need for associated shaping, polishing and measurement methods. One effect of this directly affected us: an influx of visitors seeking basic advice and training. Several unlimited-circulation Technical Notes and Papers have documented some of the developing technology but a more compendious work has become necessary.

Two approaches were suggested: either a review of the field of cutting and polishing equipments and their associated techniques; or a description of what has been developed and is in use at RSRE in processing tens of thousands of faces per annum on established and research materials. The latter approach was decided upon since it has (for us) the merit that we can write about personal experience secure in the knowledge that everything described has been put into practice often on a pre-production/production scale. The snag with limiting the text to first-hand authority is its specificity and we have been unable to comment on many machines, instruments and methods which are undoubtedly equally efficacious.

Conventional polishing methods can achieve much in highly skilled hands but the change from familiar glasses to the less familiar elements and compounds often requires a re-thinking of traditional expertise which can be time-consuming and frustrating. Moreover, there are few positive criteria to guide the technician for, with these new devices, some of the questions that arise are: how flat is economically necessary, how specular should the finish be for a given operating wavelength and how damage-free should the surfaces be? In the field of instrument design it is axiomatic that no component should be over-toleranced and unnecessary precision is usually obvious: except, possibly, in the preparation of solid-state laser rods, surface flatness and finish tolerances generally are more indeterminate and the elastic instruction "best possible" can as easily result in much too good as in not good enough. Accurate measurement is essential for most components: it is as basic to technology as Mathematics is to the Sciences. The instruments and methods detailed in Chapter 6 are therefore applicable not only to, but beyond, the field of this book.

The preparation of specular surfaces tends to encourage adherents to a single type of polishing matrix—a pitch-user views other matrices with suspicion; the soft-metal enthusiast is unwilling to vary his techniques; and devotees of cloth-facings ignore all others. Thus they may expect a polisher to give optimum results on a component for which it is strictly unsuitable. This is equally applicable to classes of abrasives and etch-slurries. It is, for example, like a miller-operator who will tax his machine to carry out a process which could be better performed on a centre-lathe (when both are available), and might even cut away large volumes of material which a bandsaw could remove more readily. A hand-worker might extend the analogy since an expert with hacksaw and file can produce some shapes which stretch a very complicated numerical control machine to its limits. However, precise single surfaces are often required quickly and if the selection of a matrix or combination of matrices together with mechanical and/or hand-polishing methods can best achieve these, then familiarity with all of them must be worthwhile. Although many of the machines and jigs described are now commercially available, sufficient detail has been given in the text and illustrations to enable small laboratory and college workshops to construct them.

Our belief, therefore, is in a flexible approach via the employment of a variety of manipulative and mechanical techniques. Managers tend to promote one or the other without considering a blend—possibly because they fear a clash between the hand-worker who painstakingly acquires his skills over a period of years and resents 'instant' machinery; and the machinist whose more speedily attained expertise causes him to avoid (and impugn) the kinaesthetic learning processes of the craftsman. Entirely mechanical methods, however large and costly, are often as ineffective as entirely hand processes when the protracted delivery of a critical new opto-electronic component delays a project.

There has been one major unsatisfying factor throughout most of the programme: too often problems have been solved semi-scientifically since the pressures of work and shortages of qualified staff have denied the opportunity for more rigorous investigation and explanation.

One caution: the expositions may be simple but their executions seldom are.

Acknowledgments

We wish to acknowledge our debt to the inadvertently unacknowledged: all those who may find details herein of their published and unpublished hardware and ideas. For their initial persuasion and subsequent encouragement to persevere with this book we thank many colleagues, in particular Mr S B Marsh and Dr P A Forrester. To all those who brought us their problems, we are grateful, for it is in the search for solutions that technology often advances.

Geoff Fynn has not forgotten and now wishes to acknowledge his indebtedness to the Royal Society Mond Laboratory in Cambridge where, many years ago, he received his initial training and acquired an enduring engineering/science philosophy.

The greater part of the manuscript in its draft and final forms was typed by Mary Thomas, Diane Jones and Elizabeth Fynn during their leisure hours. To them, and to our wives who coped with months of artistic abstraction, we are inenarrably grateful.

Metals Research Ltd kindly provided figures 1.19, 1.20, 1.48, 1.49, 1.51, 1.52, 1.53, 1.54, 2.45, 2.46, 2.47, 2.50, 2.62 and 7.2.

The book is published by permission of the Director RSRE: Crown Copyright HMSO.

1 Cutting

'is it original
it was once i answered truthfully
and may be again' (don marquis)

Cutting may be taken to mean the subdivision of a solid into accurately dimensioned pieces. With electronic materials both slicing and dicing operations have to be considered which invariably require an instrument less usual than a knife or a hacksaw, as the following listed techniques indicate. They have been applied with varying degrees of success by workers in the field:

(i) sawing with both fixed and free abrasive;
(ii) cleaving, scribe-and-break, fracture;
(iii) microtoming;
(iv) thermal cracking, for example hot wire, flame;
(v) chemical attack, such as acid/base etching;
(vi) gas-pressured abrasive particle stream;
(vii) high-speed liquid droplet impingement;
(viii) ultrasonic shaping;
(ix) electrodischarge machining (EDM);
(x) electrochemical machining (ECM);
(xi) various vacuum techniques such as ion beam etching, electron beam machining;
(xii) laser beam machining.

The list is long and not exhaustive, but of the techniques those in (i) most easily cater for a wide range of sizes, hardness and ohmic properties. The last two, for example, are applied specifically as dicing and shaping operations on thin components, enabling extremely small kerf losses and very precise linear dimensions to be obtained. This chapter deals with sawing as it has developed and been applied in our hands at the Royal Signals and Radar Establishment (RSRE).

1.1 General Considerations

Consider the different hardnesses of materials: where they are soft the problem is to cut delicately enough so that they are minimally damaged and the cutting-abrasive life is extremely long; hard materials may withstand harsh cutting conditions, but the abrasive wears away rapidly. Ideally, therefore, the cutter has to be designed to cope with materials ranging from the soft to the very hard. The extended Mohs's scratch hardness figures, (table 1.1) are an assessment of the abrasive characteristics of a range of solids. Broadly, a material can be scratched by any of a higher Mohs's value and remain unmarked when abraded by one lower in the order. In spite of the fact that crystallographic orientation makes some of the values given for other materials more qualitative than quantitative, they are a very useful starting point. Table 1.2 gives, in the second column, the values for electronic materials cut. The question of the relative hardness of a specimen is probably the first one which requires an answer; other factors that have to be considered are displayed in table 1.3. The principal parameters that can be juggled to achieve good results are cutting load, speed, abrasive size, slot width and lubricant.

Table 1.1 Mohs's scratch hardness.

Mineralogical name	Typical chemical form	Formula	Value
Talc	Hydrous magnesium silicate	$Mg_3Si_4O_{10}(OH)_2$	1
Gypsum	Hydrated calcium sulphate	$CaSO_4.2H_2O$	2
Fingernail†			
Calcite	Calcium carbonate	$CaCO_3$	3
Penny piece†			
Fluorite	Calcium fluoride	CaF_2	4
Apatite	Calcium fluorophosphate	$3Ca_3P_2O_8CaF_2$	5
Knife blade†			
Window glass†			
Orthoclase	Potassium aluminium silicate	$KAlSi_3O_8$	6
Steel file†			
Rutile†	Titanium dioxide	TiO_2	
Quartz	Silicon dioxide	SiO_2	7
Topaz	Aluminium fluorosilicate	$Al_2F_2SiO_4$	8
Spinel†	Magnesium aluminate	$MgAl_2O_4$	
Corundum	Aluminium oxide	α-Al_2O_3	9
Silicon carbide†		SiC	
Boron carbide†		B_4C_3	
Diamond	Carbon	C	10

†These interpolated substances give additional qualitative reference values.

Table 1.2 Typical data for low-speed peripheral wheel cutting.

Material sawn	Mohs's value	Diamond blade data: Type used	Diameter and width (mm)	Grit size	Typical cutting speed (rpm)	Cutting pressure (g)	Typical depth of cut (mm)	Approx. cutting duration (min)
ADP	~2·5	Sintered	125 × 0·4	160	~200	50	50	25
Anthracene trans-stilbene	~3	Nickel-bond	100 × 0·3	280	100	10	50	5
Barium titanate		Nickel-bond	100 × 0·3	200	75	30	5	1
Boron carbide	>9	Sintered	75 × 0·25	200	400	250	10	45
Borosilicate glasses	~6·5	Sintered	100 × 0·3	170	250	250	8	1
Calcium fluoride	4	Nickel-bond	100 × 0·3	280	100	60	35	10
Calcium molybdate	~5	Nickel-bond	100 × 0·3	280	200	25	10	5
Calcium tungstate	~4	Nickel-bond	100 × 0·3	280	100	30	12	5
Cerium fluoride	~4	Nickel-bond	100 × 0·3	280	150	25	20	4
Ferrites	~7	Sintered	125 × 0·5	200	200	200	25	<10
Gallium arsenide	~3	Nickel-bond	100 × 0·3	280	175	60	20	<10
Gallium phosphide	~4·5	Nickel-bond	100 × 0·3	280	200	75	20	<10
Germanium	~5	Nickel-bond	100 × 0·3	280	300	30	6	5
High alumina ceramics	~9	Sintered	75 × 0·25	300	200	60	0·5	2
Hypodermic needles	~5·5	Nickel-bond	100 × 0·3	280	75	50	0·5	1
Indium antimonide	~3·5	Nickel-bond	100 × 0·3	280	250	25	20	10
Indium antimonide	~3·5	Nickel-bond	75 × 0·25	400	75	5	0·05	<2
KD*P	~2·5	Nickel-bond	100 × 0·3	280	200	10	10	10

Table 1.2—*Continued*

Material sawn	Mohs's value	Diamond blade data: Type used	Diameter and width (mm)	Grit size	Typical cutting speed (rpm)	Cutting pressure (g)	Typical depth of cut (mm)	Approx. cutting duration (min)
Lanthanum bi-fluoride	~4	Nickel-bond	100 × 0·3	280	150	25	20	4
Lanthanum hexa-boride	>9	Sintered	100 × 0·3	200	300	750	25	<20
Lead telluride	~2·5	Nickel-bond	100 × 0·3	280	100	10	15	<15
Lithium niobate	<5	Nickel-bond	100 × 0·3	280	200	60	5	< 5
Miniature glass galvo mirrors	~6	Nickel-bond	75 × 0·25	400	75	5	0·1	< 1
Oilstone	~9	Sintered	125 × 0·4	170	200	50	25	~10
Neodymium doped glass	~6	Nickel-bond	100 × 0·3	280	200	30	12	2
Periclase (MgO)	~6	Nickel-bond	100 × 0·3	280	250	250	10	4
Plastics	—	Nickel-bond	100 × 0·3	280	300	80	25	20
Proustite	~2·5	Nickel-bond	100 × 0·3	280	150	20	10	5
Quartz	7	Sintered	100 × 0·3	170	200	250	6	< 1
Resinated glass fibre	—	Sintered	100 × 0·3	220	400	100	12	10
Ruby/sapphire	9	Sintered	125 × 0·4	160	300	350	20	<20
Rutile	~6·5	Nickel-bond	100 × 0·3	280	250	200	5	<2
Silicon	~7	Nickel-bond	100 × 0·3	280	400	200	25	8
Silicon carbide	>9	Sintered	75 × 0·25	200	400	250	10	4
Spinel	8	Nickel-bond	100 × 0·3	280	250	250	10	4
Strontium titanate	~7	Sintered	100 × 0·3	170	250	200	12	10
Tungsten	(tough)	Sintered	100 × 0·3	170	150	250	6	20
Tungsten carbide	(tough)	Sintered	75 × 0·25	300	300	20	5	20
Uranium carbide	~3	Nickel-bond	75 × 0·25	400	200	10	3	3
Yttrium aluminium garnet	~8	Nickel-bond	100 × 0·3	280	300	250	10	4
Yttrium lithium fluoride	~3·5	Nickel-bond	100 × 0·3	280	200	250	40	30
Zinc selenide	~3	Nickel-bond	100 × 0·3	280	200	30	14	10
Zinc sulphide	~2·5	Nickel-bond	100 × 0·3	280	200	20	14	10

Table 1.3

For the material	For the blade
Low damage (both mechanical and thermal);	Low damage;
Minimum kerf loss;	Low wear;
Parallelism;	Cost;
Flatness;	Replacement problems;
Cut-off time;	Good housekeeping
Crystallographic orientation;	
Safety (of personnel);	
Cross-contamination;	
Adaptability (eg as slicer and dicer)	

Sawing, by whatever means, involves the interaction of a cutter and a specimen: motion is necessary, although which of the two members is dynamic and which static seems to be optional. In the interaction, disruptive forces are inevitable and it is this state, relative to the strength of the material being cut, which has a profound influence upon the resulting surface-finish and work-damage. Clearly, enough force has to be applied within the cutting zone to remove material, otherwise no progress would be made. There must then be an area of shear or disruption at which material is torn from the wanted faces. Force equations may be derived (more readily for free abrasives) but they vary at least as often as the specimen varies, or the particle size and sharpness changes. It is the domain of the empiricist, where the application of experimental bias (AEB) is not easy to supplant.

In the case of a diamond wheel (a fixed-abrasive mode) its edge, when examined closely, will be seen to be irregular. Even a perfectly circular blade must be asymmetric to the extent that the irregularly shaped diamond abrasive particles project from the edge. Thus, in the typical situation of a wheel cutting through a small workpiece, a roughly toothed wheel engages in a short cut. Such an operation requires that either the worktable moves progressively into the saw blade or vice versa. When the work is solidly held and fed against the cutting edge it is the refinement of this movement which contributes much to the quality of the sawn surfaces. Since the engagement of finely irregular cutters with small brittle crystals at any dynamic loading is the cause of a measure of damage (and from our experience there is no reason to doubt this) some form of pressure-limiter is desirable.

Superficially, a resilient mounting of one or both members should serve, but a blade needs to be held in a rigid relationship to a shaft-housing for precise motion which leaves little scope for any resilience there. Wire saws, whether fed with loose abrasive or using a diamond-coated wire, are exceptional. Even though the wire is tensioned to maintain positional accuracy it is flexible enough to act as a force-limiter. The price that has to be paid is the difficulty of achieving very flat and parallel slices from anything other than small diameter crystals. However, in sawing arrangements like a rotating disc or a strip of metal used edgewise, as in reciprocating-blade saws, rigidity is fundamental. Disc saws are capable of dealing more quickly and more accurately with a wider range of materials than wire saws, therefore the inbuilt rigidity of the cutter is a necessary evil which has to be included in the design rather than left out. The only manoeuvre remaining is to provide the work with a flexible mounting, but here again precision and resilience do not go together.

Early lapidary's saws, drawn and described concisely by Charles Holtzapffel (1850) would have cut many of the delicate materials which proved difficult in the nineteen-fifties—difficult principally because saws were used at high rotational speeds, intended for such robust materials as quartz, silicon and germanium. A lapidary's saw or crane (figure 1.1) was undoubtedly modelled on the lines of similar previous machines. The disc rotated around the vertical axis and was driven fairly slowly by hand. The work was pivoted on a vertical shaft, too, and any required cutting load was applied by gravity, using a cord over a pulley and a suitable weight. Two useful pressure-limiting features were incorporated: a freely moving, light workfeed arm combined with an effective loading mechanism. The low mass of the feed arrangement together with a slow rotation meant that the inevitable blade irregularities could be 'followed' by small movements of the work. Though high-speed annular saws have continued to develop and their performance on fragile materials has improved, the slow-speed peripheral-diamond saw (SSPDS) which bears a marked resemblance to the lapidary's crane is as much the general workhorse now as it rapidly became some fifteen years ago. It would be pointless to suggest that it is the definitive crystal slicer, but crystal slicing forms only a part of overall cutting work.

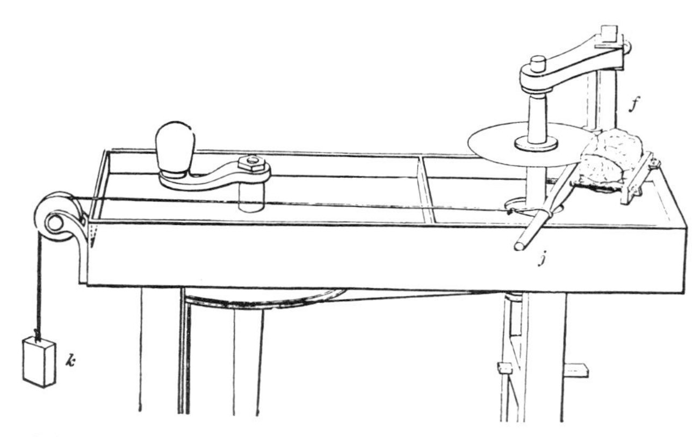

1.1 Lapidary's crane.

1.2 Prototype Slow-Speed Peripheral-Diamond Saw and its Consequences

This form of saw, therefore, where a rotating disc with its outside edge either impregnated or coated with diamond powder to abrade a slot, was our first instrument designed to deal with the complete hardness range. Although others more detailed have evolved, its development has in no way outdated it. An article (Fynn and Powell 1965) written some years after the saw's construction is quoted, since it underlines the value of a simple effective tool as a means of solving a large number of enduring cutting problems. In addition, the changes which have been made to later models can be more readily appreciated.

The saw came into being at the Royal Signals and Radar Establishment, Malvern, to satisfy a demand for the precise cutting of very brittle semiconducting compounds. Conventional methods, which encompassed wire sawing, air-abrasive and ultrasonic cutting, were all too imprecise. High-speed diamond blade sawing resulted in fragments rather than accurately dimensioned pieces; this was due to local overheating of the specimen and positive feed contact combined, giving high-frequency stresses far beyond the elastic limit of many of the compounds. Some laser materials, too, have this fragility—calcium fluoride, for example—and some, like ruby and sapphire on the other hand, soon glaze and warp internal edge blades rotating at high speeds, but can be readily dealt with on the slow-speed saw with little evidence of blade wear and distortion.

Basically, the problems resolved themselves into satisfying the following mechanical conditions: the diamond blade must resolve slowly and truly and be adjustable in speed from zero rev/min; the workpiece feed mechanism must have a low inertial mass and the cutting pressure must be variable. With these in mind and with the constant need for versatility the instrument evolved to the following pattern.

1.2.1 Mechanism

Figure 1.2(*a*) is a schematic plan of the complete experimental mechanism. Figure 1.2(*b*) shows the complete prototype saw. Because the initial limitation of the true running of the saw is inherent in the accuracy of its components, all turning must be concentric and all faces and shoulders square with their axes. An aluminium baseplate A, approximately 25 × 15 × 4 cm, carries the entire machine. An aluminium alloy block B contains two ballraces, a medium-duty journal D adjacent to the saw blade and remote from it and a light-duty race E. The pulley C is positioned in a slot in the bearing housing. The saw-shaft F requires a saw-supporting cheek G which must be square with the rotational axis of the shaft to minimise axial excursion at the periphery of the blade. For this reason the shaft and cheek are centre-turned. End float is excluded by adjusting the shoulder depth of the remote end cover H, thus giving a slight axial pre-load to the ballraces. However, a wavy washer would give the same result, used in conjunction with angular contact races.

The saw disc I is a diamond wheel nominally of 0·5 in diameter centre-hole, 7·5 cm outside diameter and 0·025 cm thick. It is located on a spigot J and clamped between the shaft cheek G and a free cheek K by a machine-cut screw L and washer M. Some of the different types and sizes of discs in use are described later. The wide range of materials dealt with requires a corresponding speed adjustment, so drive to the saw is provided by an 'O' ring belt N from a 0·1 bhp shuntwound 24 V DC motor O, the speed of which is controlled by varying the armature current. Belts can be replaced by removing the spindle nut P and the pulley locking screw Q and sliding the spindle free.

It is an essential feature of the workfeed mechanism that the worktable R be axially parallel with, and at right angles to, the saw disc. It pivots on the axis of the feed micrometer S and must remain true during the subsequent cutting movement and the requisite incremental changes. Because some of the materials sawn are adversely affected by any 'bumpiness' in feed mechanism a damping head T is used to iron out intermittent contact. The table assembly has very low inertia so that the work mounted on it backs away from any irregularities in the shape of the saw. Thus the excessive stresses which high spots on the blade can cause—if the feed motion were uniform and unyielding—are avoided. In effect, the work material itself controls the rate at which it is plunged into the saw, a situation directly the reverse of that normally used. The damping mechanism consists of a closed cylinder with two ball races U, able to rotate concentrically at the end of the micrometer spindle. A further spacing cylinder V reduces the internal radial gap to 0·075 mm and a aluminium stearate grease is forced into the head through the nipple W. The worktable and counterweight arms, X and Y respectively, are attached diametrically to the head and are at right angles to S. The counterweights Z are adjusted to provide the necessary variations in cutting pressure.

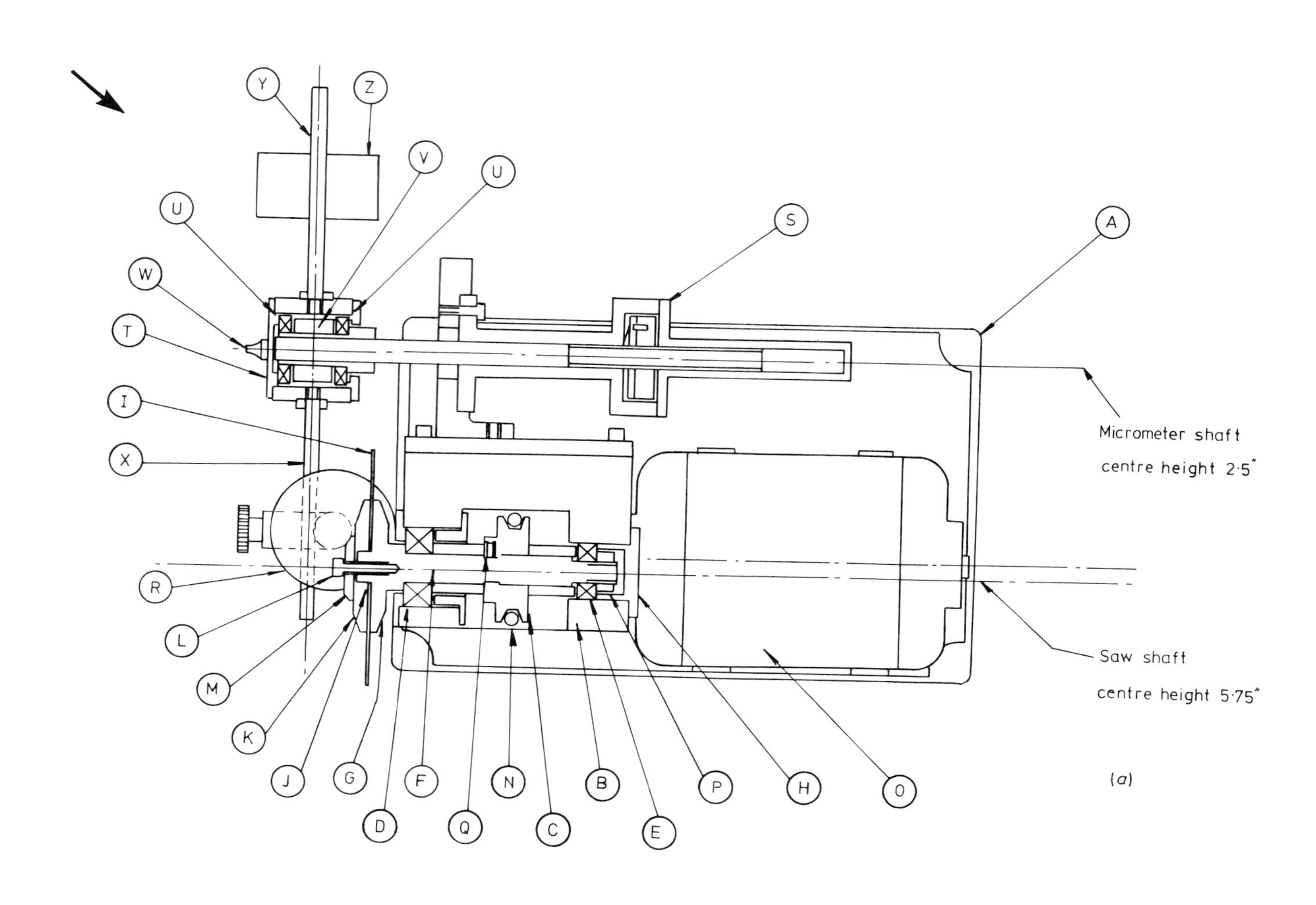

1.2 (*a*) Plan view of the complete mechanism of the prototype peripheral saw: the arrow (top left) indicates the angle from which the photograph of the prototype saw (*b*) was taken.

1.2.2 Mounting

Ideally, specimens should be wax-mounted (a full description of this process is given in chapter 4) on a material of matching thermal coefficient of expansion and of similar hardness, but since this is not often possible, glass is commonly used instead. In this way damage to the turntable is avoided. The choice of table is indicated by the operation to be carried out. A simple aluminium table, adjustable in three planes but light and structurally stiff, is most useful in the cutting of both delicate specimens and heavy crystals where any tendency to chatter might result in chipping or cleaving. A turntable is used with specimens which require accurately related angular sawing, and cylindrical work can be vee-block mounted. The work-tables are described in §1.6.3.

1.2.3 Diamond discs

Those used originally were of the Neven metal-bonded diamond type (Impregnated Diamond Products Ltd) and they have not yet worn out nor distorted in operation even after regular use over several years. Occasional operator carelessness has resulted in bump-damage to the blade and this has been remedied by trueing (sideways) against a dial test indicator (DTI) (the general use and metrological range of these instruments is described in chapter 6). When used on extremely brittle semiconducting compounds it has been found advantageous to true the wheel radially on the machine by a spark erosion technique. This consists of rotating the edge of the saw at about 200 rev/min close to a slowly revolving cathodic brass cylinder, pumping a dielectric fluid to the discharge region and spark-machining till the discharge is continuous.

Latterly, however, carbon-steel centre electro-metallised discs have been tried and have proved excellent in performance and resistance to mechanical damage. They are not, of course, amenable to spark trueing and deteriorate fairly rapidly when used on materials of Mohs's scratch hardness value greater than or equal to 9, but the manufacturing process permits the use of finer diamond in a higher concentration—a combination which itself results in keener cutting of the more difficult substances. Economically, too, these latter wheels are proving themselves to be a good investment. Table 1.2 gives some indication of the range of discs used.

It is possible when cutting to have too light a pressure for abrasion and thus merely burnish the materials: a useful indication of sawing progress can be obtained by arranging for a DTI to be actuated by the counterweight arm. This technique has the advantage of showing any blade eccentricities at slow speed (figure 1.3 shows the test in operation on a Macrotome saw—manufactured by Malvern Instruments and marketed by Metals Research Ltd).

1.3 Monitoring sawing progression with a dial test indicator.

1.2.4 Lubricant

Because the saw was originally used for shaping compounds thought to be marginally affected by water, a non-aqueous lubricant was considered to be advisable. Thus a 50/50 vol/vol ethane diol/methylated spirit solution has been used which has proved to be efficient with a variety of materials, since, moreover, it does not readily attack many mounting waxes and pressure-sensitive adhesives. However, 'Transcut 27' in water is satisfactory with specimens unaffected by aqueous lubricants.

1.2.5 Conclusions

The facility of starting the saw from rest with extremely light cutting pressures is probably a major factor in avoiding damage. Moreover, the non-positive work-feed mechanism results in a levelling of peak pressures and is, in effect, a safety valve for both discs and materials. Because there is no measurable rise in temperature of the workpiece during operation, saws and specimens are not susceptible to direct and indirect thermal damage. Economically, the saw is relatively cheap to make, easy to use and maintain, is adaptable and does not need constant attention in operation. Absence of dust or spray from the sawing process reduces the hazard to the operator. The sawn surface of a 3·6 cm diameter plane is flat to within ±15 μm, and the surface finish is good enough to make any more lapping minimal or unnecessary. Figures 1.4 and 1.5 are typical surface profile measurements made on ruby and calcium fluoride. Figure 1.6 shows the edge chipping on indium antimonide.

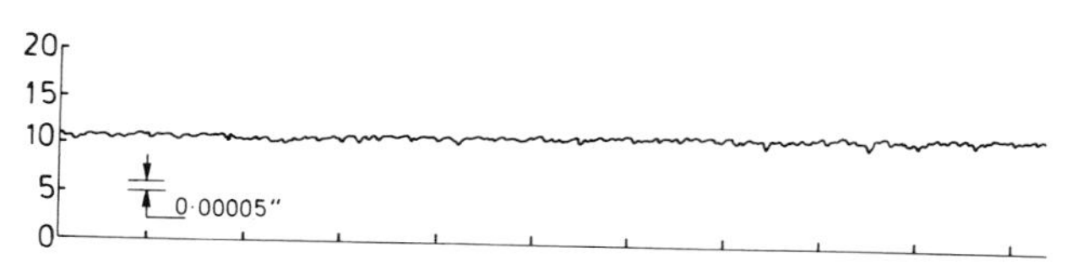

1.4 Talysurf record of sawn ruby surface.

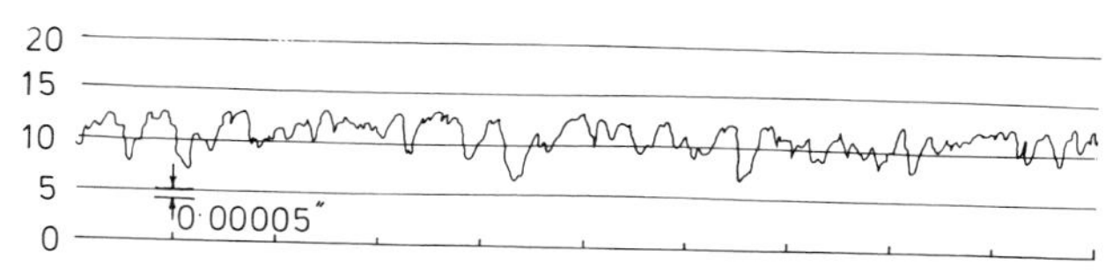

1.5 Talysurf record of sawn calcium fluoride surface.

1.6 Photomicrograph of slots sawn through 0·004 in thick indium antimonide (uprotected surface). Note: saw blade nominally 0·010 in; wider slot is produced with blade 0·010 in; narrower slot with blade cheeked to within 0·125 in of its cutting edge.

1.2.6 Consequences

The sawing carried out proved that the original philosophy was sound and retarded any further development for several years whilst data accumulated. However, limitations in its use were discovered, too, and a reappraisal of the mechanism became necessary. What had been learnt? The adoption of slow speeds simplified design and gave improved results; the damping mechanism itself was effective and cheap; and the worktables catered for most of the specimens. These ingredients were therefore considered to be essential in any modified or new instrument, but other factors had become apparent and merited attention.

(1) Even though the micrometer thread positional accuracy was better than 5 μm in 2·54 cm (~1:5000) and the arc of the cutting plane was similarly precise, cuts made through the larger diameter crystals tended to wander. Thus a departure from plane on a surface of a 2 cm diameter slice could be about 8 μm instead of less than 4 μm. A blade is normally stiffened by cheeks to minimise wander, but the greater the necessary protrusion of unsupported disc, the greater the possibility of curved sawing, especially of the harder materials. Increasing the thickness of the peripheral wheel once more restores the stiffness, but increases kerf loss.

Annular blade saws, where the disc is stretched across a chuck, and cutting carried out by abrasive on an internal edge, solve these problems (at least for materials <9 on the extended Mohs's scale)—and pose a few more. Section 1.3 deals with the genesis of the low-speed annular saw.

(2) The majority of users were not primarily concerned with fast cut-off times on large quantities of material, and this would have been a negation of the technique in any case. However, sawing many slices or dice in one operation speeds-up the process without violating the principle, and there were successful attempts at multi-blade peripheral sawing, more especially of dice (see §1.5). Large diameter crystals needed some less expensive arrangement than a bank of ganged diamond wheels, and a bench-top loose abrasive reciprocating-work saw evolved as a result (see §1.4).

(3) Sawing some materials relative to a preferred crystallographic plane became increasingly necessary. It could be achieved either by mounting the crystal on a goniometric worktable and setting up in accordance with x-ray data; or by establishing a face on a saw near to the Laué back-scatter equipment then re-mounting the work relative to this face and carrying

out repetitive slicing elsewhere. In the latter case, one goniometer and associated saw suffices and the cutting quality can be coarse; whereas the former technique calls for several gonimeters, preferably small ones, and needs a sawing operation which has to contend with the significant additional mass. Section 1.6.1 details several techniques (in addition to the standard method) which have been used on a variety of low-speed saws for re-establishing crystallographic orientation. The standard method initially used on annular saws in noted in §1.3; on the reciprocating-work saw in §1.4.

(4) One of the problems in sawing cylindrical (or irregularly shaped) specimens was that the counterweight load varied little throughout the duration of the cutting operation, but the length of cut varied greatly. For a cylinder, the pressure is a maximum at the beginning and end of the cycle and a minimum at the centre. A short counter-balance arm, set so that its horizontal position coincides with the centre of the specimen, reduces the effect slightly; and bolstering the crystal so that it becomes effectively square can be a satisfactory, if somewhat clumsy compromise. Several methods have been devised to improve loading generally and they are described in §1.6.2.

(5) The method of changing the position of the specimen parallel to the saw axis by means of the micrometer thread and nut had proved itself to be accurate but tedious in operation. Some quick-release mechanism for the arm and damping-head which allowed approximate positioning was desirable as well. In consequence it became easier to mount the saw-head on a lead-screw driven carriage in order to obtain precise positioning relative to the specimen.

(6) Low-speed sawing is a plunge-cut technique—that is, the crystal is passed through the plane of the saw in an arc and the shape of the cut is a consequence of the blade diameter. When long cuts have to be made a series of plunges ('nibbles') facilitates the operation, and the design of a twin-boom table made this possible see (§1.6.3).

(7) Superficially, a sliding table would satisfy the requirements of (6) above, as well as providing uniform depth of slot, but such a mechanism transgresses the principle of accurate yet flexible sawing which the more delicate materials require. Some workers needed long cuts of uniform depth in robust crystals and a sliding table has been found to be effective (§1.6.3).

(8) One of the more demanding features of the prototype saw was the difficulty of manipulating the worktable and the micrometer nut without accidentally contacting the saw blade. A damping-head lock-nut was needed, with the reservation that it should not interfere in any way with the dynamic function. A balanced damping-head assembly where work and counterweight maintain the arm horizontal, should change position when a 5 g weight is added to the worktable.

(9) Since the saws proved themselves to be undiscriminating in the material they cut, an adjustable end-stop was essential. On one occasion, an unattended saw continued unnoticed through a ruby boule, glass mounting strip and small goniometer plus its various metal components and emerged undamaged. It was an unsolicited testimonial to the saw's versatility. Section 1.5.5 deals with a damping-head lock and various end-stops.

(10) The O-rings used as transmission belts had the merit (for us at least) of immediate availability, but they were inconsistent in durability. Some lasted for several months while others cracked and broke very quickly. They can be mended with cyanoacrylate resins, but small vee-belts have an obvious advantage, and these have been used on later models and the commercially available versions which have resulted from the prototype and its lessons. Low-speed saws manufactured by Malvern Instruments and marketed by Metals Research Ltd, range from the simple peripheral Macrotome II, through the dual purpose 20 cm peripheral/annular Microslice II to the automated 30 cm dual purpose Microslice IV.

1.3 Low-Speed Annular Blade Sawing

Although low-speed peripheral sawing has been applied very successfully to the general field of electronic materials ranging from the nitrides (<10 on Mohs's scale of hardness) to some of the more recent semiconducting compounds (~ 2) such as gallium arsenide and lead telluride, where thin slices of weak materials are required with minimal kerf loss, low-speed annular blade sawing (Fynn and Powell 1970) offers important economic advantages. For convenience, the equipment and technique used for gallium arsenide (GaAs) will be described as an example of the general treatment.

1.3.1 Converted high-speed saw

The first attempts at cutting slices of GaAs below 0·75 mm failed on a conventional high-speed saw (the Capco Q.35) in its standard state. It was thought that this was due to the inertia of the workfeed table and heat/high-frequency stresses induced by the relatively high rotational speed of the blade. Since the easier modification to make was to reduce the rotational speed, a 0·1 bhp DC motor was coupled to the rear of the saw-shaft via a light-duty vee-belt. The motor was provided with the simple speed control which allowed a start from rest (see §1.2.1). At that stage the heavy workfeed table was still in use and some slices 0·5 mm in thickness with a surface finish of 0·75 μm centre line average (CLA) were obtained at a spindle speed of about 170 rpm. Attempts to reduce slice thicknesses below this figure usually resulted in fragmentation. Moreover, a slight bumpiness in the ball-bearing slideways of the machine, when combined with the necessarily light feed pressure, caused the mechanism to stop periodically even during successful cuts and thus produced ribs on the faces of the specimens. A close inspection of the slideways showed that lowering of the hood had caused a slight hammering of the ball-bearings on the ways, and this was responsible for the intermittent progress. It seemed safe to assume that a new slideway and balls might overcome the difficulty, but even then we would have been working at the fine limits of the mechanism.

A second modification was therefore indicated: the use of a standard damping-head (see § 1.2.1). Though not considered to be an ideal solution, it produced a great improvement. Figure 1.7 shows the modified Q.35 with its damped see-saw worktable in position clamped to the locked table of the saw. The essential mechanical requirement in the modification was that the axis of rotation of the new see-saw arrangement had to be parallel to the axis of the saw-shaft to 5 μm per 2·54 cm. With the work then mounted on a light arm, the feed direction was vertical and the pressure against the diamond edge controlled either by counterweight or by spring loading (see §1.6.2). This experimental hybrid has cut thousands of slices of GaAs and lithium niobate in particular with a single electrodeposited diamond blade (Impregnated Diamond Products Ltd—200/240 BS mesh grit) and a kerf loss of less than 200 μm. A typical counterweight load for slicing GaAs boules of 25 mm diameter is 50 g; slice thicknesses have been reduced to 200 μm when required, and, for some special glasses, to 100 μm. It is difficult to give the cutting duration, since this is roughly in proportion to the delicacy of the operation: a very thin slice requires progressively more gentle handling. Since we are often dealing with relatively expensive materials, time is seldom as economically important as material saved, and, on the whole, sawing processes are much easier to implement than subsequent lapping processes. It may be worth mentioning that the time required to saw silicon would be about an order greater than that of current high-speed sawing methods. However, the slice thickness obtainable, plus freedom from deep damage to the work, might well balance time expended in subsequent lapping processes.

1.7 Annular saw modified with damping-head/workarm assembly, operated at low speed

1.3.2 Modified low-speed saw

Since we had borrowed two features of the low-speed saw to improve the performance of a high-speed one, the next logical step was to convert an existing prototype Macrotome II so that internal diameter sawing was retained without losing the facility for peripheral sawing. This was done by fitting a reduced-size stretched-blade chuck in place of the normal saw and by raising the centre of the damping-head shaft by 9 cm with a simple jig-bored block (figure 1.8). The latter modification is necessary because the work contacts the cutting edge at the top rather than at the bottom of the diamond disc.

1.8 Jig-bored block attached to Macrotome II, for use in the annular mode.

Successful high-speed annular sawing depends to a great extent upon dynamic balancing of the chuck to minimise resonances, and precise centring of the blade—that is, the inside edge should describe as true a circle as possible during rotation on the saw-shaft—preferably to better than 25 μm. Changing to a very much reduced rotational speed brings the advantages of relaxing the need for accurate balancing whilst allowing run-out of the cutting edge to a greater extent. Although it is always advisable to obtain as precise a circle as possible, discs showing excursions up to 100 μm have cut delicate crystals at low speeds with no apparent deterioration in performance.

Tensioning an annular disc by clamping it between two concentric rings then stretching this assembly across the mouth of a chuck is a critical technique. Unequal stressing, caused partly by uneven clamping—grip/slip—can result in a non-circular cutting edge in addition to off-centre rotation, while the anisotropic blade material compounds the difficulty. Some experiments with blade tensioning were undertaken, ranging from the lightly stressed state through to fracture point.

Because of its low kerf loss (~150 μm), an Impregnated Diamond Products blade, type FX diamond on a 50 μm thick stainless steel base material, was mounted concentrically between two flat aluminium alloy rings and held in place with an epoxy resin as the bonding mechanism, then stretched lightly over the greased, radiused lip of a chuck (see figure 1.9). In use, it was probably the most durable of all the arrangements; however, replacement of the blade when it finally became necessary was not easily effected. The idea of a package assembly with dispensable rings was rejected as probably uneconomic, but it would be difficult to argue the case convincingly either way.

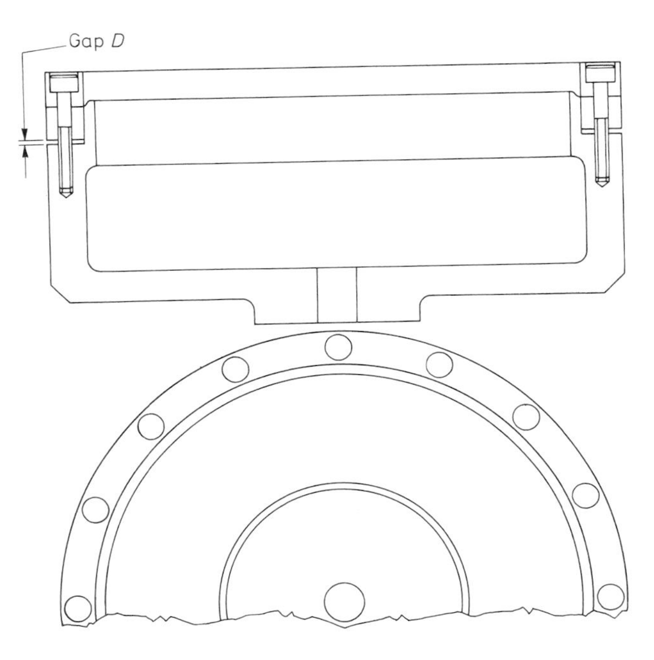

1.9 Diagram of the annular chuck with stretch-rim.

The next step, therefore, was to provide the clamping rings with interlocking toothed faces, conveniently provided by a thread-chaser (~16 threads per inch). A liberal greasing with a molybdenum disulphide paste was carried out since it minimised the risk of rupturing the blade when it was squeezed and stretched between the rings (figure 1.10).

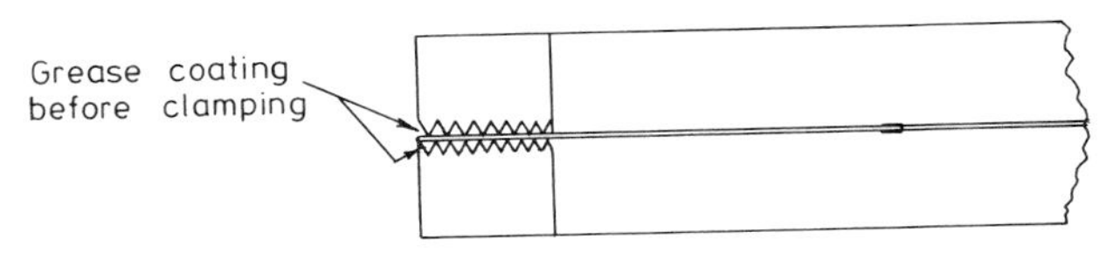

1.10 Form of clamping ring-faces.

This initial stretching during changing increased the inside diameter of the blade typically by about 300 µm. Some sawing was carried out with no additional tensioning, but the cutting edge deflected fairly readily under medium loads and tended to shear-off slices.

The clamped blade was tensioned further in the conventional way by pulling it down over the lip of the chuck with sequential tightening of the retaining bolts. A typical graph (figure 1.11) drawn of the increase in the internal diameter (ID) of the 50 µm thick blade versus the decrease in gap width (*D* in figure 1.9) showed an initial sharp change which decreased then steadily increased again until yield and fracture were reached. The stiffness of the blade was measured by using a Post Office tension gauge No. 8 to deflect the blade through 50 µm, in conjunction with a DTI (figure 1.12). The effective area of the tension gauge plunger tip was 0·07 cm^2, applied at a point 3 mm from the cutting edge. The data obtained are shown graphically in figure 1.13.

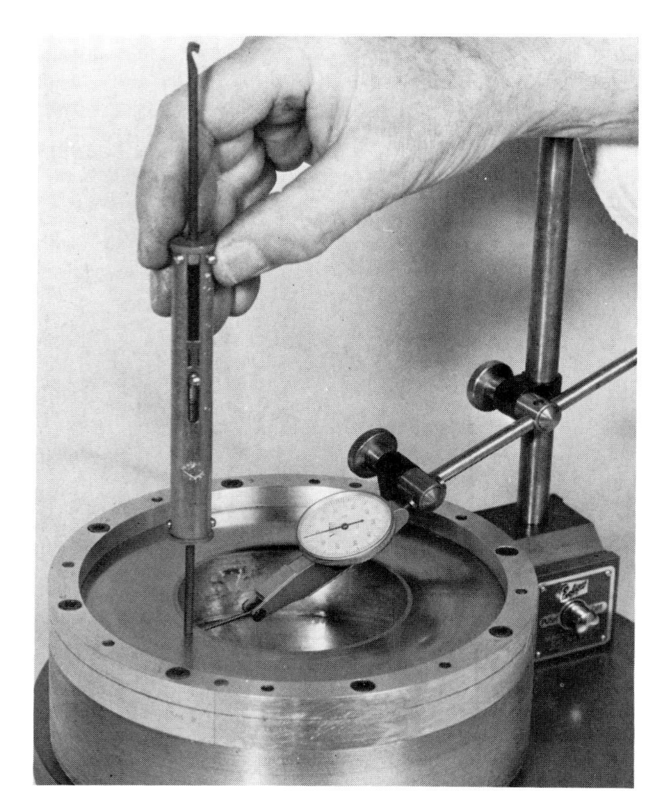

1.12 Post-office tension gauge and dial test indicator in use to measure blade stiffness.

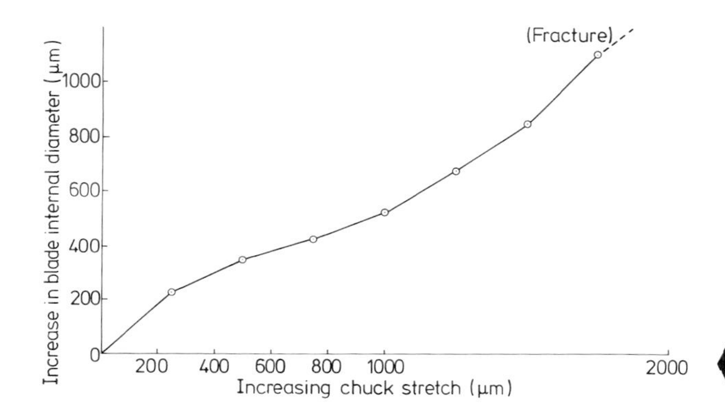

1.11 Graph of blade changing internal diameter with increasing stretch.

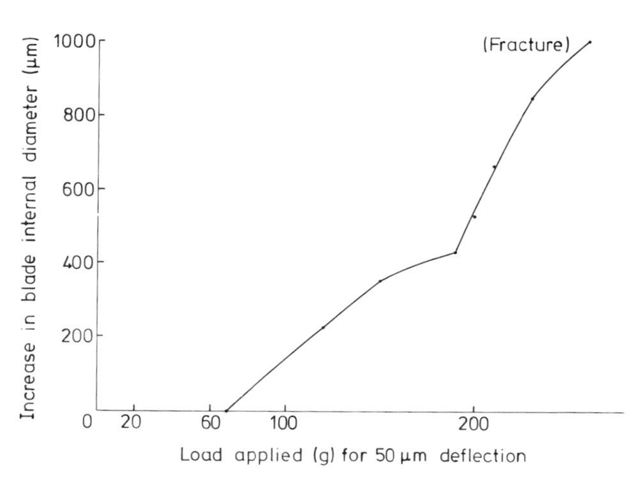

1.13 Graph of deflection of blade-edge versus applied load.

It was concluded that blade may not have an optimum stretch diameter/stiffness point but that there was a stretch-plateau (on this particular material at least) where stiffness did not significantly improve with further tensioning. Since the closer the blade is to the yield point the more prone it is to damage—and consequent ribs or pips—it is advisable to stretch one until the plateau is reached and stop there. There is the added advantage that the electroplated nickel which retains the diamond abrasive is under less stress too. From the graphs it can be seen that pulling the blade down over the chuck rim by about 800 μm is most effective with the blades that we normally use. If a blade tensioned in this way is subsequently removed from the chuck, (still in its clamping rings) it regains its unchucked internal dimension.

In performance, the modified Macrotome II was identical to the hydrid saw. It was, of course, slightly more limited in the diameter (32 mm) and length (50 mm) of crystals it would cut, but its lightness resulted in portability and its power consumption was 70 W for average cutting.

1.3.3 Sawing techniques

The general procedures and maxims for peripheral sawing are equally applicable to stressed-blade work. The specimen is mounted on a glass strip, long enough to extend completely within the chuck cavity and Tan wax (see chapter 4) is used as a cement with which to affix the boule. It may be used, too, as an encapsulant for some weak materials. Obviously, it is possible to attach extensions to the normal worktable and arrange that the latter does not enter the chuck; but when material is being sawn with related sides it is an advantage to use the rotational axis of the specimen table (or turntable) in order to avoid having to unfasten, realign and refasten the boule on its extending finger. The standard practice is to arrange the work so that the end sawn is remote from the machine—that is, the entire crystal is inside the chuck at the beginning. This permits the operator to remove and inspect a slice more conveniently after a cut has been taken. Since here one has the classical case of sawing off the branch on which one is sitting, an end-stop is usually arranged to limit the travel of the counterweight arm to the required arc (see §1.6.4).

If the crystal is being sawn relative to a crystallographic axis, it is the standard practice to provide it with a reference face on an x-ray room saw. This axis can then be re-established by aligning the prepared flat surface on the front of the boule with the sawing plane. A convenient way of achieving this is to remove the saw retaining bolt from the shaft, and, in its place, mount a DTI. Readings from the stylus movements are then taken along two arcs, approximately at 90° to each other, one made by swinging the DTI on the saw spindle, and the second by moving the work on the damping-head spindle (figure 1.14). The various degrees of freedom allowed for in the worktable construction can be used in order to achieve the required orientation. Other methods of alignment are dealt with in §1.6.1. A check on the thickness of the first sawn slice is a good test of the accuracy of realignment.

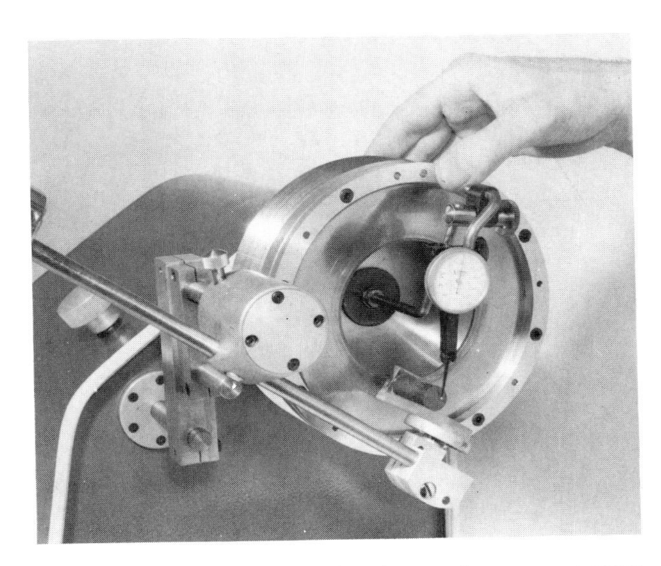

1.14 Re-establishing orientation of a crystal on an annular saw.

The boule mounting system is important and should have the following features: (i) rigidity, (ii) a base which can be sawn without damage to the blade and (iii), a cement that does not 'load' the saw—that is, fill in the areas between the protruding diamond abrasive particles and thus become progressively ineffective. In addition, an encapsulating layer can be effective in reducing edge chipping, and, where single slices need to be removed seriatim as sawing progresses, an arrangement which provides a cement-filled space between boule and base is desirable. The latter system also allows the whole boule to be sawn and left in place, too.

A system which satisfies these requirements is shown in figure 1.15. It is in effect, a fabricated glass angle-plate, made from 6 mm thick plate and cemented together with an epoxy resin. At the first stage of the mounting process, a small flat is hand ground on the end of the work remote from the aligned face. This flat is cemented to the vertical face of the glass angle-bracket, usually with an epoxy resin. In order that the boule can be spaced from the platform, it is packed up with several thicknesses of cardboard which also hold it in position while the resin cures. Quick-set room temperature epoxides expedite the process, whereas the slower-setting resins can be accelerated with an oven-cure at 100 °C. However, since the assembly must be heated to this temperature in order to fill the interspace when the card has been removed (and coat the boule if desired), the time saved is debatable. In either case, due regard has to be paid finally to susceptibility to thermal shock. By these means the whole of the boule may be cut with each slice retained in place, firmly held in the bed of wax at the bottom and protected around its periphery. Alternatively, each slice may be removed with a hot wire cutter (figure 1.16).

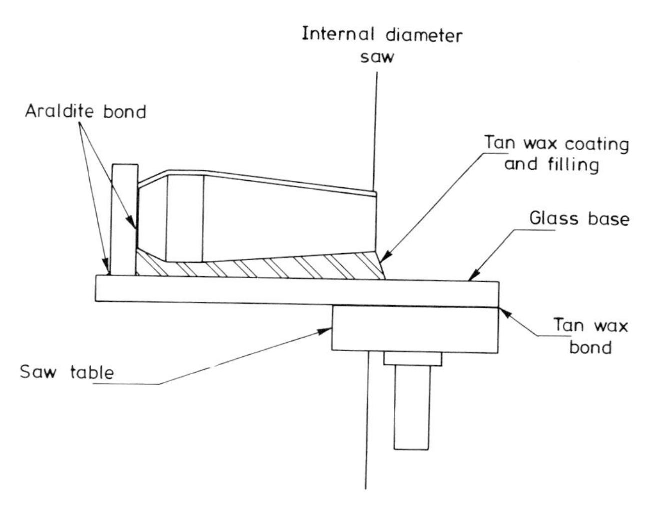

1.15 Boule mounting system for internal diameter sawing.

The techniques so far described are satisfactory for a wide range of delicate materials (see table 1.4) but there are those which do not give repeatable results below 500 μm in thickness. Notable examples are lead telluride (PbTe) and indium arsenide (InAs), and in order to obtain slices of these materials down to 200 μm it is advisable to give individual face support to each of them. This is done by using an 18 mm length of 12 × 12 mm, 18 swg, nickel plated copper-angle waxed to the face of the boule (figure 1.17). It is achieved by warming the angle section and melting on a liberal coating of a wax with a melting point of about 60 °C (see chapter 4). The hot waxed angle is then pressed lightly against the exposed end of the work and allowed to cool, bonding the slice over a greatly increased area on the glass base. A saw cut is made through the work and into the glass, and the angle-plate plus slice is removed by applying a pulse of heat from a small soldering iron which melts the wax locally (figure 1.18). The slice itself can be allowed to slide off the metal angle-bracket under its own weight when warmed in an oven or over a hot-plate. The vicinity of the boule end is quickly cleaned—preferably with brushed-on solvent and a dish to catch the fluid—and the angle-plate rewaxed in position. Aerosol degreasing can be used, but always with a directional pipe which restricts the solvent to a small area and not as a dispersed cold mist. On those occasions when extremely temperature-sensitive materials have to be thin-sliced, temporary support is given to the specimen by means of a thick grease: or an embedding resin such as acrulite, or polyester–glass/slate filled resin. The slices are separable by soaking in dichloromethane to remove the polyester: the polymethylmethacrylate can be treated by immersion in agitated chloroform. A film of water covering the surface of volatile solvents retards their evaporation.

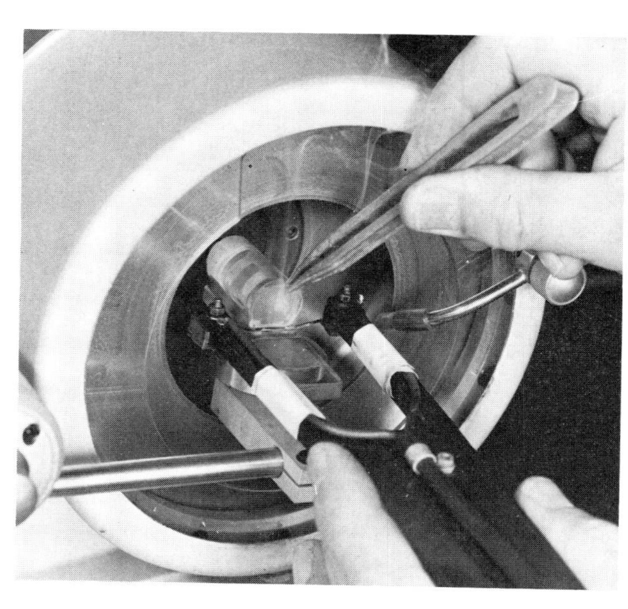

1.16 Hot wire cutter for slice removal during the cutting process

Table 1.4 Typical data for low-speed annular blade cutting

Material cut	Mohs's value	Blade type (grit)	Cutting speed (rpm)	Cutting pressure (g cm^{-1} cut length)	Kerf loss (μm)	Surface finish (microinches CLA)
ADP, KDP	~2·5	FX	150	30	150	12
Barium strontium niobate (BSN)	~5	280	250	50	175	12
Brass		FX	150	50	150	6
Cadmium mercury telluride	~2·5	320	125	10	150	20
Chalcogenous glasses	~3	FX	150	30	150	~ 3
Crown glass	~6	FX	200	50	150	8
Gallium arsenide	~3·5	280	180	50	175	12
Gallium gadolinium garnet (3Gs)	~7	FX	200	50	150	8
Gallium phosphide	~4	FX	180	50	150	12
Germanium	~5	280	250	50	175	12
Indium arsenide	<3	FX	150	30	150	12
Indium antimonide	~3·5	320	150	20	150	12
Indium phosphide	~3	FX	150	30	150	12
Lead germanate	~3	FX	150	50	150	20
Lead telluride	~2·5	320	150	20	150	6
Lithium fluoride	~2·5	FX	125	20	150	16
Lithium niobate	<5	280	200	50	175	8
Proustite	~2·5	280	150	20	175	16
Quartz	7	280	200	50	175	8
Rutile	~6·5	280	200	50	175	8
Silicon	7	FX	200	50	150	8
Spinel	8	280	200	50	175	8
Tri-glycine sulphate	~2·5	FX	150	30	150	12
Yttrium aluminium garnet (YAG)	8	280	200	50	150	8
Yttrium iron garnet (YIG)	8	280	200	50	150	8

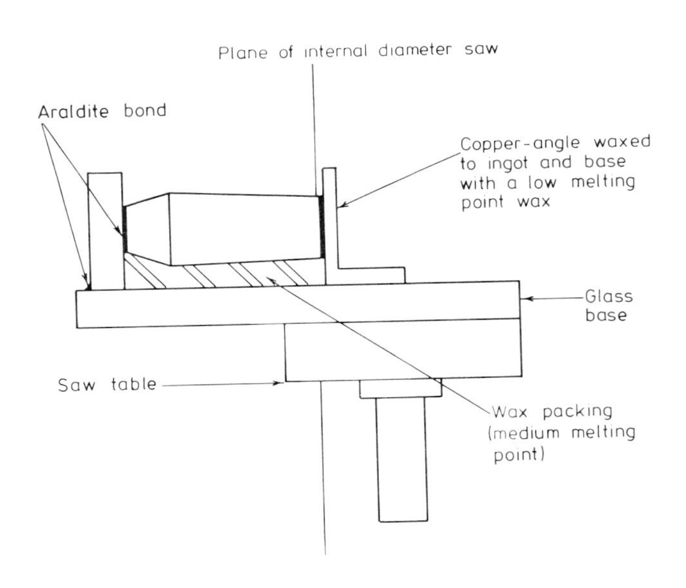

1.17 Face support bracket for delicate materials.

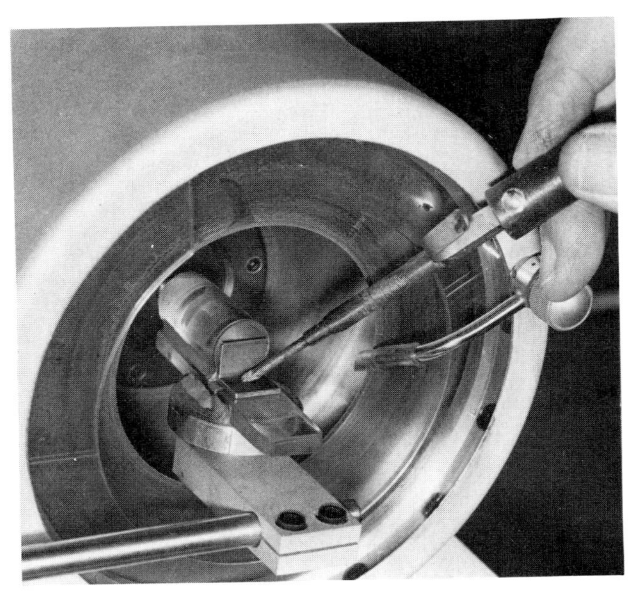

1.18 Small soldering iron in use for removing an individual slice and supporting bracket.

1.3.4 Microslices III and IV

Both peripheral and annular blade versions of the low-speed saws so far dealt with have been manually operable only. Commercialisation inevitably suggested a model that could be either manually operated or automated and the Microslice III (annular only, figure 1.19), then, latterly, the Microslice IV (annular and peripheral, figure 1.20) were developed by their manufacturers to satisfy a demand. A brief note of the capacity of the Microslice IV, and of the automatic mode of operation as we have used it, is a guide to effectiveness on repetitive slicing schedules.

1.19 Microslice III automated low-speed annular saw.

1.20 Microslice IV automated peripheral/annular saw, shown in the peripheral mode.

The saw chuck's rotational speed is continuously variable between zero and 550 rpm. Electro-metallised diamond blades, 30 cm outside diameter and 10 cm inside diameter, are used for annular cutting, while the peripheral mode can accept discs up to 17·5 cm diameter. The maximum depth of cut for a cylindrical specimen whose axis is in the plane of the saw-spindle is about 7·9 cm. Crystals with off-axis orientation that often need to be skew-mounted for their slicing, and non-circular specimens, reduce the available maximum depth of cut. Since the depth of the chuck is 6·8 cm it accepts material up to this length within the cavity for the standard in-to-out cutting procedure (cf §1.3.3). For sequential automatic slicing which begins with the work outside the chuck cavity and ends with the sliced boule inside, it should be remembered that the operator can program a greater cut-length than the chuck depth will take—that is, the total set travel could be 8·7 cm into the 6·8 cm deep chuck.

Automatic indexing of the work is carried out in 10 μm steps up to a maximum of 0·999 cm per cycle (0·999 mm on the Microslice III) with the number of cutting cycles, equal to slices, variable between 1 and 99 (1 and 999 on the Microslice III).

When the saw is operated manually, the feed rate of the work into the sawing edge is controlled by adjusting the counterweight. The positive drive automatic mechanism can be varied so that the work advances between about 6 and 50 $\mu m\, s^{-1}$. Resilient electromechanical constant-loading methods and devices which have been developed for very demanding cutting operations on all the saws are detailed in §1.6.2. The method of re-establishing a crystallographic axis from a reference face is described under the general techniques in §1.6.1.

The operations in an automatic control sequence are:

(i) the work to be sawn is balanced by the counterweight which is then moved towards the damping-head until the specimen is the heavier by about 100 g;

(ii) the fast approach switch is selected and the work electrically driven to within about 1 mm of the edge of the blade, then the feed-rate potentiometer adjusted to give the required movement for the material being cut;

(iii) when the cut has been completed, the adjustable depth-switch reverses the motor drive acting on the arm, allowing the work to return slowly to its starting position (i.e. withdrawing it from the blade);

(iv) the in-feed automatically indexes the pre-set distance moved and records a completed cycle;
(v) the counterbalanced arm begins the next cutting cycle;
(vi) when the programmed sequence of operations has been carried out the work is finally withdrawn from the blade and the machine—with the exception of the lubricant control—is switched off. Switching the saw to its auto-hold mode permits the operator to interrupt a cutting cycle immediately after the withdrawal stage (iii) above.

Slices may then be removed (e.g. by the hot wire cutter, §1.3.3) without interruption of the programmed sequence.

1.3.5 Conclusions and discussions

Low-speed annular sawing is not suggested as a revolutionary competitor to the standard high-speed technology, especially in the field of robust semiconducting elements like silicon and germanium. Its chief value lies in coping with materials which are often so weak that they might be termed potentially friable. However, they can be safely cut with surface flatness better than $2\,\mu m\,cm^{-1}$ and with uniform flatness to the same tolerance; with minimal damage and a surface which reflects the diamond particle size in the wheel; with little observable wear on the saw (even when using blades coated with circa 400 BS mesh diamond grit which give kerf losses of about $125\,\mu m$); and in very thin sections. There is an economic case for considering techniques where the work is not subjected to the rigours of high-speed sawing. Certainly devices are getting smaller and often thinner and in this age of mechanical mutation the nutcracker may have to usurp the sledgehammer.

It may appear, too, that low-speed peripheral sawing is uneconomical and as far as a repetitive slicing programme is concerned this may be true; but a slice of some research materials may cost several hundreds of pounds and the increase in yield, from decreases in both breakages and thickness, may be an additional 50%. For general purpose sawing and sample numbers, however, the more robust peripheral instrument is essential.

There is at present no facility for ganging annular blades as there can be for peripheral discs (see §1.5). Moreover, no satisfactory sintered annular disc is yet forthcoming for long-life slicing of materials as hard as or harder than sapphire, though electrometalled blades will perform satisfactorily over a limited period when operated at low speeds.

1.4 Reciprocating-Work Saw

Despite the success of the low-speed annular saw and its commercial development in the Microslice range—including the automated versions—we still encounter materials which need a great deal of careful, time consuming, individual attention for their repetitive slicing. The ganging of peripheral blades has been used specifically for dicing, (see §1.5), and has not been pursued as a multi-slicing technique. The low mass of a wire in sawing, in conjunction with loose abrasive, promises the most damage-free mechanical cutting technique, but the tendency of the wire to wander and wear are major snags. Exchanging the wire for steel tape, tensioned so that its long narrow edge travels accurately perpendicularly through the work, decreases the imprecision and increases the blade life. Multiply the number of tapes, space them uniformly and a reasonably precise multi-slicing method is available for most of the softer electronic materials. The Norton Vacuum Equipment Division's Model 262 saw is a notable example of a reciprocating-blade machine—that is, the work is fed into a shuttle of tensioned blades either supplied with abrasive slurry or coated with diamond in electroplated metal. We have used a modified version for cutting large acoustic surface wave materials (see §7.2) such as lithium niobate and quartz, (figure 1.21). There is, of course, the cautionary point that all one's eggs are in the basket with this technique. A disadvantage too, where heavy reciprocating masses are involved, is that it is easy to contaminate soft substances with ingrained hard particles. However, an important reason existed for using a similar reciprocating saw, but redesigning it as a low-mass bench-top model; the request from users of semiconducting and electroluminescent compounds that cross-contamination between materials at all stages, including sawing, should be avoided. In essence, that required either a complete saw, or a sawing mechanism plus lubricant

1.21 Norton Model 262 wafering machine; workarm changed to be low mass, down-fed.

system for each material. Moreover, because of the toxic nature of much of the swarf produced, minimal quantities of lubricant containing this detritus would be dealt with more easily. A recirculated charge of abrasive slurry in a small tank can be more readily handled than cleaning out a large capacity sump; and the same tank, complete with blade pack(s) provided for each material can be cheaper and easier to change than chucks complete with tensioned blades.

The objectives, and design considerations arising therefrom are,

(i) low reciprocating mass and variable, low reciprocation rate to minimise damage to the work. These point to:

(i) (*a*) reciprocation of work and workarm rather than blades; since the mass of a frame tensioning the blades is usually the greater its motion would be incompatible with the stub-axle method envisaged;
  (*b*) the system of work-load catering for delicate cutting, preferably on a principle similar to that used on the low-speed diamond saws;
  (*c*) continuously variable stroke-length adjustment;
  (*d*) reciprocation rate between inching speeds and about 400 strokes/min.

(ii) Small volumes of abrasive slurries to facilitate cleaning of components and disposal of contaminated waste; helped by:
  (*a*) rapid recirculation of a relatively small volume ($\sim 500\ \text{cm}^3$) of abrasive slurry;
  (*b*) a plug-in pump so that maintenance and cleaning can be easily effected:
  (*c*) interchangeable tank and blade-pack systems.

(iii) An inexpensive, durable mechanism; therefore:
  (*a*) precise sliding ways are essential and their primary accuracy should be bought in standard form rather than machined specially in-house;
  (*b*) protection from abrasives must be provided for the sliding ways by separating the reciprocating carriage from the cutting mechanism.

(iv) Work and worktable must be readily accessible and this is achieved by having:
  (*a*) the work fed progressively down through the blades rather than being up-fed;
  (*b*) a reorientation facility available, therefore a tilt and rotate mounting system is needed;
  (*c*) a rapid clamp/unclamp worktable.

A prototype reciprocating-work saw, christened 'Abraslice' has been made and performs well enough for the design to be refined: the Abraslice II has resulted (produced commercially by Malvern Instruments).

1.4.1 Mechanism

A 240 V DC shuntwound 0·1 hp motor A (figures 1.22–25) forms the power unit for the saw. Its field windings are fed with the full bridge-rectified voltage and its armature receives the rectified output from a variable transformer, (figure 1.25). The 10:1 reduction gearbox B is coupled to the pulley wheel and reciprocating carriage E via a vee-belt C giving 20:1 overall reduction of the motor speed. A tensioning screw A_2 may be used to correct undue slackness of C caused by its stretching during constant use, but the belt should not be overtightened because this places an unnecessary radial load on the ballraces of the pulley wheel rotational shaft.

The carriage E moves along accurately parallel tracks F_1 and F_2, formed by precise, hardened-steel cylinders (>620 VPN) 2·54 cm diameter. Three ball bushings give a semi-kinematic translation along the bearing cylinders, one running around F_1 and two round F_2. A scotch crank converts rotary to reciprocating motion. The stroke adjusting bolt D is used to vary the length of stroke between 5 and 15 cm.

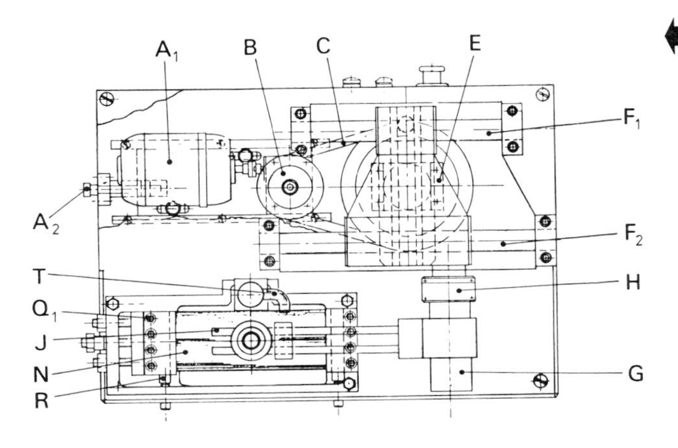

1.22 Abraslice II plan view with the top-cover partly removed to show the drive mechanism.

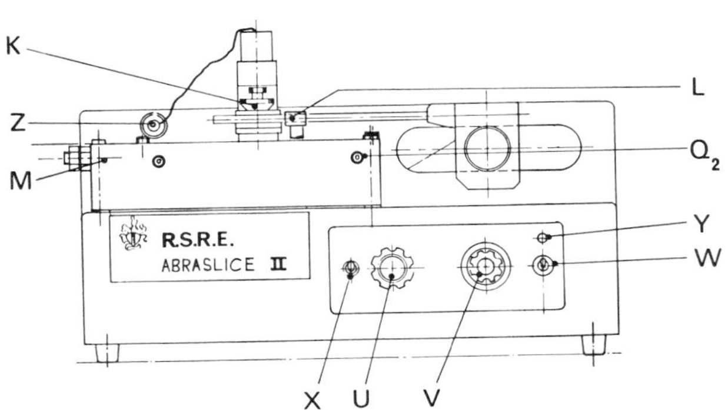

1.23 Abraslice II: front elevation.

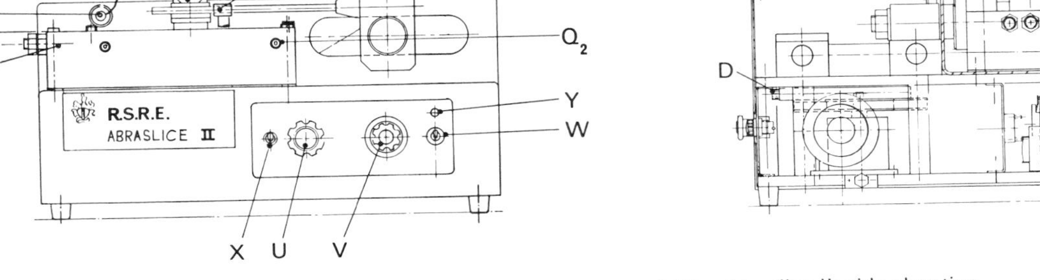

1.24 Abraslice II: side elevation.

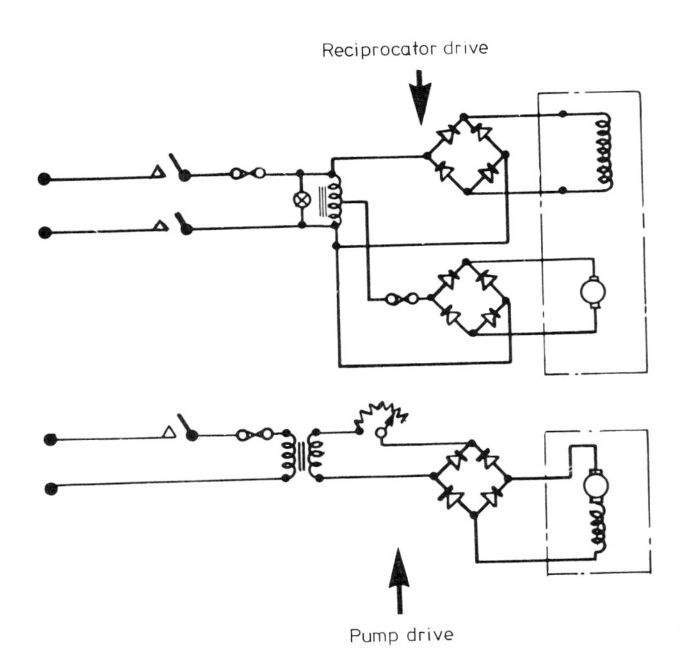

1.25 Circuit diagram for drive and pump.

Figure 1.26 shows the crank (or yoke) in greater detail: it is inserted between the linear carriage E, and the larger pulley; the upper face of which is slotted diametrally to form a slideway in which a sub-carriage runs. The leadscrew, in the form of a cap-headed bolt D engages in a female thread in the sub-carriage. The latter carries a ballrace which forms the crankpin of the scotch yoke, its radial position being controlled by rotating the leadscrew, clockwise to decrease the length of the stroke. Access to D is obtained through a grommeted hole in the rear panel of the machine when the reciprocating mechanism is at one mid point of its travel.

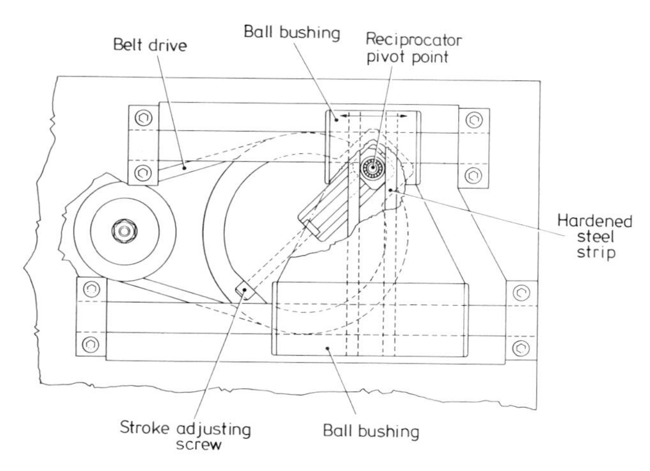

1.26 Scotch yoke assembly, converting rotary-to-linear motion.

The ballrace crankpin engages in a slot in the lower face of the reciprocating carriageway E, lined with hardened steel strips which form a close running fit on the outside diameter of the ballrace.

The workarm pivot shaft, in the form of a stub-axle G protrudes from the linear carriage and carries the twin-boom workarm J which pivots on two light duty torque-tube ballraces, widely spaced on the axle and given some measure of damping by being packed with a heavy duty chassis grease. For convenience in setting up a specimen relative to the blade pack, the bearing housing which carries the workarm can be unclamped and slid along the stub axle. Off-loading of the sawing pressure on the work is provided by a spring tension unit H. Thus the standard sawing load is the dead-weight of arm, table, mounting strip and specimen. It can be reduced by increasing the tension in the spring unit which can effectively balance the mass of the arm, worktable and specimen. Sawing pressure may be obtained by adding weights near to the worktable K. The boom-clamp L improves the stiffness of the work-arm, and is intended to do so especially if K should be removed. The worktable is described in detail in §1.6.3. The tank M, made from aluminium and anodised to minimise corrosion, has a safe abrasive-slurry capacity of about 500 cm^3 and is held in place by three bolts which allow sufficient angular adjustment to facilitate alignment of a blade pack N with the axis of reciprocation. A maximum of two blade packs, each 5 cm wide and 20 cm open length (that is, the distance between the opposing sets of precise spacers), can be clamped in the tank, initially by four bolts Q_1 at each end. Similar forcing screws Q_2, one at each end, clamp the blades together during tensioning, which is achieved by tightening the three nuts P. The packing blocks R are pressed against the blade-pack ends by the forcing screws, and, in the event of a single pack being used, a block of sufficient thickness is added.

A 24 V DC centrifugal pump S recirculates the abrasive slurry from the tank bottom and over the blades via the delivery pipe T. Section 1.4.2 deals in more detail with this type of pump since it has been used successfully for many years for a variety of operations such as recirculating electrolytes, lubricants and polishing slurries (chapter 2). Control of the pump delivery rate up to about 1 l min^{-1} has been obtained by placing a rheostat in the rectified voltage (see figure 1.25) adjusted by the knob U. The reciprocator speed control V can vary the rate from inching to about 300 strokes/min (a speed which is almost impracticably high because of splashing). W and X are the motor and pump on–off switches respectively and Y an indicator lamp for the motor–gearbox on–off condition. Figure 1.27 shows the complete saw.

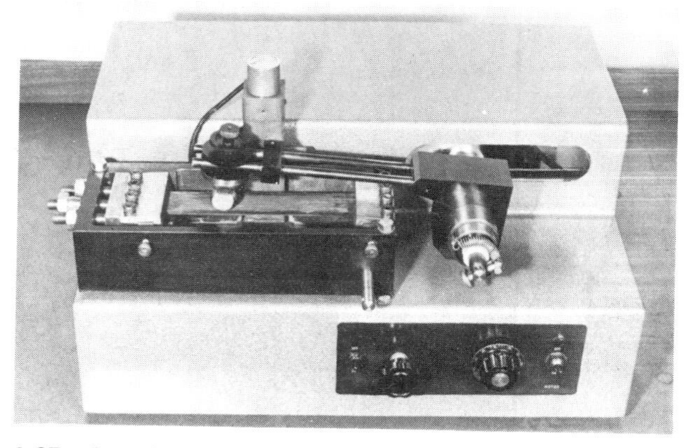

1.27 Complete reciprocating-work saw, with spring tension unit shown at outer end of pivot-arm shaft.

1.4.2 Centrifugal pumps

For any equipment in constant use, the maximum service obtained from it for the minimum attention given to it is a measure of its efficiency—efficiency which, even so, can be initially expensive because of stringent durability requirements. Happily, the centrifugal pumps are cheaply made with generous tolerances allowable on all fits: models have run continuously for months with a bearing clearance at the impeller end of several millimetres. In fact, anything initially precise rapidly becomes statically slacker when abrasive slurries are recirculated. The problem of wear is sidestepped to some extent in peristaltic pumps which can cope very efficiently with fluids of various viscosities and chemical reactivities, but their flow-pipes are relatively short-lived with lapping and polishing fluids. We constructed a centrifugal pump in about 1956 (figure 1.28) with an impeller of Perspex and bearing bushes made from synthetic resin bonded fabric, which has been in constant use since then, recirculating acid–copper plating solution.

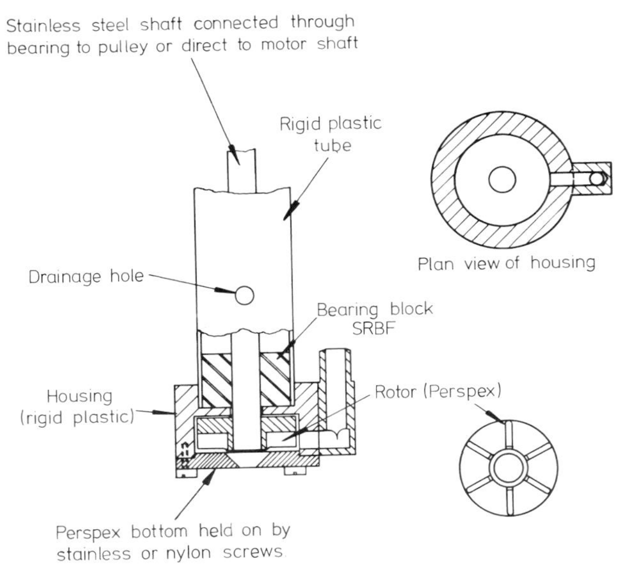

1.28 General type of centrifugal pump for slurries and electrolytes.

Driven by a fractional horsepower electric motor, it has served as the prototype for all the pumps described herein. Small DC brush motors are safe and very well tried and their ultimate renewal is easily carried out. They do not function indefinitely if the liquid floods into them by rising up past the bottom bearing. It is for this reason that overflow drain holes are provided in the tubular body, which should never be allowed to block up (e.g. with soaked tissue) nor be plugged.

Figure 1.29 illustrates the version of the pump used on the Abraslice II: its square section departs from the cylindrical (i.e. lathe oriented) design, mainly so that it can be clipped firmly in position over a recess in the tank wall without protruding into the blade-pack area. A copious flow of slurry must be available, both to keep abrasive particles well agitated and in suspension, and to minimise any tendency for detritus to clog the blade interspaces at each end.

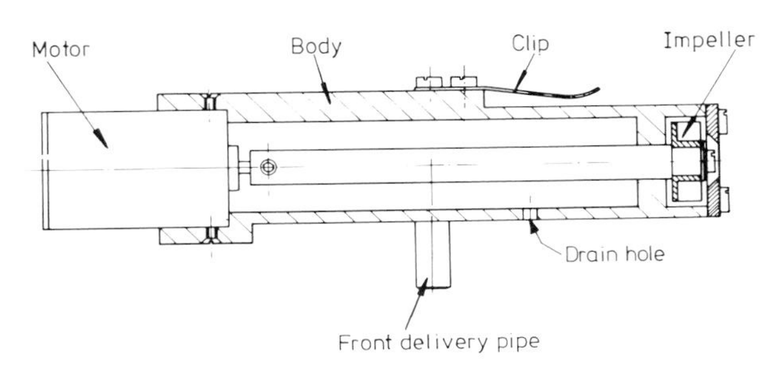

1.29 Abraslice II pump.

Thick, semi-solid slurry can impose sufficient drag on slices during cutting to shear them off either from the bulk crystal or at the glue-line. The delivery pipe copes fairly successfully with one blade pack, but the output may be increased by drilling a large hole diagonally through the corner of the pump body about 1 cm above the blade-top line.

Figure 1.28 shows a Perspex impeller that can be simplified in order to cheapen the design further. A tri-lobar disc is equally effective since the three apices (at 120°) centralise the rotating impeller as the bearing wears, yet continues to sweep-out an adequate volume of slurry.

1.4.3 Setting-up procedure

The blade packs that are normally used on the Abraslice II are available from Varian Associates Ltd, made from approximately 0·152 mm steel tape, very accurately spaced in order to cut uniformly thin slices.

Various spacers/blades are obtainable made up to about 3·75 cm widths overall—for example, 0·5 mm slice thickness (when the abrasive particles have removed an aditional 25 µm or so from the slice) from a 56-blade pack. It is possible to increase the slice thickness obtainable from a given pack by carefully snipping out blades. The pack (or packs) can be mounted and tensioned in the tank before it is bolted to the saw, or it can be clamped temporarily in place while mounting is carried out.

(1) Pack(s), spacers and clamp-plates are positioned on the clean platforms at each end of the tank.
(2) The clamping jaws are engaged with the steps ground for the purpose on the blade-pack ends, and the screws lightly tightened down.
(3) The forcing screws are turned until the assembly registers against the opposite wall of the tank, then tightened to about 700 gm ($\sim$5 lb ft).
(4) An open-ended spanner is needed to tighten the tensioning nuts evenly until the blade-pack length has increased 0·5 mm. as shown by a DTI contacting the vertical face of the movable clamping plate (figure 1.30).

When an annular chuck or a peripheral blade is changed on a low-speed diamond saw, the accuracy of re-registration is that of the original precision of the abutment faces and their subsequent freedom from burrs and dirt. The optimum co-planarity of the pivoting work-arm and the stretch-rim of a chuck can be obtained by mounting a sharp tool on the worktable and taking a facing cut across the rim while it is rotating. However, the practice is not usual on low-speed saws because they are less demanding than the high-speed versions. On the Abraslice, the changing of tank and blade pack is needed, and any system of precise re-location relative to the plane of reciprocation could be difficult to effect cheaply. It would require closely dimensioned tolerances on tanks, packing pieces, clamps and blade packs (and their mating surfaces) that are unnecessary. In the sawing plane itself—that is, the movement of the specimens down through the blades—the location is not too critical since a 100 µm tilt of the tank across its 10 cm width results in a blade tilt of about 5 µm. Tilt of the tank end-to-end is not significant. It is the realignment of the blades parallel to the reciprocating axis which has to be achieved.

The three tank-clamping bolts that are provided allow a few degrees of lateral motion to the tank. A simple means of realignment is carried out by mounting a DTI (see chapter 6) on the workarm and using it to indicate when there are similar readings at each end of the outermost blade as the tank is gently tapped into position (figure 1.31). The operation should not take longer than a few minutes to carry out and results in overall deviations from parallel of less than 25 µm without difficulty. A slightly more cumbersome method is to mount a microscope on the arm and view its graticule whilst aligning a blade. The tank is finally clamped in place.

1.30 Dial test indicator in use to show the extent to which blades are tensioned.

1.31 Method of tank alignment: dial test indicator mounted on the workarm and traversed whilst registering against a cutting blade.

It should be added that uniformity of slice thickness in many of the more robust materials like gallium phosphide is apparently not affected by several minutes of non-parallelism between blade and reciprocator mechanism: the blades appear to bend in and out of the slots as they are formed without widening them or rounding-off their extremities. Packs have been deliberately offset in an attempt to cut wider slots for a given blade separation and thus thinner slices, but without effect. Softer materials do not, however, perform in the same way and vary in flatness and uniformity.

At the end of the sawing operation, the blades must cut through the work, cement and into the supporting structure. The standard practice is followed of sticking a disposable platform on the twin-boom worktable (§1.6.3) and affixing the crystal to it. The general methods of mounting specimens are dealt with in chapter 4, and, for thick slices of robust materials where the available glueing area is large, Tan wax (a barium sulphate filled shellac) has proved satisfactory. When softer materials are sawn, however, especially into thin slices (~250 μm) the bond is critical for two competing reasons: since the work passes down through the blades during sawing, each slice finally relies upon the cement layer's cohesive and adhesive strength for its virtually rigid retention, which suggests a thick embedding layer; whereas abrading a slot through a cement that is not as hard as the specimen causes a change in sawing rate which can be minimised by a very thin glue-line. Any cement fillet can initiate the flexing of the back of the blade and causes steps on the slices. The thinnest possible filled cement is thus preferred while the sawing platform attached to the worktable should be of a similar sawing characteristic.

Good adhesion is obtained by lapping a flat strip, about 3 mm wide along the periphery of the boule in order to increase the available bonding area and allow a uniform adhesive layer to be obtained. Two flats of different widths are often worked on crystals in order to assist in identification of A and B faces after sawing, so that the wider is available for mounting. When the crystal has an off-square oriented end face, which has to be re-established, the flat along the side should be ground so that the cylindrical axis will be parallel to the tops of the blades (across the pack). This is most easily arranged by registering the end against a square metal block (e.g. an angle block) on the cast-iron lapping plate, rotating the crystal until it is parallel to the lap surface then working both components together, (figure 1.32(*a*)); or a pressure limiting jig such as the spring-hinge model shown in figure 1.32(*b*) can be used. A small amount of epoxy resin is loaded with alumina, a filament of it applied to the plane surface on the crystal, (figure 1.33) which is then pressed firmly in place on the platform. If too much resin is used, any which squeezes out of the join should be brushed away while it is still liquid. Cold-cure epoxides are normally used, of both the slow and rapid-curing types. The former allow a longer time for positioning the crystal, whereas the latter speed-up the operation but permit little time for adjustment. Their behaviours in solvents (for slice demounting and cleaning) can vary since chlorinated hydrocarbons such as dichloromethane which attack many slow-curing epoxides are less effective on rapid-curing ones; warm acetone dissolves some of these gradually.

1.32 (*a*) Lapping a flat on the side of an off-axis oriented boule.
(*b*) Lapping a flat with a pressure-limiting jig.

1.33 Thin, loaded glue-line for minimum interference with cut-through.

1.34 Re-alignment of crystal face.

When the resin bond has cured the assembly is mounted on the sawing arm and then slid along until, in conjunction with stroke adjustment, the maximum stroke-length position is found. A DTI can be bolted to the tank side (figure 1.34) and used to align the crystal end-face. Alternatively, the DTI may be clamped with a magnetic base to a heavy steel plate, resting across the tank top. The orientation face is made parallel to the vertical side of the blade pack by slackening the arm retaining bolt on the stub-axle pivot bearing and registering the stylus of the clock gauge against the work whilst the arm is raised and lowered. When the reading is uniform throughout the travel—that is, the diameter of the crystal—the retaining bolt is retightened. For adjustment in the horizontal direction, the arm is reciprocated and the worktable rotation used to give zero deflection of the stylus across the oriented face.

Because multiple slicing increases the contact area between blades and crystal, light sawing pressures are more readily provided on the Abraslice than on the low-speed saws. The process of abrasion here is more closely analogous to free-lapping: the material removal rates obtained from a specific size of abrasive particle (see chapter 2) and applied load give a reasonable guide to safe (minimal work damage) sawing pressures. Methods which result in nearly uniform cutting rates, detailed in §1.6.2, are not readily applicable to a reciprocating arm—with the exception of making all work sensibly rectangular with additional packing pieces.

Consider a typical electronic material with a Mohs's scratch hardness value less than 5: the general lapping pressure of about 100 g cm^{-2} gives a working load on each 150 μm wide blade, acting over a 1 cm cut length, of about 1·5 g. Materials greater than 5 become progressively more wear resistant and the sawing pressure on them has to be increased. Finally the abrasion characteristic of the work becomes similar to that of the blades and the process, with free abrasive at least, becomes impracticable.

Load control by means of a short, heavy counterbalancing weight has the disadvantage of adding to the mass of the reciprocating mechanism; a longer arm (more equal to the length of the work-arm) can reduce the counterweight mass but results in a space-consuming mechanism. The tension spring unit overcomes either snag. It is adjusted by releasing the pivot clamp-bolt, and, holding the tension unit stationary, rotating the pivot-shaft bearing housing until the spring approximately balances the work-arm assembly and specimen. The unit can be set so that the dead weight left unbalanced forms the sawing load, but it is more convenient to arrange close counterbalancing of the arm and table assembly than add the necessary weight on the twin-boom arm in addition to the work. Figure 1.35 shows the way to measure the sawing pressure with a spring balance.

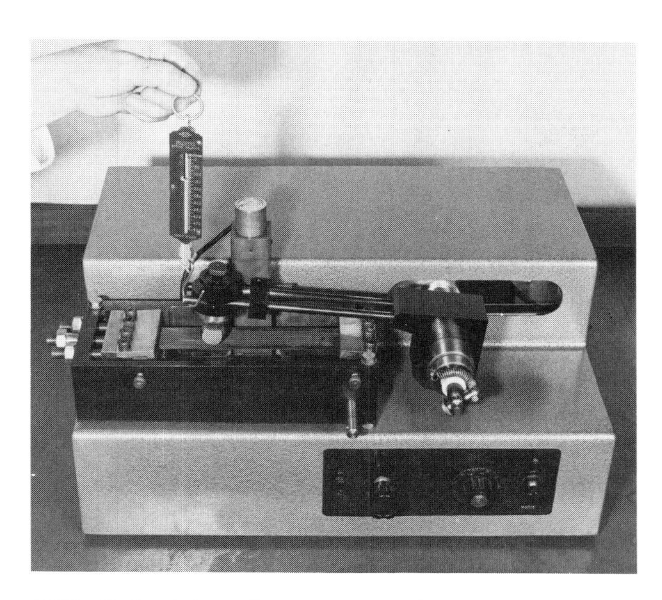

1.35 Spring balance in use to test cutting pressure.

1.4.4 Operation

Very soft materials are the most demanding on all saws, therefore the description of a typical cutting operation here relates to the extreme case of thin slices of cadmium telluride. For thicker slices of tougher materials, many of the conditions are less critical and can be relaxed accordingly.

The small volume of recirculated slurry which is used makes it essential that as many abrasive particles as possible are available for cutting. Despite the accepted use of light oils, because of their viscosity and lubricity, as the liquid component of abrasive slurries, we have concentrated on water-based and water-miscible fluids. The greater ease with which cleaning operations can be carried out encouraged the change. However, the particles tended to settle-out at first even though they were being vigorously agitated by recirculation, and a suitable liquid had to be found. Density and viscosity were considered to be the important variables. Density, generally, is the mass of 1 cm^3 of a species; in a liquid it is dependent upon temperature and pressure, both of which influence the free volume. Viscosity, again broadly, is a measure of resistance to fluid flow caused by intermolecular attraction. The performance of a species like glycerol, with a marked degree of association, (hydrogen-bonding between molecules so that they form larger groups) is influenced by its elasticity, as well as bulk and sheer viscosities. The engineer, dealing with lubricants, employs kinetic viscosity (viscosity divided by density); but once water is involved to any considerable extent, dissociation readily takes place even when thickening additives are used, and kinetic viscosity as a figure for comparison with hydrocarbon liquids has little meaning.

Initially, several thixotropicising agents were tried—for example, cellulose paperhanging glues—but, predictably, they suffered from the disadvantage that constant agitation broke-up the long-chain molecules and reduced their effectiveness. High molecular weight polymers like polyethylene glycol which can be bought in the range 200 g/mole to 6000 g/mole have been tried, but again in practice even those of molecular weight 400 and with viscosity $\eta \simeq 7$ cS tend to drag slices whilst losing active abrasive particles by sedimentation. The most promising fluid seemed to be one which has succeeded in ultra-fine polishing, used with 0·05 μm alumina: ethane diol (ethylene glycol), but some settling-out during lengthy sawing operations persisted.

We returned to the consideration of density, ρ, but this time that of the abrasive. For alumina (α-Al_2O_3)ρ is about 4 whereas for silicon carbide ρ is about 3·2; other promising solids, especially for the softer crystals, are powdered glasses (2.3) and boron carbide (2.5) while boron nitride ($\sim$2) should be best of all. In use, silicon carbide 800 grit stays almost completely in suspension in ethane diol, cuts freely and shows little evidence of sedimentation even when left to stand for several days. Boron carbide, 600 grit, shows no tendency to settle, but is fairly expensive.

The slurry is mixed in a 1 l beaker, 100 g SiC (usually 800 grit) in 500 cm^3 either of technical grade ethane diol or of the antifreeze solutions which incorporate sodium benzoate and sodium nitrite as anti-corrosion agents. After the slurry has been stirred thoroughly it is poured into the tank and the pump started. It is pumped sufficiently fast to give about two changes a minute (i.e. $\sim$11 min^{-1}) with the flow over the blade-pack top. The sweeping action of the crystal itself during cutting spreads out the effluent, though when a few thick slices are being cut a weir may be made out of cardboard sheet and placed across the blades at one end in order to prevent the slurry falling into the first few blade interspaces. In this special case, the flow is directed on to the weir and floods down from it and over the blades.

A slow, low-load start is always advisable on the small surface area contacted at the beginning of the operation, therefore the arm is reciprocated at less than 100 reciprocations per minute with no additional weight loading. This assumes that the arm has been set-up so that the spring tension unit balances its dead weight, then the weight of the crystal and platform form the initial sawing load. When the first grooves have been abraded—after, say, ten minutes—the additional weight can be added near the table in order to give the calculated amount, less than 1·5 g per blade for 1·2 cm diameter cadmium telluride.

The typical sawing progress of most materials may be followed by applying a dial gauge to a point under the sawing arm and recording the travel at regular intervals. A graph of cutting progression versus time usually has the form shown in figure 1.36,

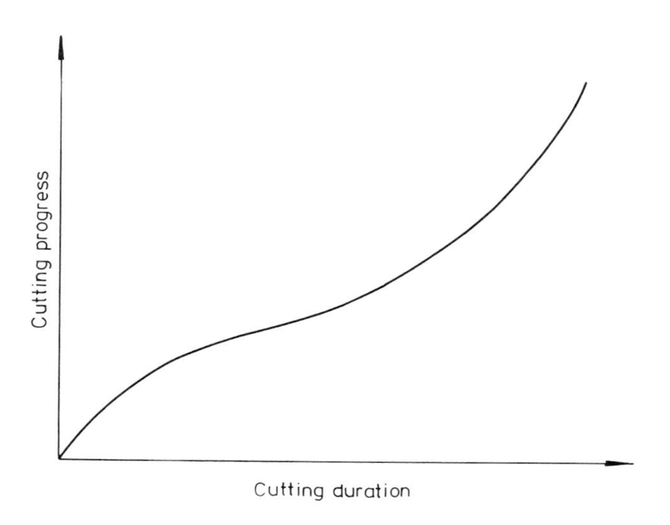

1.36 Graph of the form of normal cutting progress against cutting duration for a cylindrical specimen.

although a material like lead telluride may behave differently if the initial weight is too great. Such a compound becomes filled with abrasive very readily and slices less quickly since there is a proportion of the loose grit sliding on fixed grit. When the cut length increases to the point where the weight per unit area is closer to the optimum lapping load for the material, the rate increases. There is a change in rate, too, when the cement-line and base material are reached: ideally, any crystal should be mounted on a platform of identical sawing characteristic but this is not often possible. The supporting material, ideally, should not be softer than the crystal: but, when slightly harder like the germanium commonly used with cadmium telluride it can retard the complete cut through.

Lower density abrasives in ethane diol enable the operator to switch off a machine and restart it during the cutting cycle without apparent damage to the material caused by solidification in the blade interspaces. Moreover, at completion of cutting the crystal can be lifted from the slurry even after several hours have elapsed—for example, if a time switch has been used to terminate an overnight operation. Some of the former mixes, both hydrocarbon and water-based, effectively set when left uncirculated for a short length of time with consequent breakages of slices when they were pulled back up through the blade pack. The cut crystal, when first removed from the saw, presents a regular profile with each slice maintaining a uniform separation from its neighbours. After washing, first in water and a detergent then, preferably, in methylated spirits, the slices often lean together and should be either removed as quickly as possible in a solvent which attacks the resin used, or immobilised by painting on melted paraffin wax. The latter technique allows the crystal to be safely stored or transported in bulk; the former separates the slices which are cleaned again and packaged sequentially, either in the multi-compartment boxes described in §1.6.5, or in individual, marked envelopes. The most delicate and valuable specimens can be immersed in wax on a substrate, as detailed in chapter 4.

Slices obtained by reciprocating-work sawing, especially of friable materials, vary in thickness more than those from low-speed diamond saws. Though any slice might well be constant to better than 5 μm cm^{-1} (1 μm is often obtained) the spread over a few extreme slices has been as great as 40 μm. Thus for a nominal 250 μm thickness, a few are about 215 μm and one or two about 255 μm. This seems to be the result of a combination of small variations in blade spacers, departures from parallel of the blades (in both planes), unequal tensioning from blade to blade and any play in the reciprocating mechanism, especially on the reversal of the stroke. The assembly of the packs and the precision of blade and spacers keeps deviation within close limits for the majority of the slices.

1.4.5 Discussion and conclusions

The surface appearance which results from abrasive sawing is that of a lapping process, uniformly grey and with a texture indicative of the grit size used. There is always the danger that such a homogeneous surface can be deceptive since it may hide damage caused by excessive loads. For example. a few blemishes on a polished surface, or slight differences in arc patterns on diamond-wheel sawn slices, are seen immediately; whereas the random, matt finish persuades the eye that uniformity equals excellence. Glint from a specimen, observed as it is turned and viewed obliquely, should be examined with great care because it may be evidence of conchoidal fractures. Micro-cracks are almost inevitable, of course, but they should be small enough to be removed by etching or etch-polishing the surface: for cadmium telluride it is the practice to remove 100 µm of material. This figure may be unnecessarily large: if the sawing conditions described are adhered to, damage has disappeared by then, since scanning electron microscope channelling patterns and Rutherford back-scatter techniques indicate little crystal disorder.

Blades thinner than 150 µm ($\sim$75 µm) have been tried in an effort to reduce kerf loss but there is evidently some abrasive starvation in the cutting region which prolongs the sawing. The provision of a series of fine grooves across the blade edges should increase the supply of slurry in the grooves, and the technique is under investigation.

Loose-abrasive sawing is a delicate cutting method which can be abused very easily. Its greatest value apparently lies in economically slicing materials of less than 5 on Mohs's scale of hardness, especially those which demand considerable attention to each slice on diamond saws. The use of water-miscible slurries and individual blade-pack/tank packages help to minimise cross-contamination.

1.5 Multi-Disc Sawing

The automated Microslice III/IV is occasionally used for small dicing operations and the Abraslice, too, functions very well in this capacity. However, the best results have been obtained with ganged discs on a Macrotome II. Commercially available electrometallic or sintered-diamond wheels can be obtained with varying grit sizes, and for delicate materials *circa* 400 mesh is satisfactory. The concentration does not appear to be critical but a low one is an obvious economic advantage. These discs can give variable cutting-slot dimensions, due to wheel run-out and slight diamond protrusions. Thus the ganging of several of these blades (figure 1.37; Powell and Fynn, 1969) cannot be expected to result in a device of, say, 0·50 mm^2 having a tolerance of better than $\pm$50 µm. Edge chipping, too, is more in evidence. Therefore ganged silicon carbide discs are used in two sawing operations to cut, for example, one hundred 0·5 mm cubes of gallium arsenide.

The discs used are the Universal Grinding Wheel Co Ltd C 320 PR/VF type, 76 mm diameter × 0·125 mm thick. Collars spacing the wheels apart can be made from any dimensionally convenient flat strip, ferrous or non-ferrous. Initially, large washers 68 mm outside diameter were made from steel shim 0·50 mm thick, but when a variety of collar thicknesses are needed a resinated glass-fibre technique may be used. Two 20 cm diameter cast-iron plates are lapped flat to within two bands, cleaned and treated with a silicone release agent. Pieces of glass cloth amounting to approximately the required thickness are placed on one plate and an epoxy resin poured on. Three sets of specially prepared

1.37 Ganged peripheral discs for multi-slicing.

spacers (or feeler gauges suitably adjusted) are arranged at 120° round the periphery and the second plate added. A slight squeeze is maintained during the curing cycle to expel excess resin and ensure flatness. With care, the resulting resinated fibreglass disc is flat and parallel to better than 10 μm and can be finally made into spacing collars by conventional machining techniques.

Although they are extremely flat, the abrasive discs vary slightly in diameter and in bore size. This results in an uneven sawing load and though the damping head resolves most of the bumps, the edges of the discs should be trued *in situ*. A 14 μm diamond lap is pressed very lightly against the rotating blades with a DTI monitoring the excursion of the worktable (and thus the disc run-out). When the DTI shows no oscillatory movement, the treatment is stopped.

At first, the slices of gallium arsenide were mounted with optical stick wax (beeswax/resin, see chapter 4) on an SRBF slab, itself affixed to the worktable of the saw. Though the first series of cuts were successful, the second resulted in some of the cubes moving due to adhesive-bond failure. Subsequently, specimens were stuck down with an epoxy resin and the surface of the slice coated with a photoresist spray at both sawing stages. This was intended to minimise chipping and provide an additional strengthening bond when the cube would otherwise have become self-supporting.

By potting the specimen in resin at the bottom of a suitable polythene container, then inverting the cured encapsulation for sawing, the SRBF block and photoresist oversprays can be dispensed with. The method results in chip-free cutting but has the disadvantage that the edges of the discs tend to accumulate resin which slows up the process and has to be removed periodically by the method used for trueing the blades.

The same strictures for sawing slices of brittle semiconducting materials apply to multi-dicing practice: a start from rest with the saw assembly lightly in contact and a lubricant film over blades and specimen. The speed is not critical and may be increased quite quickly to 300–400 rpm, the overriding limitation tending to be centrifuging of the ethane diol/methylated spirits away from the discs and from the reservoir of liquid in the interspaces. Weight adjustment is more exacting, not in its overall limits, but in the method of carrying it out. Since the first sawing contact is made on a very small area, the pressure is about 15 g; but as penetration increases, the weight is increased. This is not an operation that can be easily carried out while the discs are rotating and it is advisable to slow the motor to rest, switch-off, carefully relieve the specimen of the cutting load, increase the weight, re-position and restart the saw from rest.

That gallium arsenide cuts very readily is shown by the blackening of the lubricant. Therefore the addition of extra weight does not call for calculation as much as observation, but obviously a rough limit may be assessed from a single cut on the material. The time when additional pressure becomes important is when the edge of the blade reaches the adhesive layer and then the substrate. Unless a substrate of matching sawing characteristic is used, the cutting rate depends upon its hardness. When a harder material is used, the weight can be substantially increased without harm to the gallium arsenide. In the instance of the twelve 125 μm thick blades described here, the weight was progressively increased from less than 15 g to, finally, about 200 g. However, it is emphasised that all the factors mentioned above have to be considered and the condition of the disc edge borne in mind: in its freshly dressed state it cuts very rapidly and the rate declines with adhesive loading rather than wear.

The results obtainable from sawing on the substrate are shown in figures 1.38 and 1.39. The yield in this example was greater than 90% since no edge chipping worse than 25 μm was discernible. The slightly out-of-square shape would have been improved with a graduated turntable. Sawing a specimen completely encapsulated in epoxy resin gives dice

1.38 Gallium arsenide dice ×10.

1.39 Gallium arsenide dice ×64.

without chipping. However, more time is needed and the wheel has to be dressed at least twice during the operation. Thus the method though very successful, is probably not necessary for anything other than the most damage-free devices.

Multi-dicing of gallium arsenide is successful, and since the sawing characteristics of other compounds such as indium antimonide, gallium phosphide, cadmium telluride, etc, are similar, the process should be suitable for them, too. The low cost of the discs does not necessitate a high capital outlay, and though they are fragile when rigidly set-up, any damage is not a financial disaster—in most cases the material cost far outweighs that of the discs.

Although the method of feeding the work into the blades is the conventional plunge technique (i.e. the specimen is mounted on the see-saw arm and allowed to feed into the discs) there is no reason why a sliding table, §1.6.3, suitably delicate in operation, should not be more convenient.

In the preliminary stages, devices as narrow as 200 μm were cut without difficulty, so the technique should be applicable to smaller, thinner dice as well. Though there is no technical reason why suitable diamond slitting discs should not be used, the flatness, thinness and cheapness of silicon carbide blades favour their use.

1.6 Techniques and Accessories

Several of the techniques used and accessories have been grouped together in this section because they are relevant to more than one of the saws.

1.6.1 Crystallographic reorientation

The clock gauge technique for realigning crystal end faces (whether re-establishing a crystallographic axis from the x-ray room sawn surface or simply squaring the boule) has been described already. Saws which have workarms that are detachable and precisely replaceable (Macrotomes and Microslices) allow realignment of the crystal end face to be made at a place remote from the machine. (Though the Abraslice technically comes into this category, the standard technique has not been varied.) Since the damping-head spindle is parallel to the saw-head rotational axis, it can be held in a vee-block on a surface table which replicates its position on the saw. Vertical alignment is automatically obtained by registering the plane surface on the end of the crystal against a square block on the table, but horizontal alignment has to be achieved simultaneously by solid reference to the vee-block. Magnetic clamps simplify this transfer of orthogonality, but refinements of the method can make it less exacting.

An auto-collimator (see chapter 6) allows the use of a special cylindrical block with its end face precisely square with the bore in which the damping-head spindle is a sliding fit (figure 1.40). With the spindle held vertically in the block, the workarm pivots in the plane of the surface table. The registration end face is uppermost and has a small parallel optical flat placed on it (or better still, a semi-polished face provided). Either by use of the worktable clamps and adjustments or by the warm plastic wax technique described below, the crystal is moved until the image from the flat is identical with its returned image when placed on the surface plate itself. A DTI can be used in lieu of the optical flat and auto-collimator.

A simpler technique that may be used when an auto-collimator is unavailable employs a vee-block clamped to an angle plate (see chapter 6 for an outline

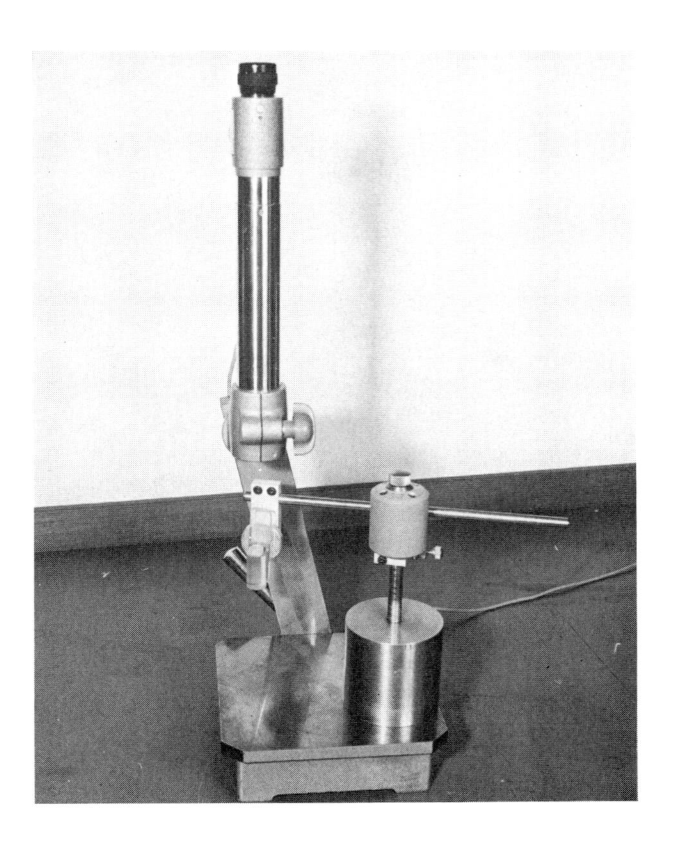

1.40 Re-establishing an oriented face, by the auto-collimator method.

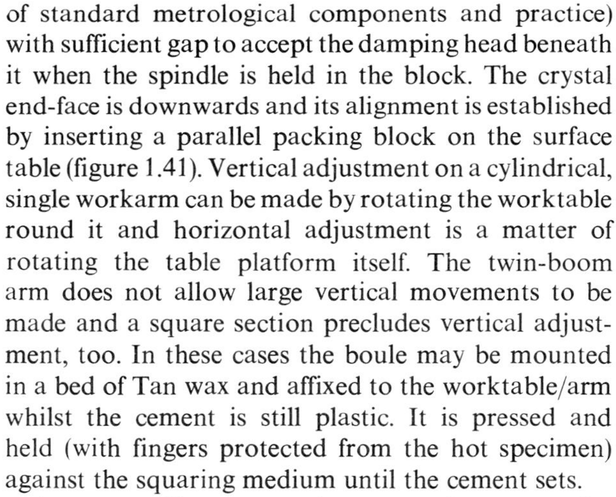

of standard metrological components and practice) with sufficient gap to accept the damping head beneath it when the spindle is held in the block. The crystal end-face is downwards and its alignment is established by inserting a parallel packing block on the surface table (figure 1.41). Vertical adjustment on a cylindrical, single workarm can be made by rotating the worktable round it and horizontal adjustment is a matter of rotating the table platform itself. The twin-boom arm does not allow large vertical movements to be made and a square section precludes vertical adjustment, too. In these cases the boule may be mounted in a bed of Tan wax and affixed to the worktable/arm whilst the cement is still plastic. It is pressed and held (with fingers protected from the hot specimen) against the squaring medium until the cement sets.

An auto-collimator can be mounted on a saw for direct realignment but the combination of an optical instrument and a dirty machine is best avoided.

1.41 Re-establishing an oriented face by contact with a surface parallel to the sawing plane, warm wax method.

1.6.2 Constant load cutting

Low-damage slicing of a delicate material, where the cut-length changes as it is being sawn, is not simply a matter of effective damping. Two further improvements that can be made are to reduce the mass of all the workarm components to a minimum (by using magnesium for tubular arms and worktable, for example) and by arranging an in-feed which caters for the changing area under the edge of the saw. Constant-load cutting refers to the latter and several methods have evolved which have been variably successful. The simplest is provided by blocking the crystal, whatever its shape, until it is approximately square overall (figure 1.42) so that the length of cut remains the same and the pressure is thus kept constant. For best results the packing shoulders should be of the same (or slightly harder) material as the crystal. Advantages are that any shape can be dealt with and the method does not involve moving parts; the disadvantages are that the mass is increased, the sawing process prolonged and setting-up is more exacting.

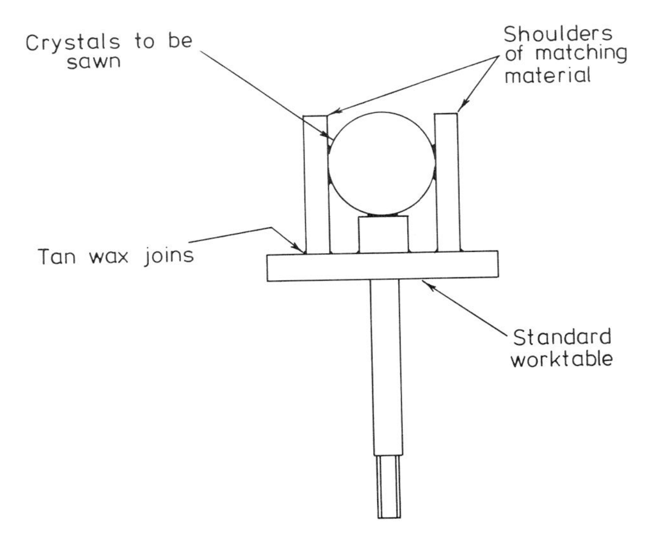

1.42 Blocking a crystal to make it effectively square as a method of making sawing pressure uniform.

When cylindrical crystals (the most prevalent form) are being cut, a sinusoidal pressure variation is suggested where the load is light at the beginning, increases to a maximum at the crystal's midpoint and then falls again to a minimum at the end. One way of achieving this is to place the workarm counterweight, in the form of a permanent magnet, between the poles of two other magnets (figure 1.43). During sawing, the magnetic counterweight moves away from the 'unlike poles' position at the start (when the weight acting to feed the boule into the blade is a few grammes greater than the magnetic attraction) to a position midway through its arc of travel. Thus the flux falls in accordance with the inverse square law, $\Phi \propto R^{-2}$, where R is the distance between poles, and the load effectively increases. As the weight continues to approach the lower pole, it is increasingly repelled and the cutting load decreases. Although the device works satisfactorily it has the disadvantages that it restricts the free movement of the arm and needs tailoring to suit the differing diameters of crystals. However, for a repetitive slicing programme carried out on uniform boules, it performs well.

Many high-speed diamond saws use constant-velocity feed arrangements that suit high-inertia work-holding mechanisms since the associated hydraulic components add little to the large mass of the whole system. In the lower-inertia workfeed associated with most low-speed sawing systems the workarm must not be too solidly coupled to a drive or escapement mechanism because a rigid feed would result, no longer allowing the work to follow the irregularities of the blade.

A mechanism which satisfies the low-inertia requirement uses the conventional workarm with or without its counterweight. A micrometer head with a ball-ended spindle drives the workarm into the specimen by means of a leafspring or rubber sponge pad (figure 1.44). Mounted beside the micrometer head is a small variable speed motor–gearbox unit (24 V DC; 500:1 reduction) the output shaft of which has a small pulley-wheel fixed to it. An O-ring belt couples this to the micrometer thimble. By the application of a suitably adjustable voltage to the motor the micrometer head is rotated and applies pressure to the workarm.

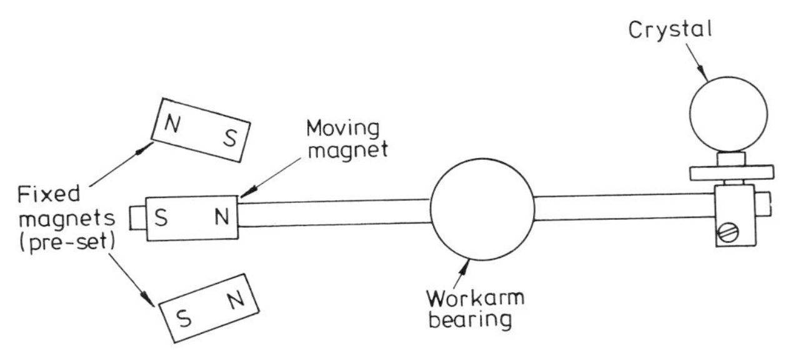

1.43 Approximate constant-load cutting for cylindrical specimens by magnetic means.

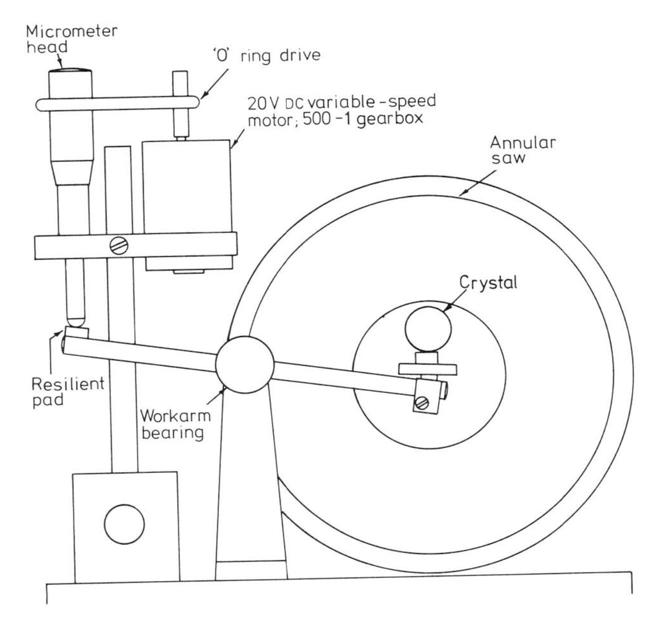

1.44 Controlled buffered down-feed drive.

The safe cutting characteristic of any material can be determined by standard sawing methods over the mid-section of a crystal as, say, mm hr^{-1}. Setting the micrometric drive (directly from the thimble/sleeve scale or with a DTI) to the rate assessed, or slightly slower, means that the crystal is sawn uniformly whatever its presented cut-area. In operation, the small movements generated by blade eccentricities and irregularities are accommodated by the sponge rubber block or leafspring. The motor has sufficient torque to provide the cutting pressure on an unbalanced arm, but if the counterweight is used initially to balance the work assembly, the resilient coupling between spindle and workarm is less compressed. A coupling could be placed between the worktable and the workarm but would then have to be both resilient and precise in movement: a spring hinge consisting of crossed leaves would be suitable. It follows that any high-speed saw can be converted to conform with low-speed practice by inserting a spring hinge of the type noted above between the worktable and arm; and by reducing the saw-spindle speed.

Although not involving constant-load techniques, one other simple method that can result in more sensitive cutting if used judiciously is to reduce the mass of the feed-arm assembly by replacing the counterweight with a long tension spring, coupled to the arm and to a point remote from it (figure 1.45). The change in tension undergone as sawing progresses and the spring slackens has less effect with a long spring. The spikey attachment which results has proved effective enough to justify a neater design in which the counterweight arm is removed and the spring wound round a grooved PTFE pulley (figure 1.46), thus making it effective in a small radius.

Summarising; all these methods lessen sawing damage on the less robust materials to some extent and allow thinner slices to be cut. Their use is largely dictated by the economics of the final device fabrication; the better the crystallographic order required, the more exacting the sawing conditions.

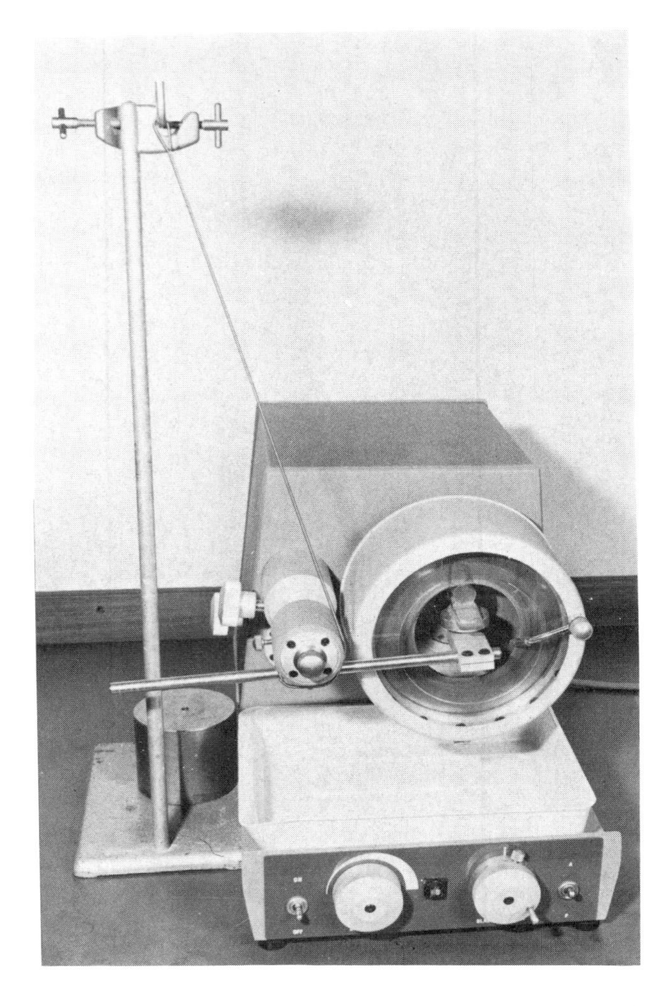

1.45 Long tension spring arrangement for application of lower-mass cutting pressure.

1.46 Long spring method, with spring wound round PTFE pulley.

1.6.3 Worktables

Some means of holding the specimen at a specified angle to the cutting disc and maintaining its position during sawing requires a worktable assembly which can be adjusted and clamped easily. Robust materials, often large and heavy, are not demanding, either in terms of worktable mass or damping; but one fundamental maxim for sawing delicate electronic materials is that the mass should be kept as low as possible for all workarm components. Of the tables which have been developed for peripheral saws, none is lighter nor more generally adaptable than the simple aluminium alloy (or magnesium) version (figure 1.47),

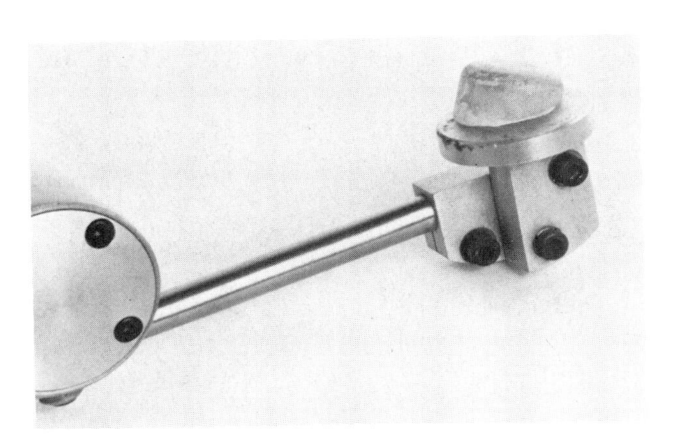

1.47 Simple sawing table.

clamped on a cylindrical arm and having most adjustments available for presenting different aspects of the work to the disc. Its workarm clamp allows unlock–set–relock movement along the arm and rotation around it; the work-platform clamp can swivel through a limited arc in the plane of the arm, while the platform itself rotates freely when unclamped. Superficially, moving the worktable along the arm permits the specimen to be set so that, in the case of a large-diameter cylinder, its centre may be fed into the disc along an arc which gives the most efficient cutting. However, swivelling the worktable via the platform clamp has a similar effect without losing the initially perpendicular arrangement of the assembly.

Annular disc cutting requires a worktable which can support a crystal within the chuck cavity (see §1.3). At its simplest, a long glass strip may be waxed to the table described above; but there are occasions when exact realignment of a crystal can make use of the rotable platform mounted at the end of an extension finger (as in figure 1.48). In those cases where reorientation is carried out with the setting-wax technique (that is, worktable complete with arm and clamping head on a remote surface table, jig and auto-collimator or dial gauge apparatus, §1.6.1) the extension finger needs no turntable.

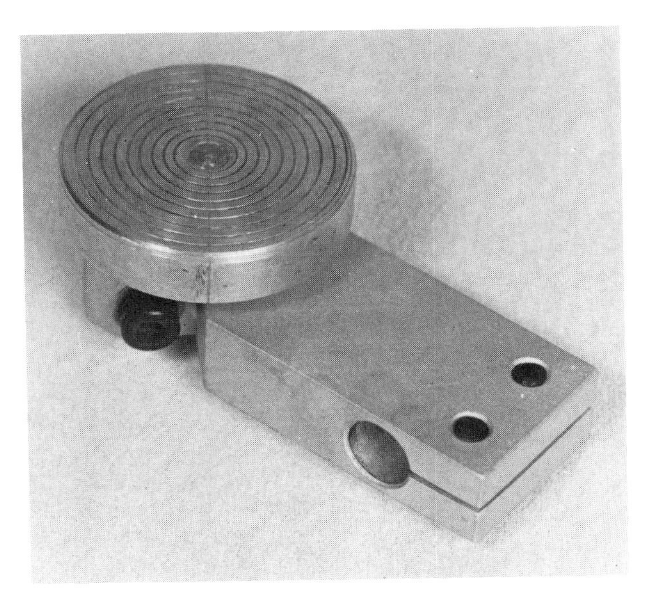

1.48 Worktable on extension finger for annular sawing.

The simple, peripheral-saw table is intended principally for plunge-cutting, where the disc makes a deep semicircular incision through the work and into the supporting platform. When specimens are large in diameter or long, too much time is spent in sawing into the supporting material. Either a sliding table is needed which, complete with drip-tray, passes beneath the blade, or a precise unlock–slide–relock mechanism which arranges for a series of shallower plunge-cuts to be made without losing alignment. The former method defeats the principle of the damping head, of course, since a sliding table has to be positively pushed or pulled at a rate which suits the blade's rotational speed. Figure 1.49 shows such an attachment which is satisfactory for sawing shallow slots in fairly robust materials (Mohs's value >4) where the depth is pre-set, the damping head locked in position and a suitable weight pulls the table along its slideways.

1.49 Sliding worktable assembly for long uniform-depth cuts.

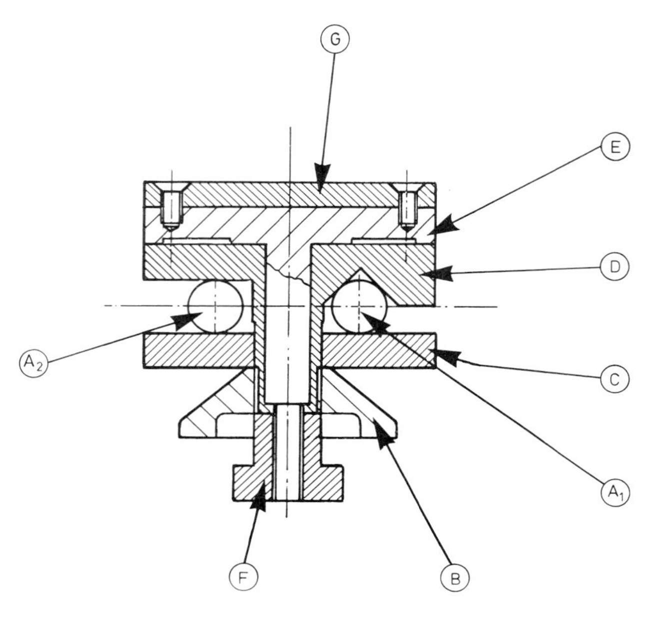

1.50 Diagram of multi-plunge twin-boom worktable.

The twin-boom assembly, as shown in figures 1.50 and 1.51, permits a series of related cuts whilst still employing the damping-head. Although the mass is relatively high because of the large worktable, twin cylinders and additional boom clamp, the table is invaluable for use with very hard substances such as sapphire and as the standard arrangement on the reciprocating-work saw (§1.5). Figure 1.50 shows its construction in detail: twin-boom arms, A_1 and A_2, formed from aluminium alloy tubing are clamped by a large knurled knob B between a pressure-plate C and the threaded base-member D which has a hole drilled centrally through its threaded shaft. Through this passes the stem of the worktable E which can be locked in place by the knurled nut F. Thus the base-member can be left in place while E is rotated or removed for specimens to be cemented to it or its supplementary platform G.

The heaviest worktable, a calibrated rotary one (figure 1.52) allows angularly related cuts to be made with an accuracy of about 6 minutes of arc ($\simeq 2$ m rad $= 20\ \mu\text{m cm}^{-1}$, see chapter 6). The design follows conventional machine tool practice, based on a 40:1 worm-gear reduction mechanism. A scale on the turntable rim divides the circle into either 10° or 10 grade intervals (where there are 400 grades in a full circle). The 10-division wormshaft dial thus

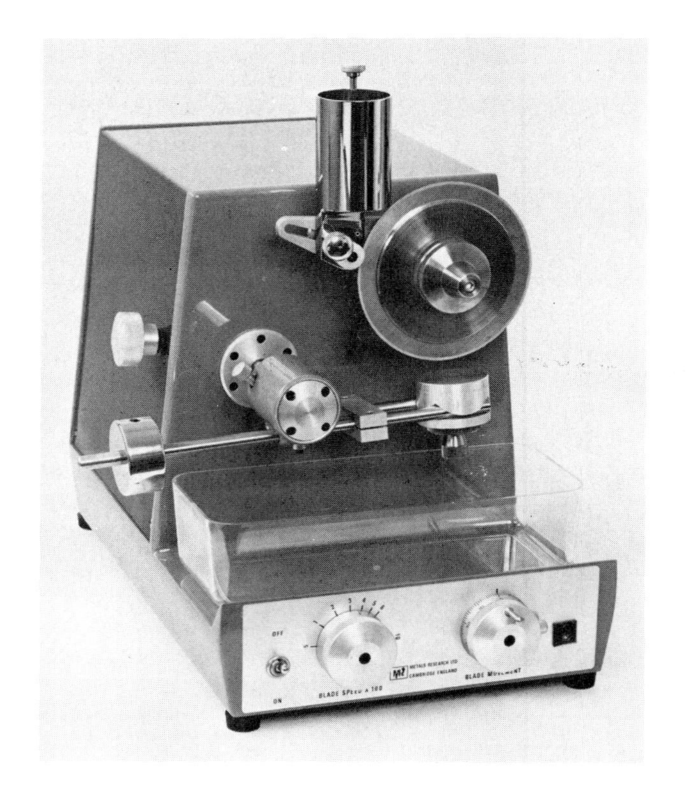

1.51 Twin-boom worktable on saw.

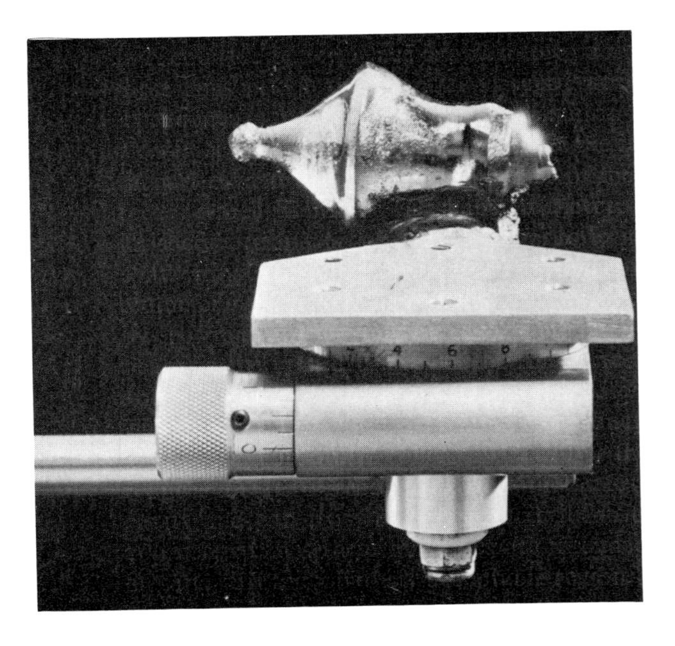

1.52 Rotary table for angularly related cuts.

1.53 Damping-head lock.

indicates single units directly and, by estimation, 0·1 of a unit. Both scales are read against one common fiduciary line. A friction plate is fitted to the worktable shaft and can be adjusted to hold the table from rotation against any lost motion (backlash) within the worm and wormwheel drive—for light sawing at least, it is not necessary to lock the table shaft.

Work is cemented to a supplementary plate which is screwed to the rotary table itself. Apart from the obvious uses of the table, such as positioning work for cutting parallel or angular faces, a number of saw cuts made at intervals of a few degrees can produce a form which is readily lapped or ground cylindrical.

1.6.4 Damping-head locks; end stops and progress gauge

Two additional features which improve disc saws are a damping-head lock, and some form of end stop which limits further workarm travel at the end of a cut. In addition, the end stop can be fitted with a microswitch which signals the end of the cut by buzzer or similar audible means. The damping-head lock, one form of which is shown in figure 1.53, prevents the arm assembly from rotating during setting-up. Any such clamp mechanism or pressure plate must release cleanly so that its action does not interfere with subsequent movement of the arm.

The constant down-feed device described in §1.6.2 makes an end stop redundant; but for the more usual counterbalance arrangements a strut may be positioned externally, on a laboratory stand-clamp for example or on a special bracket attached to the damping-head spindle (figure 1.54).

Some of the very hard materials are cut increasingly slowly by peripheral sintered-diamond wheels, apparently because they lose some edge diamonds and the soft-steel matrix is peened over the others. The effect is initiated by too low a cutting pressure and a dial gauge mounted to register on the counterweight arm as in figure 1.3 shows when cutting progress has stopped. The longer this state is allowed to continue, the more glazed the edge becomes and the cut is burnished rather than abraded. Regeneration of the edge can be achieved by allowing it to dip in a strong solution of hydrochloric acid and water which etches away some of the steel. The spark erosion technique described in §1.2 for trueing blades provides a better cutting edge. Further sawing then requires an increase in pressure within the limits set by the stiffness of the blade.

1.54 End stop to limit sawing at specimen cut-through point.

1.6.5 Slice cleaning and packaging

Chapter 4 deals in greater detail with manipulation, cements, solvents and packaging techniques. Tweezers, stainless steel or plastic, can be used on robust materials, but the delicate ones may require manoeuvring with brushes. Sawing debris and fluids must be removed from slices and specimens as soon as possible after they have been cut: by warm water/detergent solutions for water-miscible species and a trichloroethane rinse for hydrocarbons generally. Ultrasonic vibration can facilitate the cleaning process, but it should be used with care on less robust materials: whilst its application with chlorinated (or fluorinated) hydrocarbon solvents where the cement, too, is soluble may end in disaster. Subsequent demounting and cleaning requires a few rinses in a suitable warm solvent with vapour degreasing kept as an optional final stage in order to reduce the accumulation of contaminants.

Many sliced materials have to be numbered sequentially from one end of a boule and this order carefully retained. Individually marked envelopes or small plastic boxes are often used, but the more expensive materials are often transported in multi-compartment plastic boxes (figure 1.55) which cater for slices up to 26 mm in diameter, cushioned between discs of foamed polyurethane. The compartments are numbered: up to 23 for the smaller box, 38 for the larger and the lid is screwed in place, registered by dowel pins. It should always be opened carefully, since a few slices can become lightly attached to the foam discs in the lid and be inadvertently extracted with it. A useful technique for the carriage of single specimens is to immerse them in paraffin wax on a glass slide and mount this in a plastic box, held lightly in position by double-sided pressure-sensitive adhesive tape.

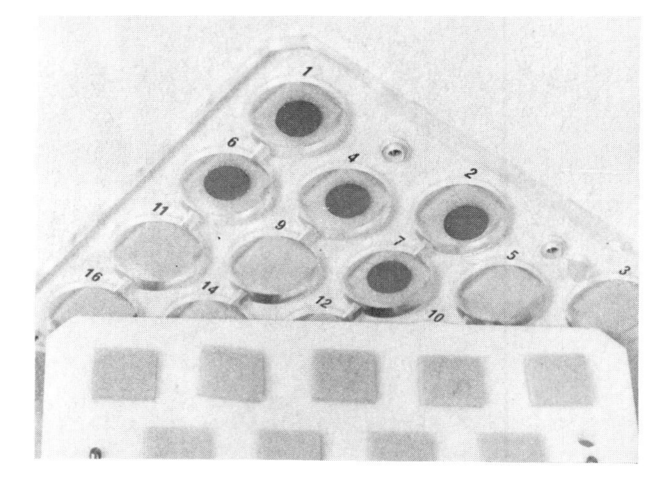

1.55 Multi-compartment boxes and foam inserts.

References

Fynn G W and Powell W J A 1965 *R and D for Industry* **40** 30–1

—— 1970 *Ind. Dia. Rev.* **30** 312–4

—— 1967 *Engineer* **22** 23

Holtzapffel C 1850 *Turning and Mechanical Manipulation III* 1311–3

Powell W J A and Fynn G W 1969 *J. Sci. Instrum.* **2** 1122–3

2 Lapping and Polishing Machines and Jigs

'Nature! whose lapidary seas
Labour a pebble without ease
Till they unto perfection bring
That miracle of polishing;' (William Watson)

2.1 Ring Lapping

Apart from some hand-lapping operations used to produce a small grey surface speedily, ring-lapping forms the major means both of stock removal and of finishing. This applies to crystals and glasses and, with suitable changes of abrasive, to ruby and sapphire as well.

The system is basically simple but an appreciation of the details of operational technique is needed to 'make' the product. At first sight the principle would seem to be foolproof. However, ring-lapping machines are commonly seen on which elementary precautions have not been taken and the results are invariably disastrous. Perhaps this is not too surprising since the operational handbooks provided with some machines do not always cover the points which we think essential for their optimum efficiency. Probably the earliest commercially available machines operating on the ring-lapping principle (referred to as tubular-lapping in Russian literature; see for example Kumanin, 1962) were those trade-named 'Lapmaster' and it is with these that we have had the most experience. There are many others similar in design, and one in particular had an interesting variation. Lap flatness control was achieved on it by means of differential weight-loading on the ring edges instead of the more usual positional variation of the control lap or conditioning ring.

2.1.1 Machines

The two machines which are described are those with which we have eighteen years accumulated experience: the 30 cm and 60 cm Lapmasters (Payne Products). The 30 cm design, (figure 2.1), has a revolving lap driven by a backing plate and a catch pin which is, in turn, motor and gearbox driven. Above the plate are three cast-iron rings, constrained by roller bars. These allow free rotation of the rings with the energy derived from the rotation of the lapping plate. The all important shape or figure of the plate is determined by the position of these rings relative to the lap centre. When, with the usual abrasive slurry, the machine is run with the rings overhanging the outer edges, the excessive work done at the periphery of the lapping plate leads to its eventually becoming convex, (figure 2.2(*a*)). The longer it is operated under these conditions, the greater the degree of convexity that will be obtained. This potential abuse may be illustrated by the fact that we have found neglected machines in which a 12 in lap was 0·010 in (0·25 mm) lower at its edge relative to its centre. Disasters such as this call for strong cures and, in this case, the plate was removed from the machine and faced in a lathe. The surface produced by a reasonably true centre-lathe should produce a departure from plane better than 0·001 in (0·025 mm)

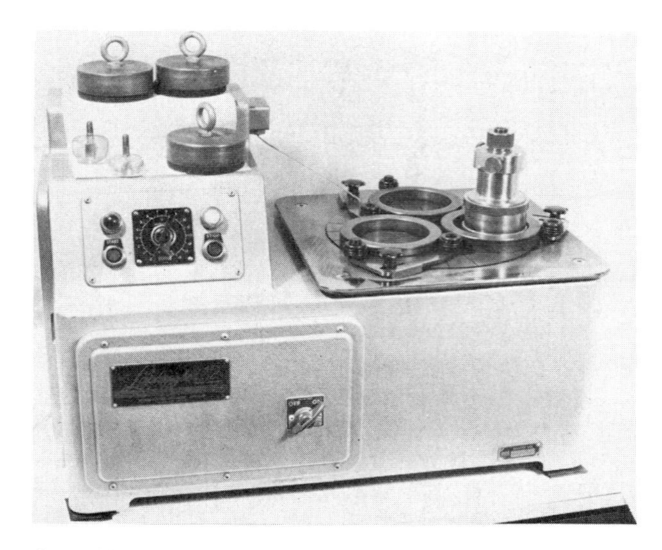

2.1 30 cm ring-lapping machine.

A concave plate surface is obtained by moving the three rings in to near tangential contact, (figure 2.2(*b*)). The concave figure forms on the plate at a much slower rate than in the case of the convex. This is in part due to the restricted diameter of the central recess which limits the overhang at the centre. If a large correction has to be made, a short double-ended shaft with a central flange (figure 2.3) is machined. Its diameter at one end must be less than the space formed by the rings at their innermost positions. The flange bears on the upper surfaces of the rings and the boss above accommodates a suitably drilled lead weight. The increased loading on the plate centre brought about in this way produces concavity rapidly—that is, it reduces a state of convexity. Although states of concavity and convexity have been described, it will be appreciated that these machines are intended to lap flat and it is the narrowly acceptable region between these alternating figures that is normally needed.

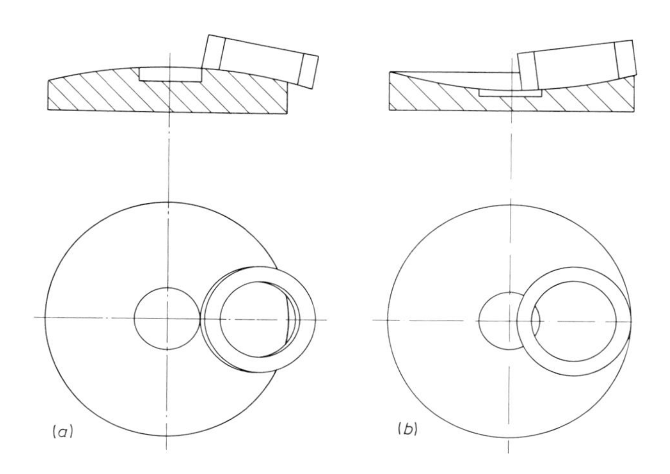

2.2 (*a*) Ring position resulting in convex lapping. (*b*) Ring position resulting in concave lapping.

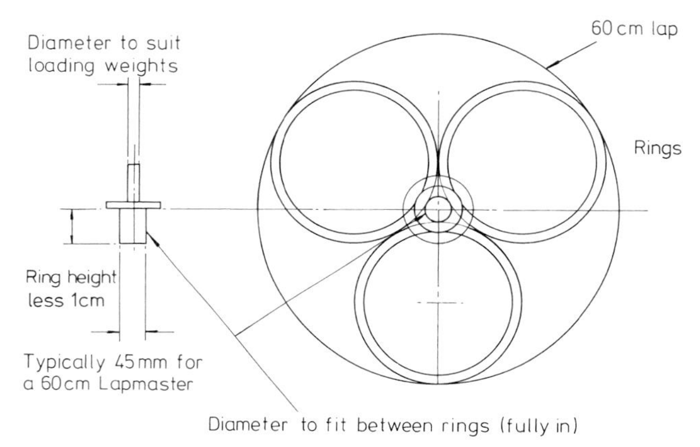

2.3 Double-ended shaft for rapid correction of convex plate.

over the diameter. The machine will condition itself satisfactorily from this state. Of course, if the shape of the newly machined plate is known its eventual flatness can be the more readily attained by corrective positioning of the rings. Rapid progress is made in this way by observing the changes in the Beilby lines and turning grooves (however faint) resulting from the lathe operation. The fact that work-damaged surfaces invariably erode much more rapidly than polished or more damage-free ones, has been noted on many occasions.

2.1.2 Flatness monitoring

It is advisable for someone to be specifically responsible for the inspection and correction of a machine by the appropriate adjustment of ring positions. A small stainless steel plate, of 12·5 cm diameter for the 60 cm machine, is kept permanently running on the machine and taken off at intervals for its flatness to be checked by interference techniques. A great deal of experience accumulates in this process and it becomes almost an intuitive forestalling of future change by observing recent trends of the test plate. On the larger machines, when the test plate does not fill the conditioning ring area, a thin Tufnol disc is used which contains an eccentric hole some 0·2 mm larger than the test plate diameter. It is as eccentric as possible without breaking through the edge of the disc. The external diameter of the Tufnol is a loose fit in the conditioning ring. In operation, the ring turns but the Tufnol disc tends to resist its efforts to rotate both disc and plate. A simple counterweight can be placed at 180° to the test plate, either on top of the Tufnol or in a hole specially machined through it (figure 2.4). In the latter case it is lapped with the test plate and can be used for checking. It may be of any convenient metal such as brass or steel.

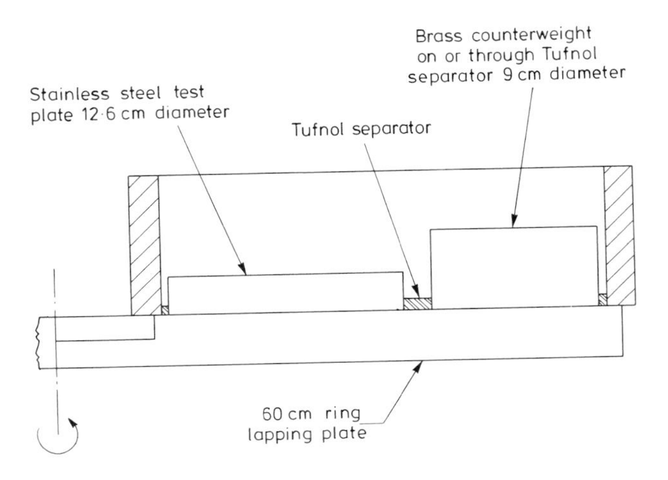

2.4 Test plate and counterweight arrangement.

A stringent test of the flatness of the machine, and hence the work it will do, is carried out by running two of the arrangements described. The test plates, preferably cast iron, are allowed to be lapped until even grey surfaces are produced on the whole face of each. They are cleaned with a solvent such as acetone and burnished together (figure 2.5). If the machine is very flat the plates take an even shine (and will show straight fringes). If the plates burnish with two bright areas in the centre, then the lapping plate of the machine is concave and the rings have to be moved out to correct this. Brightening at the peripheries indicates a convex lap and the condition is corrected as described above.

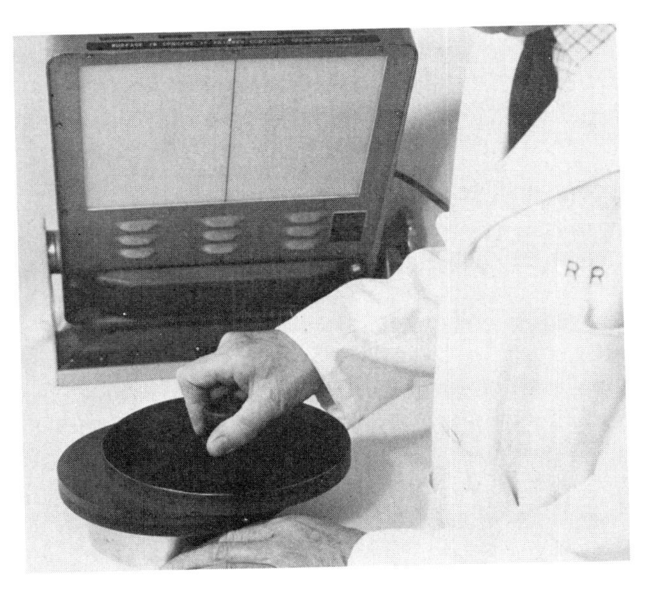

2.5 Burnishing two cast-iron laps together in order to produce specular surfaces.

The test method obviates the need for interference techniques of monitoring. It is, in effect, a three-plate system and equally as infallible as the systems used by Henry Maudslay (Gilbert, 1971), Joseph Whitworth and countless other early engineers for scraping flat cast-iron plates. Although the test is very sensitive it is obviously not quantitative, whereas systems using a reference optical flat and interference fringes are (see chapter 6)—that is, as long as the optical flat is a good one and is regularly proven with other flats. The interferoscope provides a better method because it uses a nearly parallel beam between the optical flat and the test plate (figure 2.6) and physical contact between these components is not, of course, required. The contacting system which we normally use because it is quickly carried out is typified by the Lapmaster monochromatic light arrangement (figure 2.7). It is advisable to separate the job and the flat with a piece of optical tissue large enough to extend beyond their edges and then withdraw it with great care while taking pains to keep the relative motions of the components to a minimum (figure 2.8). If the last few millimetres of tissue tear off and remain in place, it is obvious and can be carefully removed. The flat has to be viewed at as near normal incidence as possible, otherwise very flattering indications are obtained (Johnston, 1968).

For example, viewing the optical flat at 60° to the normal gives a figure half that of the correct one: two

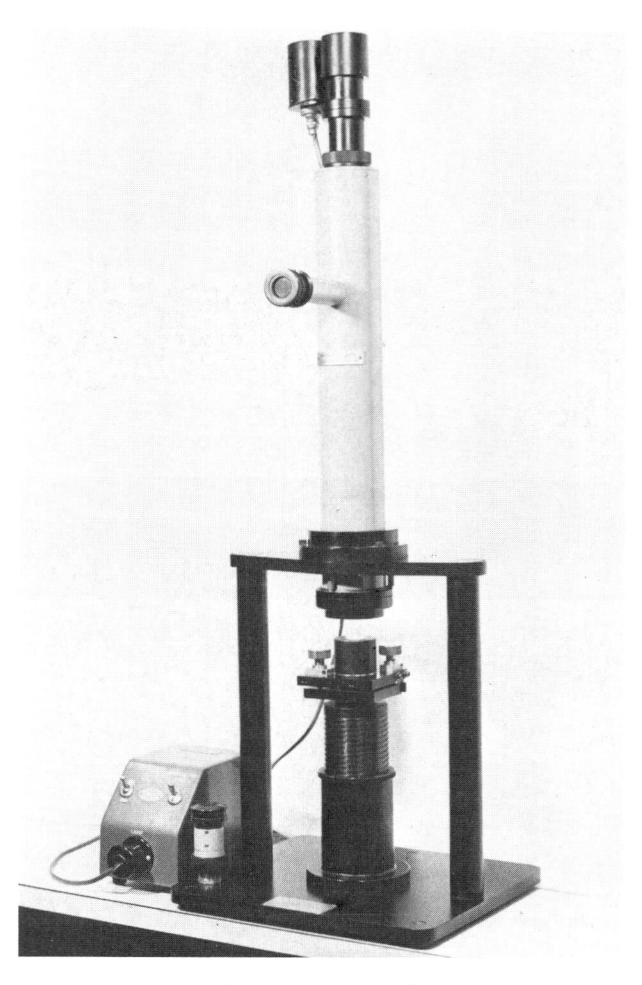

2.6 Interferoscope for proving optical flats.

2.8 Tissue removal from between specular surface and optical flat.

bands, say, instead of four. Chapter 6 deals in more detail with these pitfalls and with the interpretation of fringes generally.

The lap form can be monitored directly by placing a straight edge across it. This is the technique intended for Lapmaster owners, since a high grade straight edge was supplied with the two early machines in our possession. The method may look a little crude—in terms of optical tolerances, at least—but the test is quite sensitive when carried out with care. It is particularly useful on the 24 in (60 cm) machine where a central draw bolt is provided as a means of flexurally controlling the lap shape. One ring is removed and a diametral stripe cleaned across the lap. The straight edge is applied to the clean band, lightly held by one end and given a 1 cm movement tangentially (figure 2.9). The pivot point is ascertained from this motion; if it appears to pivot at its end, the lap is concave; if it pivots centrally then the machine is convex. Of course, the reliability of the test depends upon the quality of the straight edge and we can only quote from experience. We find them extremely good and have measured (and verified) departures from plane as small as 40×10^{-6} in in 24 in (1 μm in 60 cm)—that is: one band in 6 in.

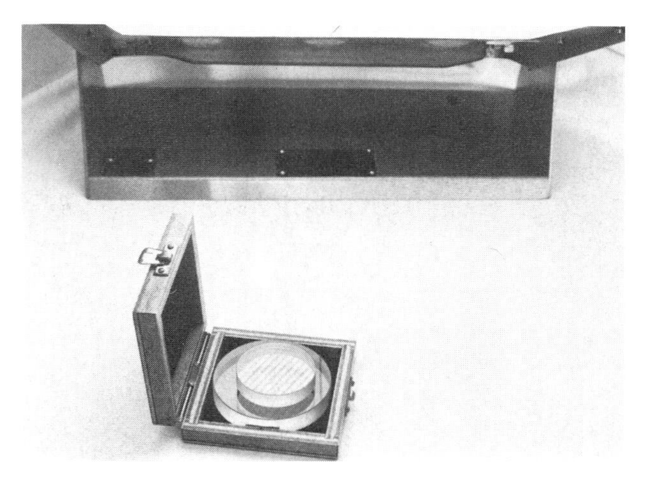

2.7 Monochromatic (helium) light source.

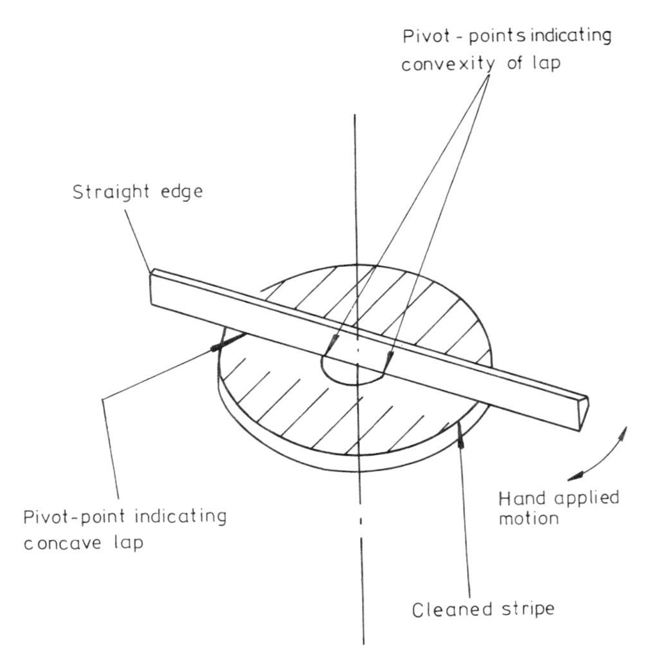

2.9 Straight-edge behaviour for various plate profiles.

2.1.3 Ring-lapping techniques

There are a few features of the lapping process which seem to result in severe turndown at the edge of the workpiece and they are not always mentioned in the manufacturers' handbooks. The commonest fault is to put small work on a large machine and allow it to run against the internal diameter of the ring, thereby causing it to precess at high speed. The remedy is to use a Tufnol or plastic disc with a hole in it to suit the work and an outside diameter which presents a slack fit in the ring, (the same arrangement as described for the test plate.) The orbiting which results sweeps the entire lap surface (within the ring) and matches the workpiece revolutions to those of the ring. The next most common fault has been termed 'tracking'. It is caused by a centrally disposed workpiece being loaded more heavily than the load applied by the conditioning ring itself, since this ceases to decide the figure of the lap when its abrasive pressure is not dominant. The annular recess which is consequently worn in the plate leads to turndown (figure 2.10). The cure is either to reduce the load on the workpiece or increase the load on the ring if the former is impossible Some measure of correction can be obtained by orbiting a small surface area, heavy workpiece on a large machine, but the practice is fundamentally unsound and must be carried out warily.

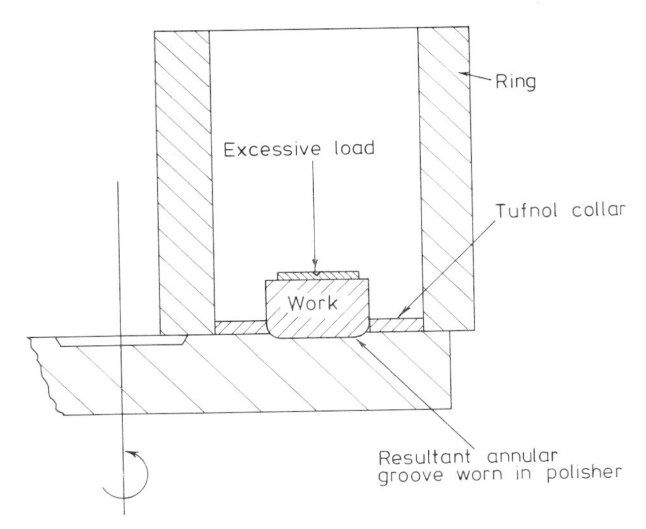

2.10 Tracking and turndown produced on a workpiece.

Another curious fault, for which the explanation remains at present only speculative, concerns abrasives. We find, for instance, that 600 carborundum cuts glass quite well using the standard lapping vehicle (Paynes No. 3 oil) but that a similar particle size of alumina does not remove glass. Yet the two worked by hand on cast iron or glass, with water as the lubricant, erode glass normally. Load seems to be important for glass lapping on ring machines. Below a certain pressure the work acquires a partial glaze and little further stock removal takes place. It is advisable to use loads of 100 g cm^{-2} or greater. There are notable exceptions to the general rule of 100 g cm^{-2}—mainly the very soft crystals, many of which work well on small or large ring-lapping machines when a suitable loading is used. (See chapter 7, tables 7.1 and 7.2: Summary of lapping and polishing data.)

Large cast-iron plates, brought to the machine for flattening from a lathe-turned state which has resulted in a concave surface, have been known to exert sufficient drag to stop the machine. Surface grinding can be a remedy but there are pitfalls since it seems to be a feature of many surface grinders that they produce two low regions where the abrasive wheel is partly off the diameter of the plate. Lapping the plate down to these low areas takes a considerable time and can be avoided by arranging packing which effectively extends its area (figure 2.11). High-grade surface grinding makes the best start from which plates may be lapped optically flat. As little as 15 min running on the 60 cm Lapmaster can be needed to bring a 20 cm cast-iron plate flat to within one or two bands ($\frac{1}{40}$ to $\frac{1}{20}\lambda_{(He)}$ cm^{-1}).

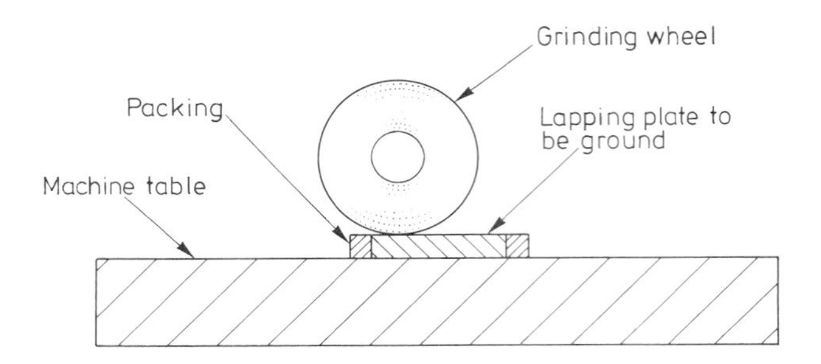

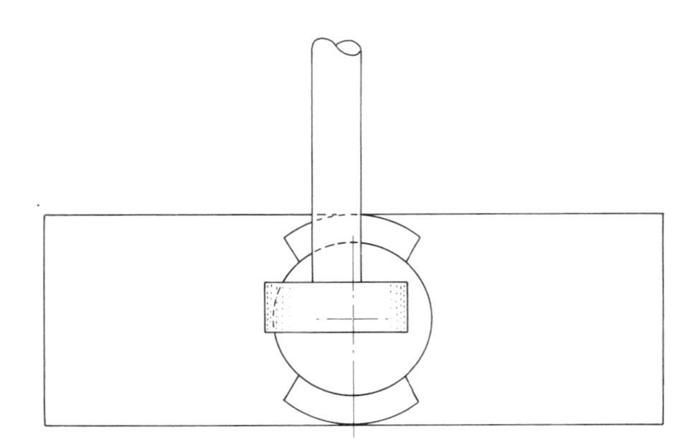

2.11 Extending effective area of laps for efficient surface grinding.

2.12 Ring-lapper loaded with low aspect work and a tall component.

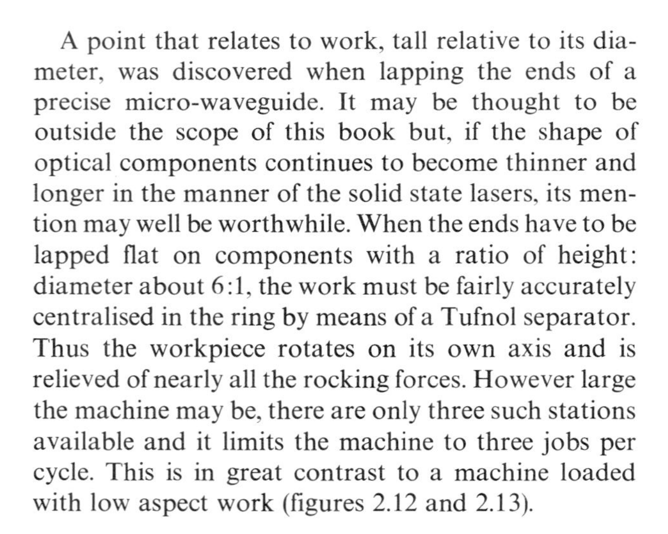

A point that relates to work, tall relative to its diameter, was discovered when lapping the ends of a precise micro-waveguide. It may be thought to be outside the scope of this book but, if the shape of optical components continues to become thinner and longer in the manner of the solid state lasers, its mention may well be worthwhile. When the ends have to be lapped flat on components with a ratio of height: diameter about 6:1, the work must be fairly accurately centralised in the ring by means of a Tufnol separator. Thus the workpiece rotates on its own axis and is relieved of nearly all the rocking forces. However large the machine may be, there are only three such stations available and it limits the machine to three jobs per cycle. This is in great contrast to a machine loaded with low aspect work (figures 2.12 and 2.13).

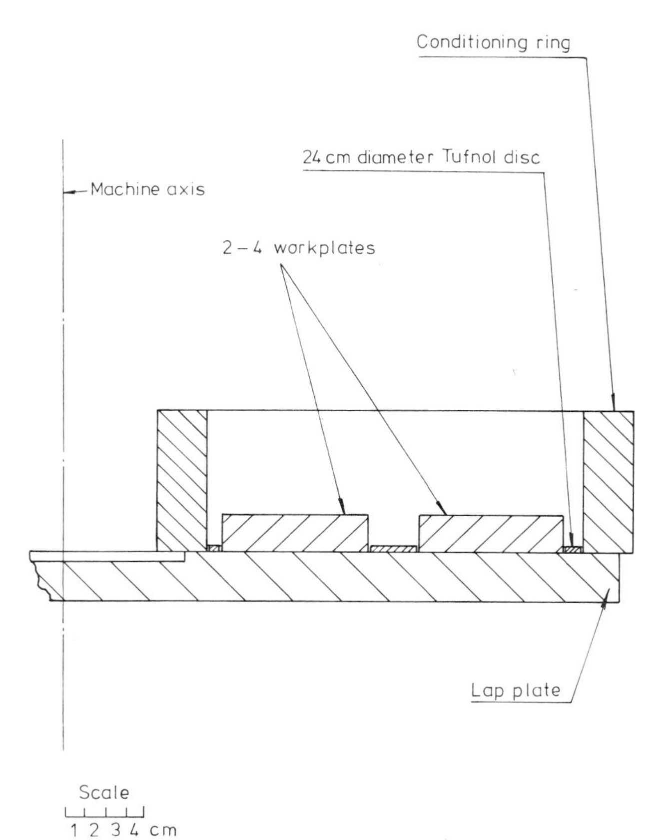

2.13 Single ring loaded with low aspect work.

2.1.4 Lapping plane parallel

A standard recommendation for the control of work-parallelism is that it should be done by offsetting pure weight loading above the workpiece. Although we make use of the technique it is less predictable than that which is called push-rod working. The well defined, and easily adjusted, force applied by a push rod is responsible for the superiority of the method. A typical arrangement is shown in figure 2.14. In essence, the work (whatever its height) is constrained by a loose fitting Tufnol ring within the conditioning ring. The method of application of pressure via the push rod is important in all but the thickest work (~ 1 cm thick, 6 cm diameter). It involves a velvet-lined Tufnol cap, fitting the work on one side and having suitable eccentric and concentric conical depressions, as seatings for the rod, on the other. When the velvet is either stuck to or laid on the well machined interior of the cap, it forms an excellent means of load application. The vast number of fibre ends, all capable of being kinked under pressure, act as efficient load equalisers. Many other materials have been tried but few seem to carry the basic accuracy of thickness of woven fabrics.

2.14 Push-rod arrangement for parallelism correction during lapping.

Foam rubbers are generally avoided because they soon start swelling at the edges due to attack from the lapping vehicle.

The progress of parallelism correction is monitored by the electromechanical methods described in chapter 6.

Although adjustable-angle jigs were developed mainly for polished work, their use for small numbers of precise components has much in its favour. For parallelism to better than one minute of arc, the procedure is the same as that for polished work: the first face has to have a slightly specular finish so that it can be aligned by optical means (see chapter 6). When a bright-line auto-collimator is used, the quality of polish on the work need be only just distinguishable from a faint grey surface. This type of auto-collimator is preferred because a special graticule is fitted to it which produces a very bright line from an effectively small surface.

If the parallelism required need not be better than one minute of arc, it can be achieved by lapping one surface of the specimen plane—carefully measuring its thickness at marked positions. After it has been mounted in the jig, a DTI can be used to establish the necessary corrective tilt of the work relative to the jig's conditioning ring. The process is made easier by placing the jig on an adjustable-angle mount in order to set its ring parallel to the reference surface (figure 2.15). It follows, of course, that in using jigs for lapped work the faces produced can have any angle up to the limit of the jig (normally $\pm 3°$).

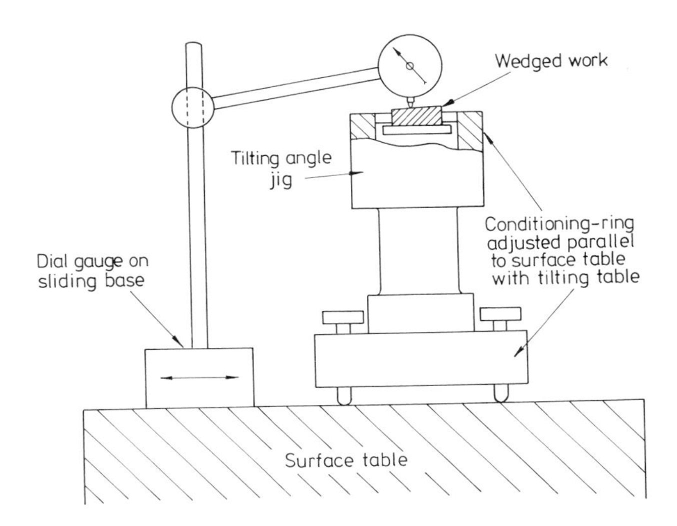

2.15 Dial test indicator measurement of specimen on jig workplate: achieving parallelism.

2.1.5 Lapping spherical surfaces

The fact that several 30 cm ring lapping machines intended for flat plate work have been brought to our notice as faulty and have been found to be some 0·25 mm lower at their peripheries at first seemed to indicate an obvious way of producing long radius work. However, this concavity equals that produced by a 45 m radius, since, from figure 2.16, $R^2 = (R - \Delta R)^2 + x^2$. Then

$$2\Delta R = (\Delta R)^2 + x^2,$$

where $(\Delta R)^2$ is, in practice, very small compared with R and the equation reduces to

$$R = x^2/2\Delta.$$

In our example, $R = (0{\cdot}15)^2/5 \times 10^{-4}\ \text{m} = 45\ \text{m}$.

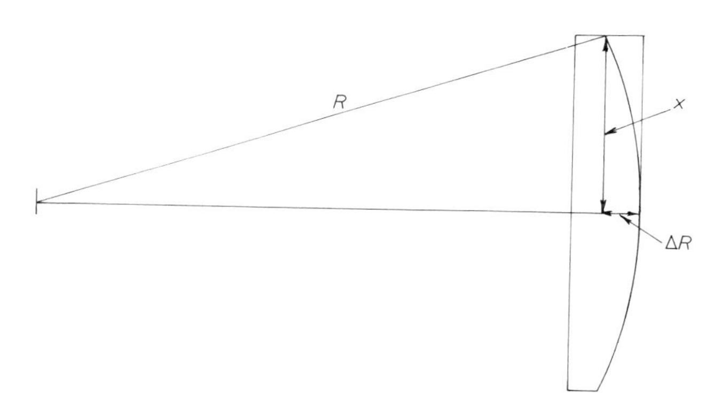

2.16 Calculation of radii for small curvatures.

In order to produce the 8 m radius mirrors which are often called for in laser optics, a 30 cm machine would require running out by 1·4 mm. Although this is within the bounds of reasonable abuse of a machine if it were intended for production or semi-production of one radius of mirror only, our very varied programme could not justify either the time required to effect the required wear or that needed to correct it. Should the need arise, the speediest way of acquiring medium radius concavity or convexity would be to machine the plate as described in §3.1.2. From the formula $\sin \alpha = D/2R$ (where D is the diameter of the cutter, R is the radius of the lap and α the angle at which the cutter axis is tilted) an 8 m radius would be produced with a cutter 45 cm diameter tilted by about 1° 40′.

The polishing machines described in §2.2 can be used equally well for lapping surfaces spherical. Specially machined plates and rings are used for medium and short radius work—say 8 m or smaller. These are made from cast iron and are described by one of the techniques included in §3.1. In line with the usual ring lapping practice, a central recess or flat is required and again the ring diameter is half, or a little less than half, that of the lap. It carried a radius equal to the plate's radius but opposite in sign. The roller bar and sweep arrangement for working shallow surfaces (figure 2.17) are similar to those described for flat plate polishing, but on deeper forms a degree of tilt to the roller bar is needed, (figure 2.18). It is easily obtained on the general purpose polishing machine (GPPM) but special arrangements would have to be made on any type of less flexible machine. For example, the Multipol requires a small modification to give it this facility (cf figure 2.48). Tools made for use on Draper-type machines (or for hand working) are

2.17 Roller-bar arrangement for lapping shallow spherical surfaces.

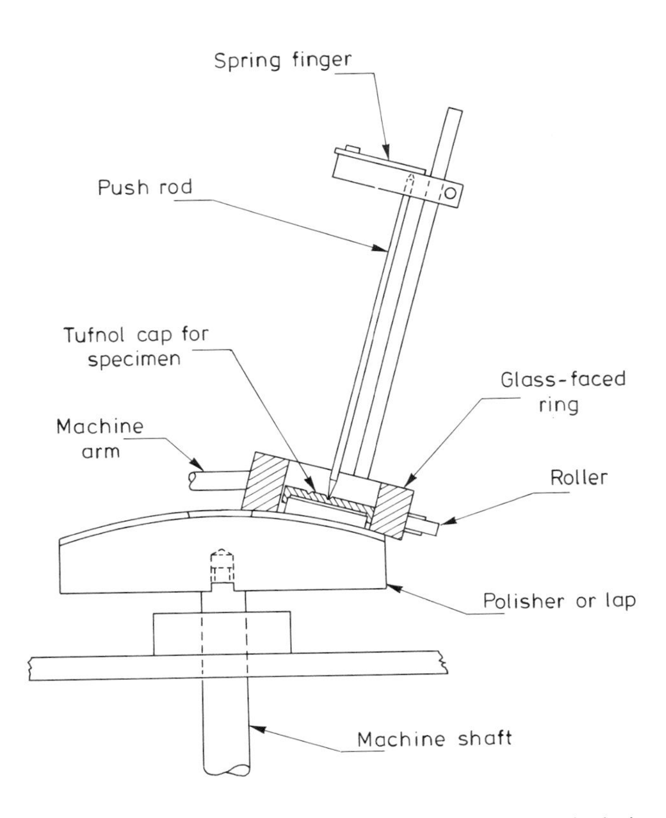

2.18 Roller-bar arrangement, tilted, for lapping deeper spherical surfaces.

2.19 Positive drive of a roller from crank mechanism.

improved by being grooved to give either conventional squares of say 6 mm side or a spiral. They then need no central recess and the surface interruption is carried to the lap or polisher centre.

In lapping (and particularly in polishing) spherical surfaces by ring methods, a general reluctance of the work to rotate smoothly is often noticed. On the older type of machine, typified by the GPPM, it is possible to arrange semi-positive drive to the inside roller of the roller-bar by connecting the stroke adjustment control to it with a light rubber driving band (figure 2.19). Although the stroke control itself orbits a little eccentrically, it ensures that the ring rotates continuously and in one direction. Such a simple arrangement is not possible with many more streamlined machines because they either have totally enclosed stroke mechanisms or no stroke at all. One remedy is to provide a separate, variable speed gearbox–motor unit to drive the inner roller. Another successful arrangement uses a system of 'O' ring edged friction drives that transfer the rotation of the lapping plate to the outer guide roller via a driving ring spaced from the plate-edge (Kasai *et al*, 1973).

Machined rings and laps are run-in with 600 carborundum and a suitable lubricant. Tests of the concave member give an indication of the radius and of the accuracy of its spherical form. A reasonably specular surface is needed on the block or ring. To produce burnishing which gives adequate reflection, the two mating members must be cleaned of all abrasive by brushing with detergent and room temperature water and the surfaces dried—again, at room temperature. They are worked together without lubricant or abrasive until brightened contact areas are seen. Surfaces obtained in this way are adequate for the operation of a knife-edge test (Strong, 1938; see chapter 6). A rougher estimate of the radius may be made by observing the image of an electric torch on the burnished lap surface. The distance at which the filament appears to be spread over the majority of the reflector gives the approximate radius. This assumes that the eye and the filament are about equidistant from the surface undergoing test close to the expected radius of the lap. Transverse movements of the eye and lamp within the radius of the mirror produce sympathetic images and movements of the eye and lamp outside the radius produce oppositely moving images. The uniformity of brightening acquired during the burnishing process gives a good idea of the accuracy of sphericity. In our experience it is certainly not worth testing by optical means until a fairly uniform reflectivity is obtained.

2.2 Polishing Machines

An initial appraisal of mechanical polishing suggested that the machine should reproduce the random motions sought after in hand processes as closely as possible. Several combinations of relative motions of polisher and work were implied:

(i) rotation of either or both,
(ii) sweep of either or both,
(iii) reversible positions (i.e. work uppermost or vice versa).

All the options tend to one end: that the motions should produce, as nearly as possible, uniform coverage and thus wear of the entire polisher surface. Where the area of a plane specimen is about 7:8 relative to that of the polisher, the departure from uniform wear is very long-term. As the area of the specimen and packing (wasters) becomes very small—for example, in a single laser rod—the effect of patterning increasingly influences polisher wear and thus specimen form. Figure 2.20 shows the different pressure areas produced when a point in lieu of a specimen is traversed across a rotating polisher. The traverse and polisher rotations are in the ratio of 5:8.

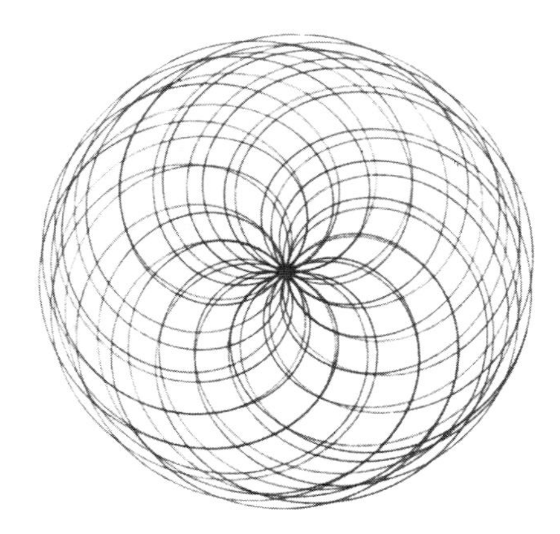

2.20 Pressure area produced by a single point on a polisher for typical stroke/rotation rates.

The imitation of hand processing gives success when the work area is much the same as that of the polisher, and for a long period in our early work on laser crystals this tenet was closely observed by making small polishers for small rods. However, when the philosophy of hand mimicry is examined with the fundamental consideration of the polisher form in mind, at its simplest and most ergonomic it calls for the dominant effect of a conditioning ring. Since the polisher, like a lap, inevitably undergoes wear, it is an advantage to have this occur on the operator's terms with a large area device loaded to override the lesser effect of the specimen. Underhill and Twyman used a ring in 1937, and though it became an important feature of their flat plate polishing machine, it apparently still perpetuated the influence of mechanised hand motions.

The principal modes of mechanical polishing may be arranged in increasing order of complexity and versatility as:

(1) non-sweep, dead weight—where the work does not change its position in operation relative to the circumference of the rotating polisher, but can gyrate if free to do so.
(2) non-sweep axially or bias loaded—as (1) but with additional load applied either with weights or with push rods and springs; the work can rotate.

The following methods are fundamentally Rosse–Draper modes and they utilise sweep and rotational freedom of the work. The poker is probably attributable to Draper (1904) and the guide ring to Rosse (1840). It is the opinion of the authors that the latter is preferable for high accuracy work since the positional control of the work is closer to the line of the polishing plane.

(3) Swept dead weight, where the work moves across the rotating polisher in an arc to a greater or lesser degree. This may be achieved by moving the work with a poker attachment applying no load or with a guide ring, which again leaves the work free to gyrate.
(4) Swept, axially or bias loaded—as (3) but with a loaded poker, or with compression applied via a leaf-spring/push-rod arrangement to the work moved by a guide ring. It must be emphasised that the poker can provide both sweep and loading, whereas the push rod provides load only and is superior in bias polishing arrangements since its remote springing allows for free rotation of the work with minimal tipping forces. The technique pre-dates machinery since Huygens (Smith, 1738) made use of a wooden pole bent between the centre of the work and either the room's ceiling or a 'strong iron spring'.
(5) The final complexity is paradoxically the simplest in operation. It is the use of a jig, unswept or swept but free to rotate, running in a roller-bar arrangement. Work may be loaded more heavily or relieved of its dead weight by a compression spring. In the latter case the wheel has turned again, in effect, to Lord Rosse's

counterweight polishing technique.

The four polishing machines that will be described are:

(1) the prototype general purpose polishing machine (GPPM),
(2) the ring chemech polishing machine (RCPM),
(3) pumped slurry pitch polisher (PSPP),
(4) the commercial version which combines most of the features of (1) and (3), the Multipol.

It has been noted in §2.1.5 that (1) and (4) can be used as precision lapping machines with no modification since they may be regarded as sophisticated single-stage ring lappers.

2.2.1 General purpose polishing machine

The prototype experimental lapping and polishing machine, later the general purpose polishing machine (GPPM, for the sake of easy reference) was developed when the need for a rapid turnover of precise laser rods became critical in the early 1960s. Since we had neither manual nor mechanical experience in optical polishing of either glasses or crystal materials, we decided to adopt the broad techniques described by Bond of the Bell Telephone Laboratories (Bond, 1962). His machine was basically of the Draper (1904) type (a design in its turn based upon one by Lord Rosse) but had been changed at the work-holding mechanism for attaining parallelism control on blocks. (See chapter 7.) For a single rod, however, the swept block was positively prevented from rotating by the way in which it was held in the sweep mechanism. Figure 2.21 shows the drive mechanism and bearing housings of the GPPM: two bearing plates A (removed) and B house a shunt-wound DC motor C with a series rheostat D in the armature circuit providing a simple speed control. This is the simpler of the two speed controls commonly used—it provides 24 V DC (either directly available or rectified AC) across the field of the motor and varies the armature current via the 12 Ω rheostat. Such a rudimentary, purely resistive, arrangement lacks torque at the lower end of the speed range and is inconvenient for pitch and cerium oxide polishing operations where friction is extremely high. The better (but dearer) system is the one described for the lap and polisher surfacing and scrolling machine in chapter 3. Polisher shaft speeds of 10 rpm may be obtained with it and the torque remains high. Loosely tensioned vee-belts are used to transmit drive from the motor to the countershaft E and on to the crank-arm shaft F. At this stage the reduction totals 16:1. A third vee-belt is used

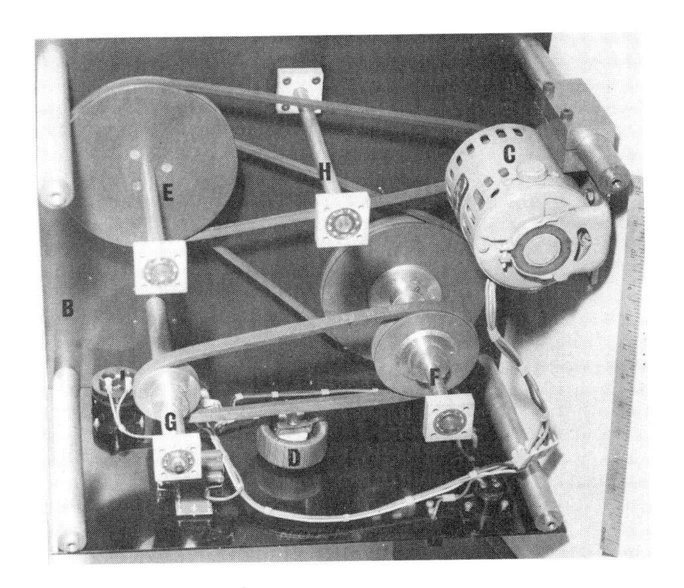

2.21 Drive mechanism on the GPPM.

between F and a lap/polisher shaft G and provides a ratio of approximately 5:8. The sweep-arm shaft H is at the rear of the machine.

Figure 2.22 shows the polishing mechanism; an adjustable crank J provides control of the stroke (i.e. the extent of the sweep-motion) of the workarm K.

2.22 Polishing mechanism on the GPPM.

The load at the polisher can be varied by sliding the weight R on the parallel arm S which is fixed at its rear end to the workarm. A polisher T is screwed to the top of the lap/polisher shaft G.

It is a feature of this machine, not yet used in practice, that the dead weight of polisher and specimen may be off-loaded by suitable positioning of the counterweight and thus it can provide all of the features enumerated in the opening paragraphs of this section.

The crank J is shown in greater detail in figure 2.23. The crank pin is adjustable in eccentricity and has a sliding block coupled to it which translates the orbital pin motion into a nearly linear sweep (arcuate in fact). The mechanism has been in constant use on a group of machines for some fifteen years and shows little sign of wear. The vee-belt driven crankshaft (1) is located with set-bolts (2) and a slotted crank plate (3). A tee-bolt (4) is free to slide along the slot and can be locked in a selected position by the control knob (9), applying its force by way of two collars (5) and (8) and the ball race (6). This provides rotational freedom to the slider block (7) at the same time as its transmits any eccentricity of the tee-bolt to the slider shaft running in the shaft bore (10). A grease-nipple is fitted approximately at the centre of the slider-shaft bore. Grease forced in at this point not only lubricates but serves to eject foreign matter, such as polishing abrasives, from within the bearing. A similar situation is found at the top ball race of the main lap/polisher shaft: located as close as it is to the underside of the tool, some contamination is inevitable and a grease nipple is fitted primarily to aid the expulsion of detritus.

The GPPM can be operated in all the main modes with their various sub-modes. At first it was utilised as a Draper machine. Thus, with an approximate diameter of the lap or polisher of eight units and for the work of seven units, a stroke one-fifth of polisher diameter effectively sweeps across and slightly beyond it at each extreme. The extent of stroke has a controlling effect on the polisher wear and consequently its shape. Later, the poker was exchanged for a Rosse-type guide-ring and the load applied with a loaded push rod or additional stuck-on weights for some shallow work. Although these modes are still in use, it is the commoner practice now to replace the guide-ring with a roller bar and convert the GPPM into a ring-polisher. The roller bar is in effect an improved guide ring since it contains the work in the polishing plane yet allows it freedom to rotate and accepts a range of diameters. The main benefit obtained from the conversion to a single-station ring machine is that the availability of sweep is particularly valuable with spiral grooved polishers and soft materials. It prolongs the polisher form and minimises the channelling effects noted in §2.1.

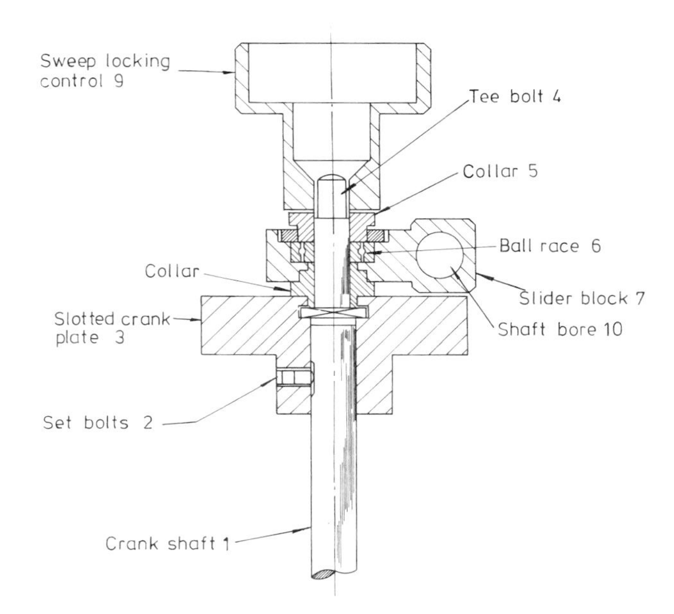

2.23 Crank mechanism: GPPM.

The sweep is controlled via the stroke control knob; the load by the counterweight in all instances where jigs (§2.3) are not used. There are two methods of applying the load of the counterweight to the work: by poker (figure 2.24) and by push rod with the work swept by a guide ring (figure 2.25). In its turn the push rod can be either solidly loaded by applied weights as in figure 2.26, or it can be spring loaded (figure 2.27).

2.24 Polishing mode, poker type.

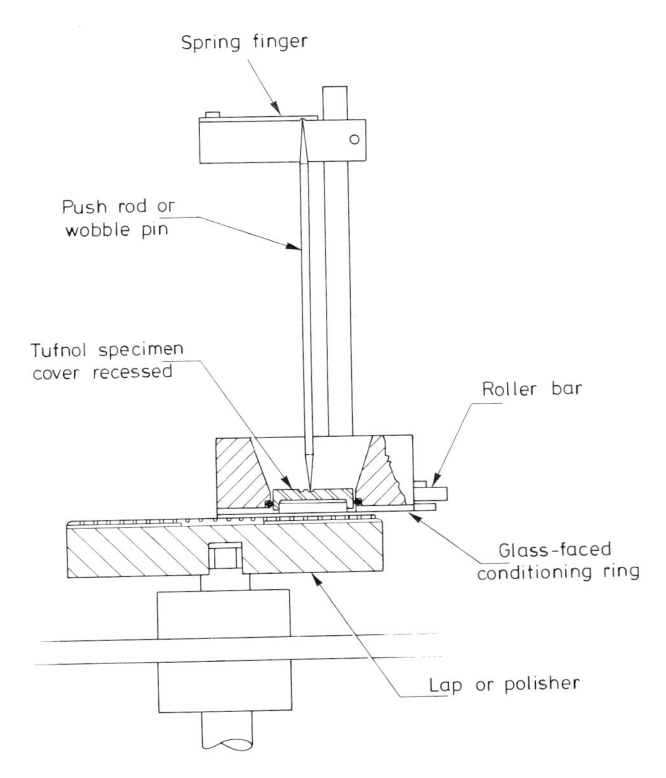

2.25 Push rod applied to a specimen running inside a guide ring.

2.27 Plain spring-loaded push-rod.

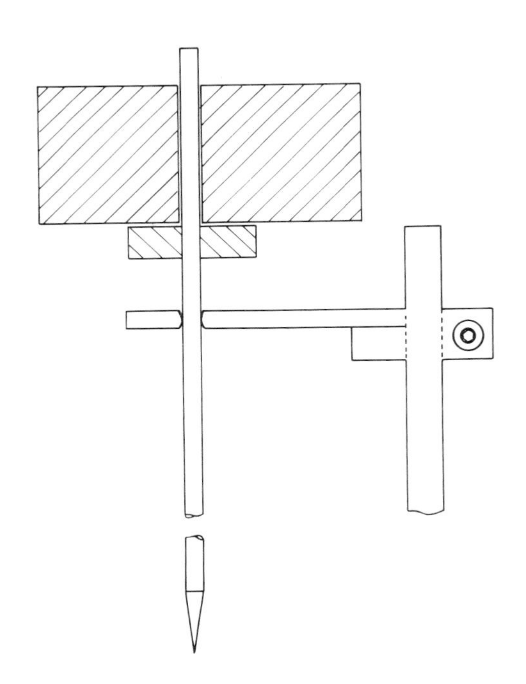

2.26 Weighted push-rod arrangement.

Some load must always be applied to a stiff rod in bias polishing, since if it were merely positioned between work and pressure point without any compressive force it would easily fall out in operation—as it has been known to do!

The spring loading at the pillar top may be achieved by a leaf spring, as shown in figure 2.27 and, in this case the polishing pressure is measured by a spring balance hooked under the spring (figure 2.28). Adjustments are made until the push rod just starts to fall away at the calculated loading. An easier method of fine adjustment is provided by a micrometer-spring abutment (figure 2.29) but a spring balance is still used in checking the load. All this assumes either that the dead weight of the work is insufficient to apply

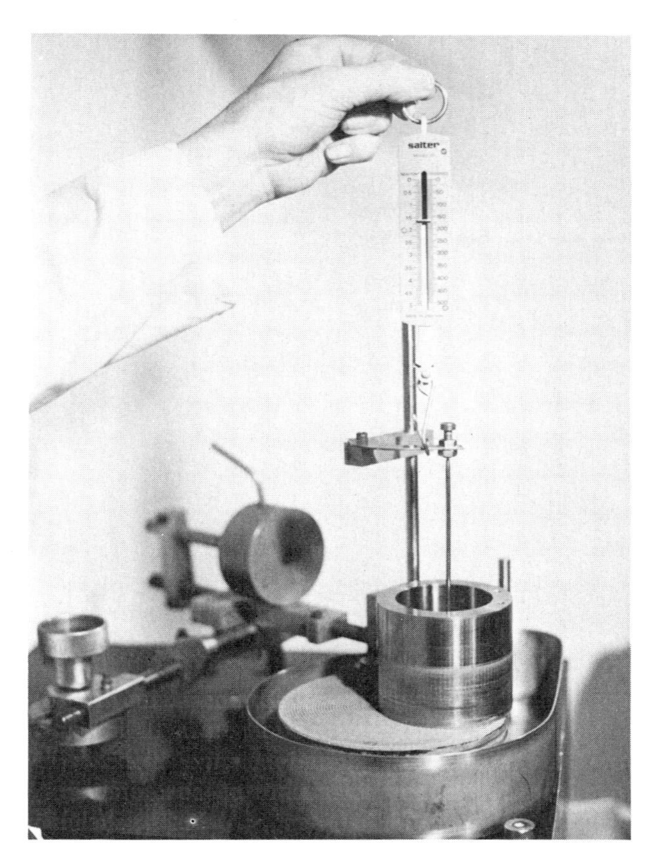

2.28 Spring balance assessment of polishing load.

adequate polishing load or that bias polishing (for parallelism correction) is required. Figure 2.30 shows a method of providing location at variably offset positions on the work for the latter purpose. Of course, a solid plate carrying thin specimens can have its own dimples, but the attachable caps have the advantage of being applicable to low or high work, be it robust or fragile. In some cases where numbers of, say, windows are polished simultaneously, additional load can be applied by sticking-on the necessary weight with double-sided pressure-sensitive adhesive tape. The whole assembly then runs within the guide ring or roller bar.

The guide ring shown in figure 2.26 has a nylon-lined inner surface which buffers it from the sharp change in contact with the work at the end of each sweep. There is, in practice, seldom any constant gyratory motion of the work within a guide ring but it undergoes intermittent rotation at each reversal of direction. As the benefits of the use of conditioning rings in many operations became apparent, the guide ring was modified to carry one by reducing its wall thickness to a minimum of about 1·5 mm and providing it with a glass-faced brass ring. It could be equally well made from stainless steel for the majority of processes.

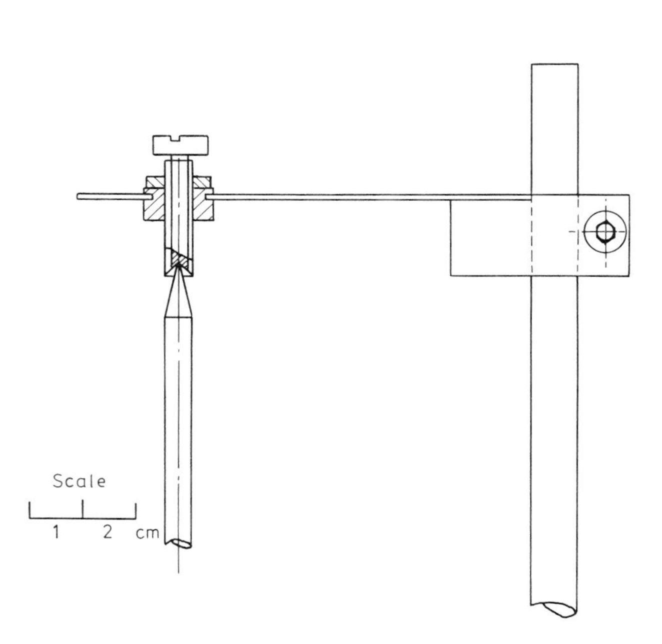

2.29 Screw-adjustment of load via push rod.

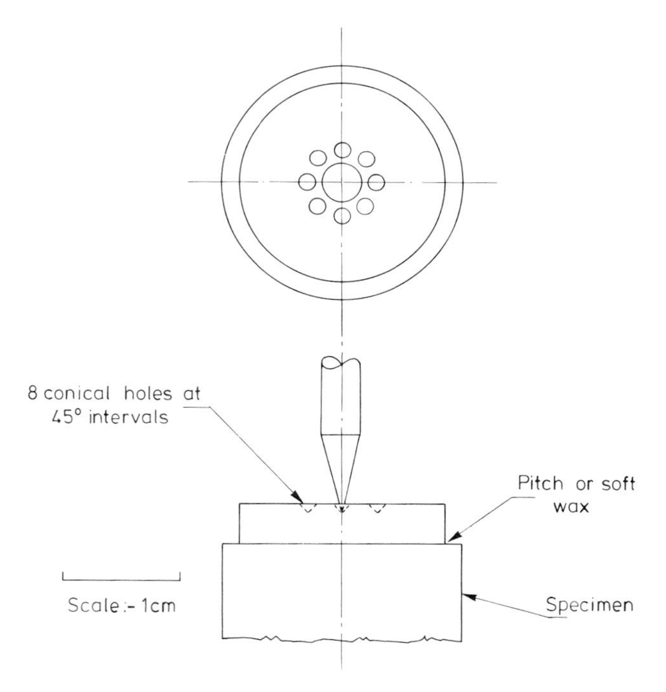

2.30 Wax-on dimple caps for parallelism correction.

The conditioning ring is fitted with an 'O' ring recessed into its internal diameter so that it can rotate freely on assembly around the guide ring. Figure 2.31 shows a typical arrangement. It is a concentric conditioning-ring method since the sweep is about the centre of the polisher. This system gives very long runs without recourse to independent polisher flattening, providing that the loading on work is within a few grammes of that of the conditioning ring. A consideration of the technique shows that it is not the optimum position for a ring, since it does not provide the overall surface control needed. When the surface of the work is less than that of the ring (the normal condition sought after in eccentric ring polishing) the work figure goes concave; conversely, when the ring pressure does not dominate, the work figure becomes convex.

Although the concentric method can be used successfully to give the desired flatness control, the eccentric ring arrangement is superior in practice. As in lapping, §2.1, it is tailored to provide one dominating load, that of the ring, and the need to equalise pressure on ring and work is avoided. Figure 2.32 illustrates one single-stage arrangement on the GPPM. It is here that the roller bar becomes important; it allows free access of a range of work diameters—plate, blocks and jigs—and permits the work to precess. It should be set in position with the lowest point nearly touching the polisher surface, unless the geometry of work or jig precludes this. It is a rule, never willingly contravened, that any movement at an angle to the polisher axis must be actuated as close to the polishing plane as can be arranged. Many of the mixed mechanical and chemical–mechanical operations performed on the GPPM have a deleterious effect on a bearing surface, and the rollers especially are prone to be affected. The best simple design so far has a PTFE (polytetrafluoroethylene) or SRBF (synthetic resin bonded fabric) roller running on a stainless steel stud (figure 2.33). An 'O' ring is placed in a groove round each roller exterior to minimise slip. The rollers have to be cleaned and oiled fairly frequently and when this is done conscientiously they are long-lived.

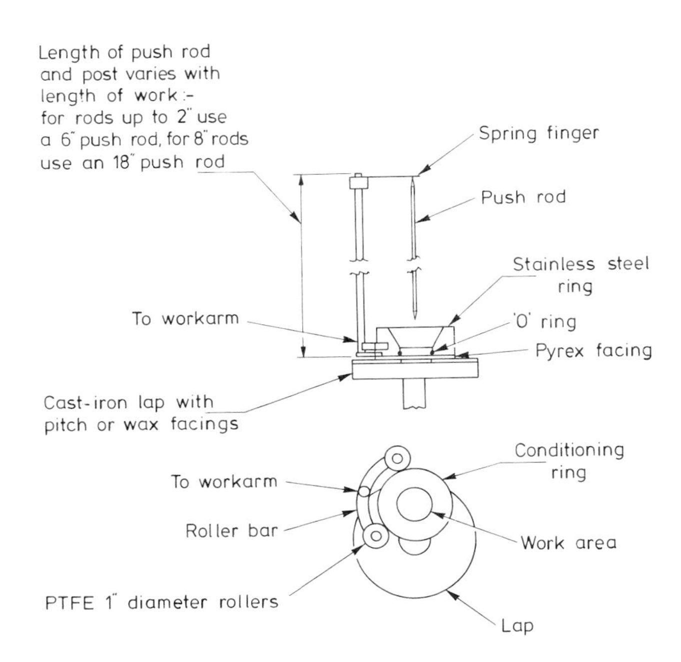

2.32 Eccentric conditioning-ring arrangement.

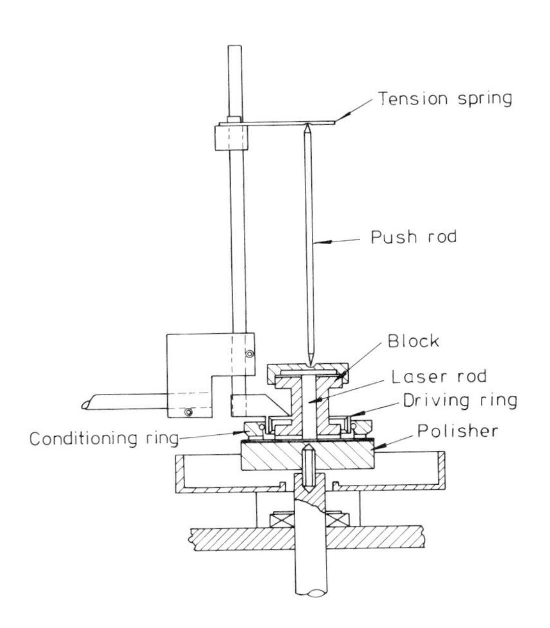

2.31 Concentric conditioning-ring arrangement.

The variety of specimens (with different areas, erosion rates, polisher matrices and abrasive slurries) which run in the roller bars, result in very different rotational tendencies. Some work turns freely whilst some stays almost stationary. The positive drive methods described in §2.1.4 may be used to correct this, but in doing so it must be noted that a diagnostic aid is lost. The behaviour of a job free to rotate within the conditioning ring and roller bar (or of a jig complete with its own conditioning ring) can often indicate polisher shortcomings. For example, convex work on a flat polisher is slow to rotate and, as its figure improves, its speed increases. The corollary is that known flat work—needing repolishing for example—

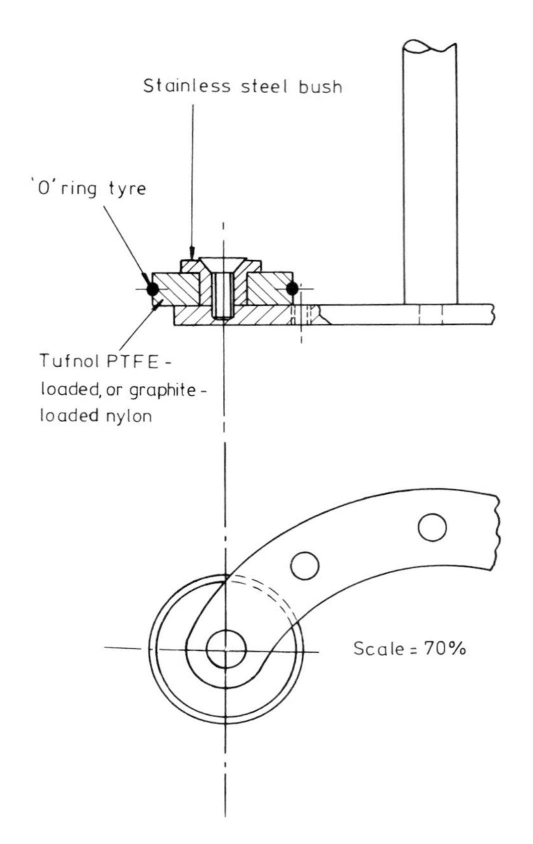

2.33 Construction of a roller.

the GPPM. Specimens polished to provide reflecting rear surfaces may be mounted in them and an auto-collimator used to establish critical alignment—indeed, much of the value of a jig is lost when optical measurement methods are unavailable.

The machine may be used for spherical work in either the poker or ring polishing modes. Figure 2.35 shows an example of the former method, with the work revolving and the polisher sweeping across it. For the latter method a range of roller-bar sizes is used. Each has a tall vertical post to carry the spring-finger. The roller bar can be inclined approximately tangentially to the polisher (see figures 2.18 and 2.36) which provides a facility not usually found on more streamlined machinery. In some spherical lapping and polishing operations where the ring mode is used, the reluctance of the work to revolve has been mentioned. Positive rotation is again easily provided by driving either of the rollers. It may be carried out with an external motor and a short round belt running in the 'O' ring groove of the bar-roller. On the GPPM it is achieved by coupling a roller by means of a lengthy PVC (polyvinylchloride) belt to the parallel section of the sweep control lock ((9) on figure 2.24)). Since both roller bar and control move sympathetically, the tension of the belt is constant.

proves a polisher to be convex if it is reluctant to rotate. Moreover, a job which tends to turn in the wrong direction (i.e. it counter-rotates) does so when the polisher has a sensibly flat surface in the main but an extended turned-down edge, amounting to a compound figure. Thus it drives the ring on its inner edge. A roller bar positioned to allow too much overhang of the work over the polisher periphery gives this effect, too.

One of the major advances in polishing surfaces plane with diameters up to 5 cm has been the adjustable-angle jigs described by Bennet and Wilson (1966). We modified their original concept by the addition of a conditioning ring instead of three feet and our prototype jig has now proliferated into those described in detail in §2.3. These devices have as integral parts of their mechanisms on and off-loading of pressure on a specimen, parallelism correction and excellent flatness control because of their conditioning rings. They are used in the eccentric mode described above: run in the roller-bar with a small degree of sweep. Figure 2.34 shows a Mk II jig (see §2.3.3) on

2.34 Polishing jig on the GPPM.

2.35 Poker mode for polishing spherical surfaces.

2.36 Tilted roller-bar polishing of spherical surfaces on the GPPM.

With the experience gained from the recirculating slurry polishing machine (§2.3.3) an additional facility became desirable on the GPPM. Figure 2.37 shows a detachable tray, complete with submersible pump pillar, which can be readily added to the machine and removed for cleaning. The small pump, described in detail in chapter 1, is powered by a 24 V DC motor and is capable of delivering up to 4 l min^{-1} of abrasive slurry. The tray can be made from 16 SWG stainless steel with its seams FDP (fusion decay proof) welded to resist oxidation and attack from etching slurries. Alternatively, it can be of welded PVC construction, but facility in fabricating plastic materials to order is less common than standard metal-working businesses. The turned-up centre section helps to protect the main-shaft bearing and to this end it matches a groove

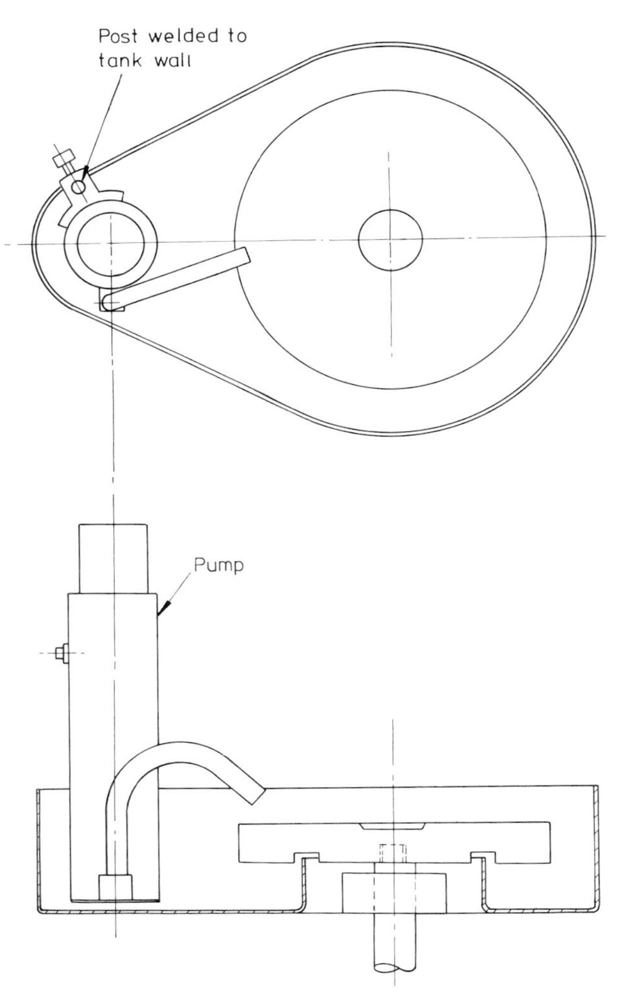

2.37 Detachable etch-resistant tray for the GPPM.

turned with a generous clearance in the underside of all flat lap and polisher bases. A feature of these trays and pumps is that they can be operated with a small amount of slurry which can be easily cleaned out and replenished. A typical arrangement on the machine with the pump in place and a 15 cm diameter polisher is shown in figure 2.38.

2.38 GPPM complete with tray, pump and 15 cm polisher.

2.2.2 The recirculating chemech polishing machine (RCPM)

Since, in the GPPM, all of the polishing modes are available, it may seem unnecessary to have designed an additional machine specifically for one operation: that of chemical–mechanical (chemech) polishing. However, it is in the quest for versatility that some of the finer guide-lines become broadened and, though the result may be an increase in usefulness, it is inevitably attended by a degrading of some function or functions. Directly, the function might be precision; indirectly it might be component wear, ease of maintenance or of cleaning-down. Concentration on a single aspect encourages the second principle of design which can be regarded (tongue in cheek) as the antithesis of built-in-obsolescence: the principle of perpetuity. (The first principle is pertinency.) It is to a great measure achieved by the use of the right materials and in making their inspection, cleaning and general maintenance as simple as possible. That probably indicates the third design principle too: popularity. The fourth principle guides us indirectly but is of paramount importance to the manufacturer—the principle of profitability.

A Cooke, Troughton and Simms machine was available for a modification and proved itself to be a valuable addition for polishing slices mounted on a 9 cm diameter block. However, its cast-brass frame was not designed to resist corrosive attack from chemech fluids and the main development in the new machine is in the change of construction materials.

The three-station polishing machine, figure 2.39, has proved very satisfactory during several years of continual use. Its three stations are arranged at 120° around a 20 cm diameter polisher B, which can be faced with a variety of matrices (the materials which hold the abrasives, chapter 3). The three roller bars C are clamped to three vertical shafts D and have adjustable locks E, which control their positions: horizontally for polisher figure control and vertically to accommodate various geometries of blocks and jigs. The roller bars are removed for polisher changing and general cleaning by the same clamping arrangement. The slurry tray F is similarly removable for cleaning when the roller bars are taken off. The entire case and tray are of welded, rigid PVC which is particularly good because of its resistance to silica sols. When splashes of these dry on, they do not adhere well and can be fairly easily flexed and washed away.

The pipe carrying the polishing fluid is kept as short as possible by using a simple centrifugal pump G made from Tufnol (see chapter 1). It delivers the slurry from the tray to the polisher through a short length of polythene tube H, at a rate of $1{\cdot}5\ \mathrm{l\ min^{-1}}$. The pump is clamped to a vertical post J which is integral with the tray wall. For safety reasons, the impeller is driven by a 24 V DC motor fed from a variable voltage DC supply.

The polisher shaft runs on ball races which are sealed at both ends and the interspace filled with grease. The upper end of the shaft is threaded 0·5 in BSW which fits existing polishers. On the lower end is a two-diameter vee pulley M (figure 2.40) which corresponds with the motor–gearbox pulley providing two speeds of 30 and 60 rpm. The motor–gearbox N is a standard 100 lb $\mathrm{in^{-1}}$ Parvalux unit with 240 V 50 Hz continuous rating. The whole unit is mounted on a pivoting frame O to facilitate speed changing and for vee-belt tensioning. Particular attention has been paid to motor ventilation to avoid heating the polisher and work area as much as possible.

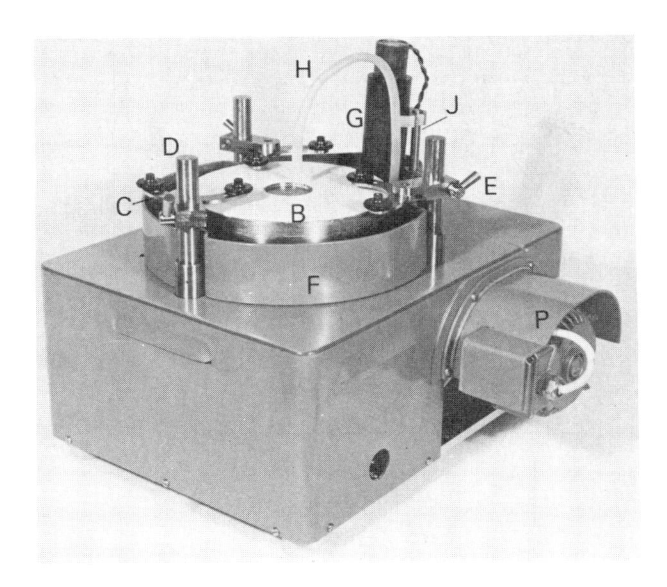

2.39 Top of RCPM: polishing mechanism.

2.40 Underside of RCPM: drive mechanism.

The motor body protrudes from the case at the rear of the machine P. In order to reduce temperature fluctuations still further, a heat exchanger can be inserted in the slurry tank or in an external storage container and water cooled as described in §2.2.3. Ease of construction has been taken into consideration in the general design of the machine. There are no castings, but plate and rod spacers are used for the main frame. The welded PVC covers and tray could be changed to Perspex or glass-reinforced plastic without detriment. Laminated wood is an excellent material for a combined frame and case when impregnated with a suitable resin (for example, an amine-hardened epoxy) or coated with a fully cross-linked, two component polyurethane varnish. It is an attractive constructional material for polishing machines in some famous optical works—reasonably so, since it has good stiffness for a given weight and a machine made from it is easily modified if requirements change. All metal parts which come into contact with the polishing slurries are made from stainless steel. The rollers on the roller bars are the same as those shown in figure 2.33.

The general rules for ring polishing are followed in practice. Most of the work done on the machines (mainly in the form of surface acoustic wave plates) is arranged annularly on heavy stainless steel blocks so that the specimens themselves function as conditioning rings. The work can approximate to a continuous surface equally well, avoiding open spaces as far as possible, particularly near the periphery of the block (figure 2.41). Some evidence of figure defects has been seen at the rotational centres of the blocks and it has become the practice now to leave that region unfilled. However, the defects that have been seen on centrally disposed work have usually been very small and are not likely to affect surface acoustic wave plates. When polishing jigs are being used on the fixed roller-bar machines, no evidence of figure defects is seen. This is possibly due to the lighter loading on the specimen, but the reason has not been positively established as yet. It is fortunate that the figure is unaffected since, for example, a laser rod may be polished on the rotational centre of a jig and ring arrangement.

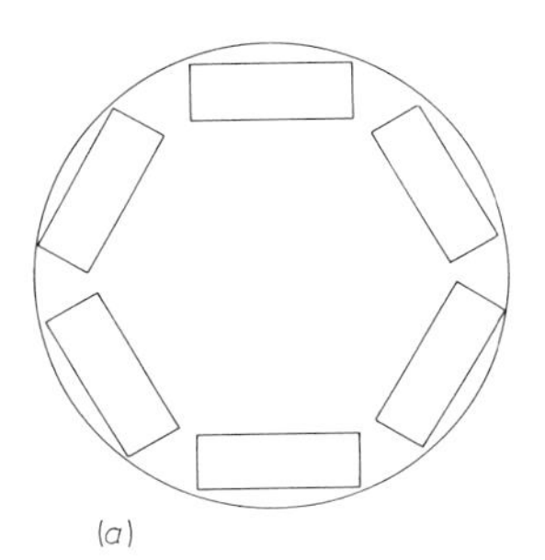

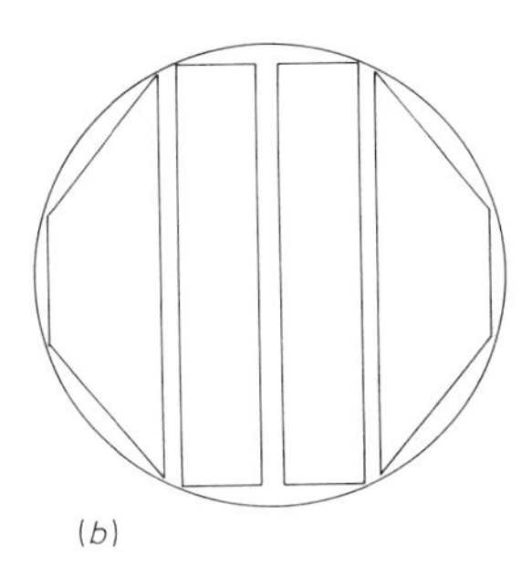

2.41 Lapping and polishing blocks; (*a*) annular, and (*b*) disc arrangement of specimens.

2.2.3 Pumped slurry pitch polisher (PSPP)

Whereas the RCPM is intended for use with a variety of chemech fluids and polisher facings, the PSPP is for use with pitch and abrasive slurries. Conditioning-ring principles are usually applied to matrices other than pitch which has been formed accurately (see chapter 3). However, there seemed to be no logical reason why it should not perform well in the ring mode, providing that the slurry feed were copious enough to ensure suspension of the abrasive and that at no time were work and ring left on a stationary machine. Failure to remove the ring at the same time as the work is taken away for testing results in the inevitable plastic deformation and the ring becomes progressively embedded in the pitch.

The PSPP is basically a ring-lapping machine converted into a pitch polisher. A 30 cm Lapmaster (of approximately 1960 manufacture) has been used for the experiment. It rotates a little too quickly for very precise polishing but it is able to cope with a large amount of medium grade work ($\lambda/6_{(He)}$ cm^{-1}) in flat glass, quartz substrate and surface acoustic wave components. The flatness limitation is not a fundamental factor but mainly due to the operating technique, as will be explained later.

It was a major operation to clean the machine for polishing purposes. It was reasoned that all traces of contamination from 600 carborundum could not be removed even with the best available cleaning and degreasing. Therefore, the parts and ways of the machine which would come in the path of the recirculating cerium oxide slurry were sprayed with PVC lacquer in the belief that any abrasive not washed away originally would be bonded in place. At this stage a rectangular rubber bung was made to fit an aperture in the machine's main bearing housing to prevent the slurry, which must circulate in a torrent, from leaking out.

A 30 cm pitch polisher was cast, as formulated by Dickinson (1968), faced and spiral grooved on a good centre-lathe. A convenient pitch is 18 threads per inch (TPI) since it allows the use of a chaser periodically to regenerate the groove in operation. It was too large in diameter to be machined on the special surfacing and grooving machine (chapter 3). A conditioning block was made from glass of 14·5 cm diameter, 2·5 cm thick and loaded with a steel weight giving approximately 50 g cm^{-2}. This is not as high as the usual loading of 100 g cm^{-2}, but such a weight was the heaviest considered reasonable for frequent manual lifting on and off the polisher. It was intended that the disc and work should be used in turn. When specimens are worked within a large ring, however, the general loading of 100 g cm^{-2} is more carefully adhered to, since the effect of the ring must dominate. The slurry recirculating system is again provided by a 24 V DC centrifugal pump (chapter 1) its intake immersed in slurry in a small brass tank placed beneath what is normally the waste abrasive outlet of the machine (see figure 2.42). Fluid is delivered to the centre of the polisher track by a suitably curved polythene pipe. The circulation rate can be set between 1 and 4 l min^{-1}, and is usually 2 l min^{-1}. Although it is much higher than the polishing requirements of the arrangement it serves to hold the abrasive in suspension and helps to equalise the temperature of the whole system.

It has already been noted in §2.2.2 that frictional and electrical heating are limiting factors with many three-station ring polishing machines. It will be clear that any polishing process which does not operate strictly at room temperature is inconvenient to use. Although working consistently at elevated temperatures should have a negligible effect on the final shape of the work, it makes testing tedious since it requires long waits in which the work and test plate reach thermal equilibrium. It is because of the unwillingness on the part of the operator to wait for temperature stabilisation that the flatness obtained is normally no better than $\lambda/6_{(He)}$ cm^{-1}. It is a question of prediction

2.42 Pumped slurry pitch polisher machine (PSPP).

rather than fact when adjusting the polishing parameters. The PSPP has been greatly improved by water cooling the slurry feed with the heat exchanger shown in figure 2.43. Tapwater is passed through the coil at a flow rate adjusted to give a slurry temperature equal to that of the room. Work has been obtained in consequence which is flat to $\lambda/60_{(\mathrm{He})}$ cm^{-1} which is an order of magnitude better than before. The operator sees a constant figure when testing and can proceed accordingly—and without delay. Many of the present-day Lapmasters are fitted with a water-cooled plate as standard. As least one record of a water-cooled pitch polishing combination should be credited to Lord Oxmantown (1828). He partly immersed his speculum mirror in a large bath of water and polished with a pitch tool uppermost.

A large number of glass and quartz specimens have been polished on the machine and it certainly has a place in the general armoury of the laboratory and small production unit. A major attraction is the speed of polishing: for instance, blocks of quartz plate ground on a 60 cm Lapmaster machine with 600 carborundum are polished in about thirty minutes to a depth well below the grey or pit level. Further improvements in surface finish can then be obtained with a few minutes' polishing on wax. It is intended that a similar machine with a large wax polisher and recirculated Linde A (0·3 μm) abrasive in water will be commissioned expressly for optimum surfaces.

2.2.4 The Multipol

Most of the principles of the GPPM have been incorporated in a commercial version: the Multipol (figure 2.44) manufactured by Malvern Instruments and marketed by Metals Research Ltd. The general appearance has been improved whilst attempting to preserve the versatility of the prototypical machine. Several sacrifices have been made but they can be overcome by small external modifications where a marginal gain may be had by their restoration. The Multipol is an enclosed machine—that is to say, its crank mechanism does not appear externally. The only control required, the sweep adjuster, is accessible by using a sliding door in the left-hand side of the machine (figure 2.45). The adjustment does not require locking as the GPPM

Polishing slury (recirculating)
Cold water input and output
Slury supply to polisher surface

2.43 Heat exchanger for ensuring cooled flow of slurry to PSPP.

2.44 Multipol.

2.45 Stroke control for Multipol.

mechanism does and there is no evidence of creep in the stroke extent during very long running periods. The general position of the workarm is set at the sweep-shaft end. An allen key is used for unlocking and adjust-it. Fine readjustment of the workarm is provided by the spring-loaded control (figure 2.46) at the rear of the arm. This gives a maximum of 4.5 cm traverse, thus more than covering the range likely to be used in ring polishing operations in order to obtain control of polisher flatness.

The drive arrangements are by a series of vee-belts. Experience with the GPPM during twelve years of continual use has proved this type of drive to be very satisfactory. The Multipol has tension adjusters (on all shafts) that can be used to regulate the belts to suit the work the machine will be expected to handle.

The motor and electrical drive arrangements are almost identical to those of the lap-polisher surfacing and scrolling machine giving a range of speeds from 5–200 rpm. High torque at the lower end of the range is valuable when finely grooved polishers and cerium oxide slurry are used to polish glass surfaces: a combination which seems to produce more drag than any other polishing process. The main shaft of the machine is threaded 0·5 in BSW to suit laps and polishers typified by the 13 cm diameter plates supplied for the Cooke, Troughton and Simms metallurgical polishing machines. These are of cast iron or bronze, relieved on the underside and light to handle; but the preferred design shown in part in figure 2.48 has perpetuated the thread size. Polishers machined from solid are stable enough for the highest accuracy work.

The stainless steel slurry tray is an integral part of the machine and thus cannot be easily removed for critical cleaning and inspection elsewhere. Even though a drain plug, shown in figure 2.47, is provided for easy removal of slurry and liquids, there is an inevitable accumulation of abrasives and debris. However, the fitting of a disposable plastic liner can remedy this.

2.46 Crank mechanism for Multipol.

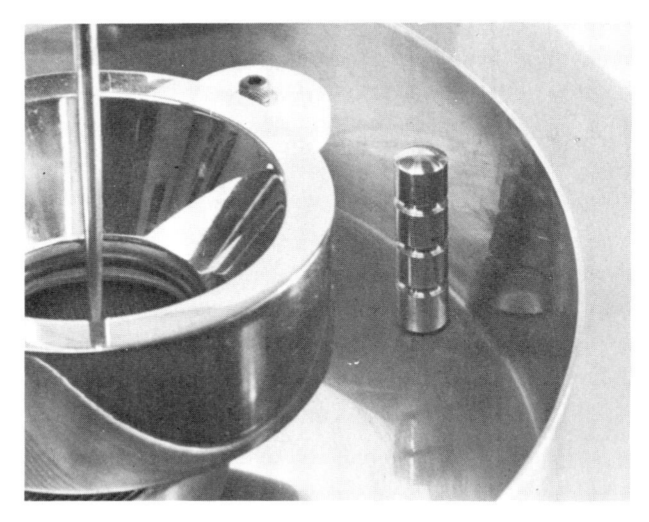

2.47 Drain plug facility for emptying the non-detachable tray on the Multipol.

The standard slurry circulating pump which is used on the saws, GPPM, RCPM and PSPP will not fit between the rim of the tray and the most commonly used 15 cm polisher bases. The generous wall thickness of the pump body makes it possible to machine two flats on it some 5 cm length while reducing the overall thickness to 3 cm. The flats are arranged to be parallel to the outlet port (figure 2.48). A leaf-spring clip, bolted to the upper part of the pump body, attaches it to the edge of the drip tray. A short length of rigid polythene tube, heated and bent to the required shape, carries the slurry to the polisher surface.

With the arrangement described it is advisable to use an abrasive suspension medium, such as the proprietary YMS (Young's miracle suspender) additive from Autoflow Engineering. Both aluminium and cerium oxides are held in suspension by this thixotropicising medium. Though desirable, it is not essential since the recirculating system can be operated by diverting a considerable proportion of the pump output to agitate the slowly moving slurry in the tray. An additional aid is provided by fastening a shallow paddle to the underside of the polisher base, extending nearly to the tray bottom. They combine to keep the slurry suitably stirred.

The main shaft is well sealed against the ingress of abrasive fluids; even so, running the system with the bearing completely submerged is potentially troublesome. A simple modification which serves as 'belt and braces' is to fit a machined nylon ring onto the bearing flange, extending up into a groove in the underside of the polishers in use (figure 2.48). The liquid level need not be as high as the upper edge of the nylon ring and, in consequence, the shaft is kept dry.

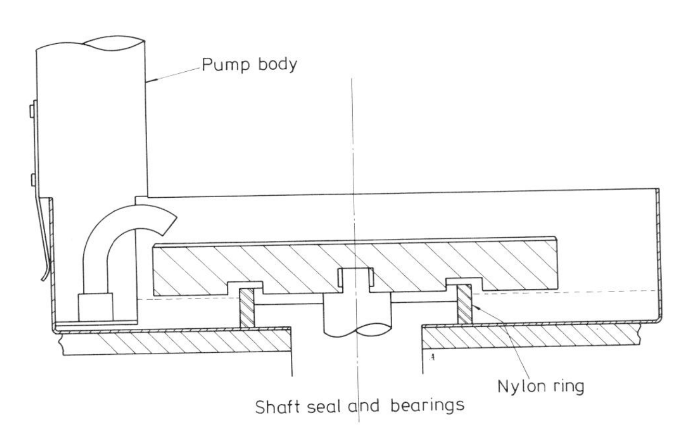

2.48 Nylon ring adaptor and clip-on pump for easy conversion of the Multipol to the fully recirculating mode.

The work head is adaptable for the various modes of polishing in poker, push-rod and roller-bar configurations. Spherical polishing in a ring requires a tilting adjustment of the roller bar not normally available on the Multipol but a simple arrangement can be made for moderate curvatures as shown in figure 2.49. Push-rod working is similar to the GPPM, with the load applied via a threaded sleeve and coil spring (figure 2.50).

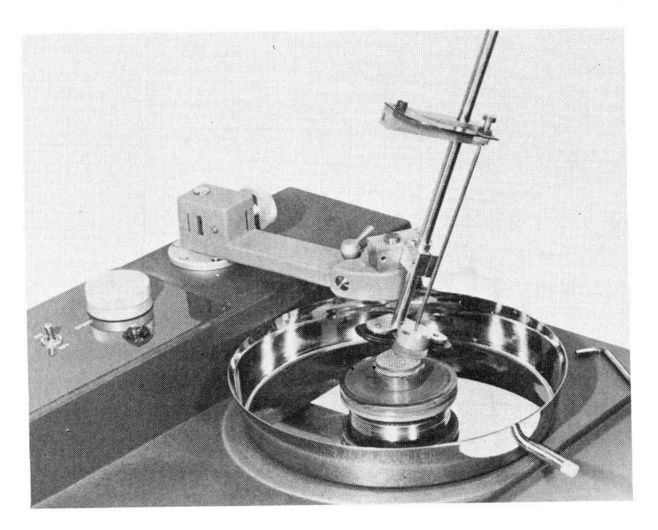

2.49 Tilted roller-bar modification to the Multipol.

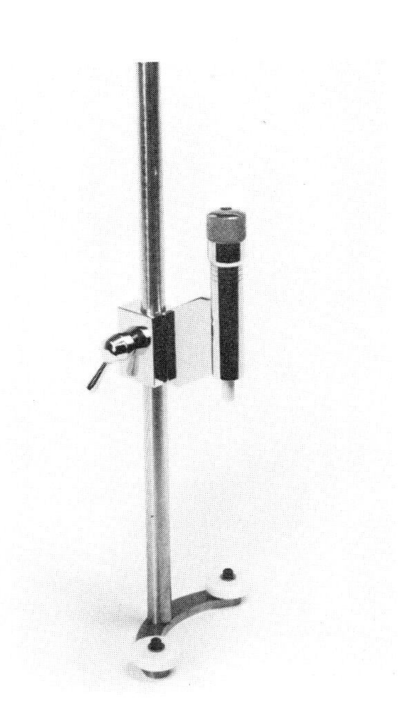

2.50 Micrometer adjustment for push-rod polishing pressure on the Multipol.

Spring abutments, or some other form of resilient loading, are always an essential part of polishing machine design, serving to equalise the pressure applied to a job which may be rising and falling slightly when running.

Blocks are, of course, used in the usual way (figure 2.51). The absence of an external rotating crank—as in the GPPM—means that any additional roller-bar wheel driving must be carried out either with a separate motor or by transferring the polisher motion to the roller.

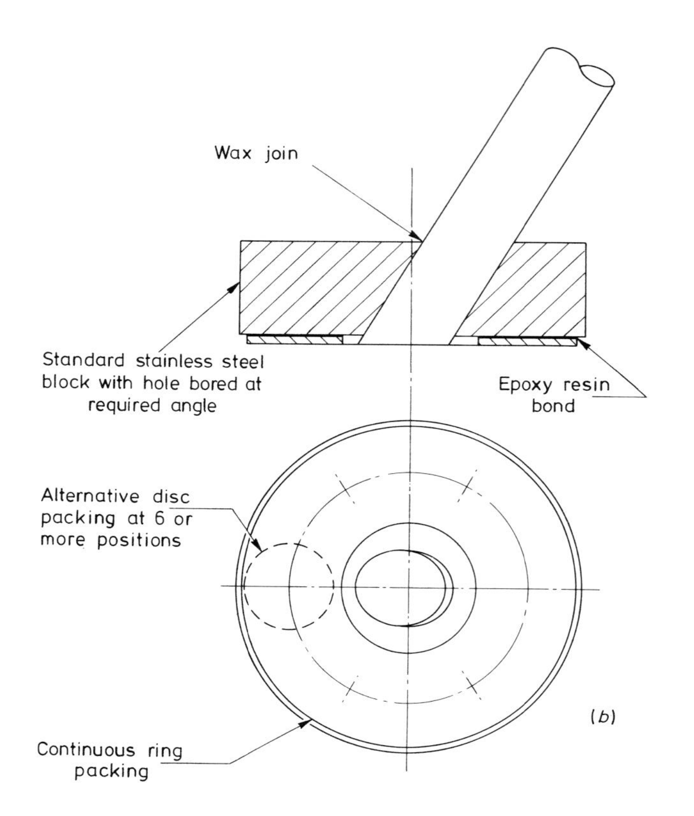

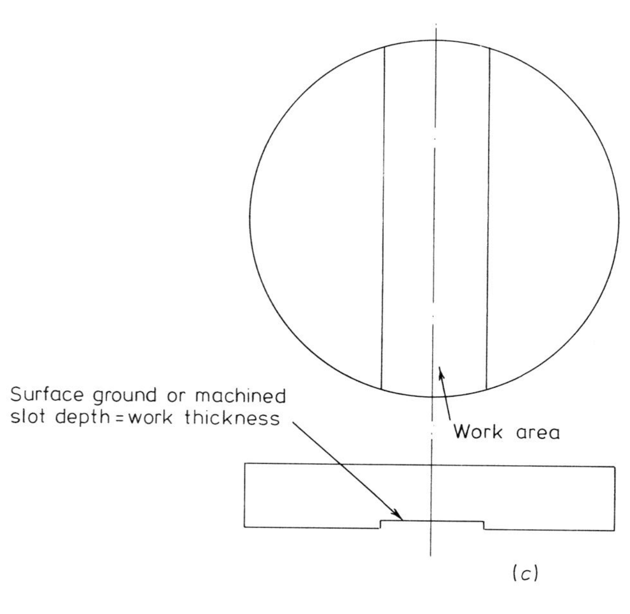

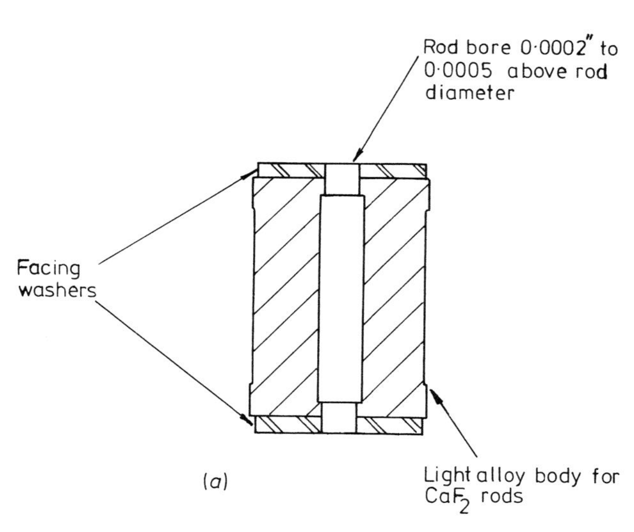

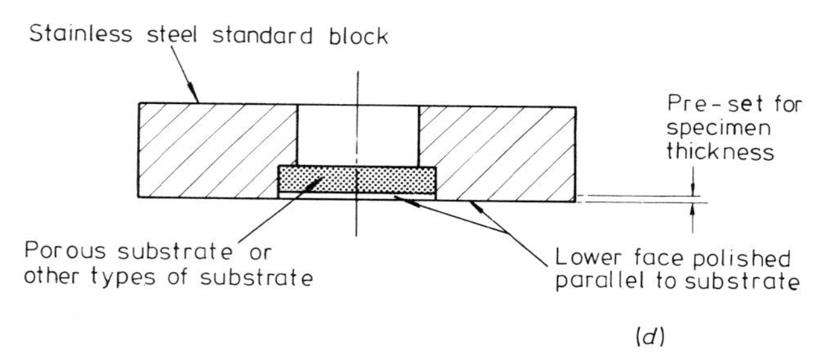

2.51 (*a*) Single rod polishing block, (*b*) block for angled ends, (*c*) slotted block, (*d*) composite block.

2.3 Lapping and Polishing Jigs

Johnson's dictionary (1848) defined a 'jiggumbob' (hence jig?) as a 'trinket; a knick-knack; a slight contrivance in machinery', and notes it as a cant word. Today's technical dictionaries might be more specific but would find it difficult to improve upon Johnson's phrases in the light of the word's general use and abuse in engineering and advertising for 'do-it-yourself' equipment. Following our practice of unilateral definition, we intend to call any adjunct (to a machine) which has an adjustable part or parts and gives precision, a jig. A piece of solid material which is plain, recessed, drilled, bored, slotted or washered to receive a specimen and hold it in a solidly established, non-adjustable position, will be termed a block. Figure 2.51 shows several types, ranging from a laser rod holder to a slice polishing block. They often provide the simplest and most effective way of repetitive production of work.

However, the development of many expensive crystalline materials precluded the use of the material itself as wasters or packing. Various attempts have been made to find them with matching erosion rates and where close matches have been found they result in good polishes and flatnesses. The differential erosion rate polishing of laser rods (Fynn and Powell, 1969) typifies the technique. Angular control, with the waster system, relies on hand, differential weight (Ramsey, 1955) or push-rod influences. Thus the need for versatile lapping and polishing jigs which could solve these problems became increasingly acute as the new, costly crystals appeared for evaluation.

The simplest forms of jigs which use a sliding plunger provide a solution to the differences in polishing rates between the specimen and control ring. Cylindrical hardened-steel rollers are often used as the sliding members since they have the advantages of precise external diameters, high surface finish and reasonably flat, square ends. In two-component jigs, only the ring bore need be precisely machined to fit the roller and, at the same setting, provide at least one end face. The operation preserves the essential 90° between bore and control surfaces (figure 2.52). Some jigs have slotted ring surfaces which allow improved abrasive feed to the work-face. However, whether they be slotted or plain faced depends on whether or not the lap or polisher is reticulated. An anti-rotation constraint between the roller and the control ring is often provided in the form of a key and keyway (figure 2.53). The former is provided by cementing a needle roller diametrically on the upper face of the sliding cylinder,

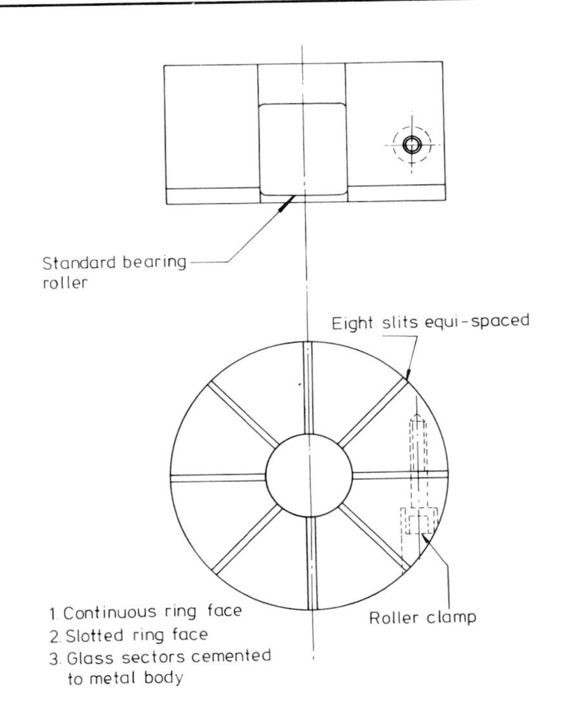

2.52 Plain and slotted-face polishing jigs.

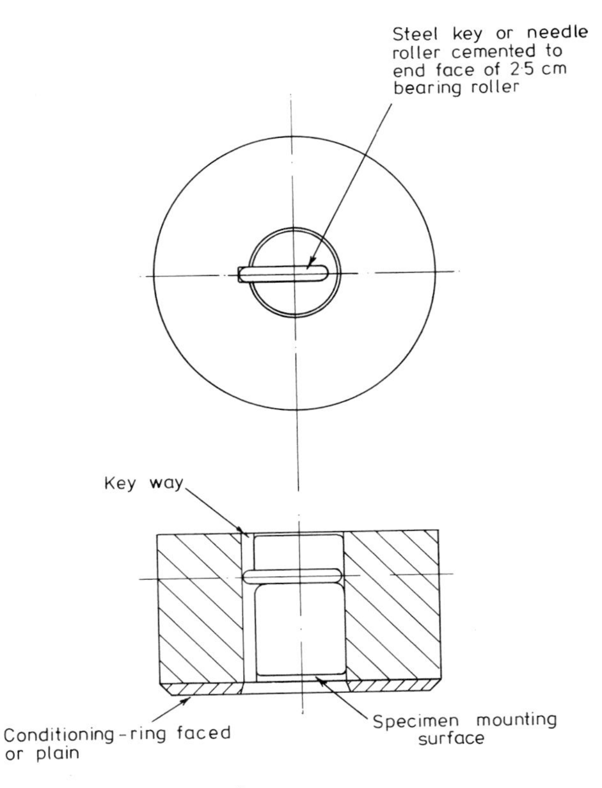

2.53 Keyway and peg jig.

the latter by a short machined slot in the ring. The roller and ring end-faces may be lapped simultaneously prior to use and preserve their co-planarity by this method. In turn, improved parallelism of specimens results, but rapid, fine control of angle is still lacking. Possibly the most significant advance in the polishing field in recent years has been in the development of adjustable-angle jigs, noted in §2.2.1. That they lend themselves readily to further modification and improvement is a tribute to their initial concept (Bennett and Wilson, 1966).

2.3.1 Prototype adjustable-angle jig

A simple adjustable-angle jig can be made by adding a bearing system, that can be tilted by means of three screws, to a conventional conditioning ring 76 mm outside diameter × 33 mm inside diameter (figure 2.54). It is regarded as a quick conversion to existing rings. The accuracy of the angle-control depends upon the stability of the three screws system and the sliding-shaft fits.

2.54 Prototype adjustable-angle jig.

There are a pin-key and keyway which prevent rotation of the shaft and a spring and adjuster nut for specimen pressure control. Specimens are mounted on detachable plates for ease of manipulation and waxing down. Both the specimen plate and shaft have central holes for optical alignment purposes—for example, with an auto-collimator—as detailed in chapter 6. The fact that the bearing system is unsealed calls for care in use in order to avoid flooding it with abrasive slurry and cleaning fluids. The jig has proved very satisfactory on ring-lapping machines where the abrasive layer is thin and similarly with soft metal matrices plus diamond polishing compounds where the surplus fluid is minimal. Large numbers of components have been produced with the jigs and accuracies of one minute of arc are readily obtained.

2.3.2 Mark I polishing jig

The lessons learnt from the prototype jig were principally those of abrasive exclusion and insufficient dead weight on the conditioning ring. Following the general rule, about 100 g cm^{-2} is the desirable pressure and the jig mass and ring area are calculated to give this in Mark I and II models. It must be remembered that the jig ring serves two purposes: it controls the lap and polisher forms and provides any required angle (within the $\pm 3°$ tilt limit) relative to the shaft. Moreover, the tilt screws themselves and the bearing system are refined for higher accuracies in flatness and parallelism to be achieved.

The jig (figure 2.55) incorporates a hollow sliding shaft (1) flanged at the lower end and with a specimen plate (2) bolted to it. The complete central assembly, including the specimen plate, can have up to 6° angular freedom by means of a 0·5 mm thick metal diaphragm (3). Three micrometer screws (4) driving via wobble pins (5) control the specimen's angle of tilt relative to the conditioning-ring face (6) which in turn defines the polishing plane. In order to accommodate specimens of varying lengths and those too large in diameter to pass freely up the central bore, the conditioning ring is adjustable axially over a length of 2·54 cm and is held firmly in position with the locking ring (7). The pressure on the specimen is controlled by adjusting the load spring (8) with the load control nut (9). Thus screwing down the nut compresses the spring, increasing its upward thrust against the outer bearing and reducing the load. Screwing the nut upwards (i.e. unscrewing it) increases the load until the spring's free length is reached, at which point the

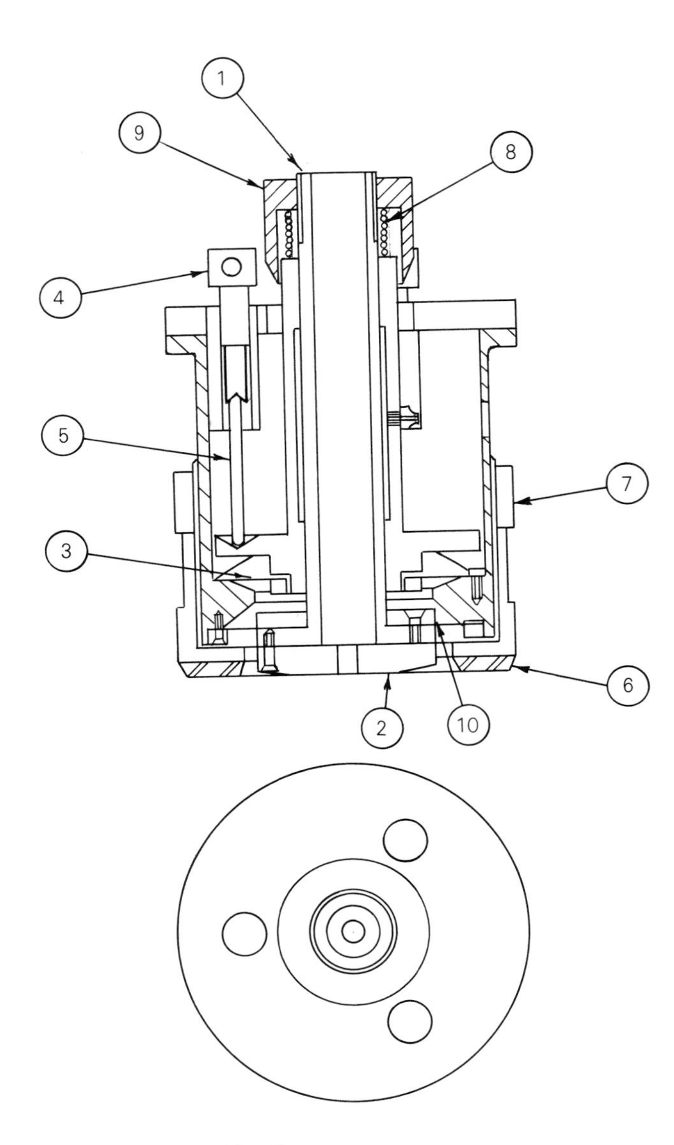

2.55 Mk I polishing jig.

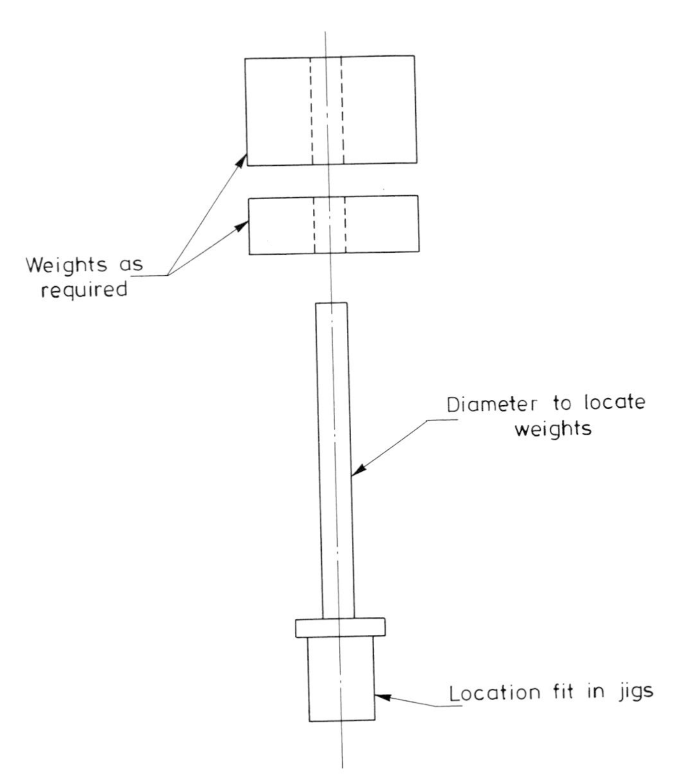

2.56 Additional weight-loading attachment for jigs.

maximum load is the dead weight of nut, shaft, plate and specimen. If a polishing pressure is needed in excess of the total mass of the central-shaft assembly, weights may be added which are located concentrically on a short stub registering in the upper end of the bore (figure 2.56). Finally, a second diaphragm (10), made of flexible plastic material, is fitted below the bearing system, bridging the movable and fixed members. It acts as a seal against abrasive contamination as well as constraining the central shaft from rotation relative to the main jig body. Because of its effective sealing from abrasive particles and liquid ingress it becomes possible to immerse the lower end of the jig in cleaning fluids and, when the robustness of the specimens allow, ultrasonic cleaning techniques can be used. The seal (10) is the one component which wears out fairly readily, and spare replacement seals should be kept in stock.

Very close fits between the sliding shaft and its bearings are required in order to obtain the best possible angular stability between the specimen and the conditioning ring. Even when such fits are machined, oil-filling of the inter-bearing cavity and its subsequent permeation of the bearings seems to be essential. It is believed that the oil film is equalised by the random side forces (of graying or polishing operations) applied to the specimen at the polishing plane and thence to the shaft. Well-oiled jigs have preserved the apparently solid relationship between specimen and ring extremely well during several years of use. Even so, it is the factor which limits the final accuracy when all other features have been optimised. While the system is in motion—that is, lapping or polishing—all is well. However, when the jig is stationary, during a period when angle-setting or correction is being carried out for example, some drift occurs—even when it is vertical. The extent of the drift depends upon factors such as temperature, existing corrective tilt, length of time taken for the resetting and inter-bearing wear. It can be demonstrated by removing the jig from a machine, rapidly cleaning it and setting it under an auto-collimator; then leaving it for a few minutes before rechecking. Drifts of fifteen seconds of arc are quite usual after use for several months and they become greater as the bearing system wears. The effect can be reduced by rapid removal, resetting and replacing on the polishing machine. Work thought to be parallel to defined limits has been subsequently proved to be in error by other means, as detailed in chapter 6. It should be noted that this relatively long term phenomenon does not affect the flatness of the specimen—at least, not to better than $\lambda/20_{He}$. In fact, no better figure would be expected if solid bonding existed between the ring (or wasters) and the specimen. Solid bonding here means normal waxing of the crystal to a rigid backing plate (e.g. figure 2.57).

The tilt screws have one trap for the unwary: it is very easy indeed to achieve angle-correction by screwing them up every time. The final result is that the diaphragm is virtually pushed out of the jig; indeed, it rapidly loses its coefficient of restitution and the adjustment becomes limited to less than 1°. A superficial appreciation of the mechanism shows how important it is to set and re-set angles as closely as possible to the flat position of the diaphragm. The snag is remedied in the Mark II jig.

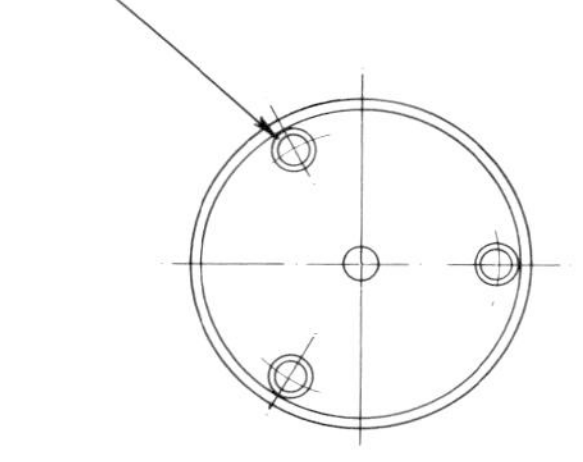

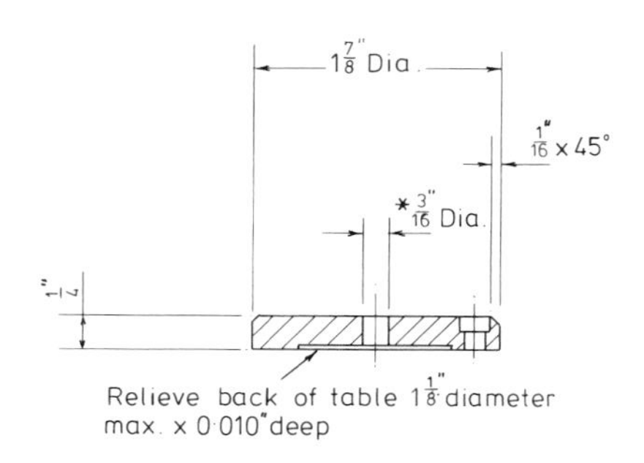

2.57 Typical specimen plate.

2.3.3 Mark II jig

It has been noted that the prototype and Mark I polishing jigs depend on the inter-bearing oil film for their degree of accuracy and thus their precision is limited by long-term drift. The Mark II jig has been developed around commercial bearings known as recirculating ball-bushings or linear bearings. These consist of hardened steel outer housings and rows of precise balls arranged parallel to the shaft axis. The balls exiting at one end of a track re-enter via return tracks. Positive contact between the fixed and moving members of a complete system is thus made by a series of multiple contact points. By means of the arrangement, oil and grease are virtually excluded from the contact area.

The shaft-fit is necessarily critical and it is achieved by grinding each shaft to match its bushings which are already pressed into their housings. An alternative system is to use adjustable bushings which may be closed slightly by using a split housing. The tolerances on the diameter of the hardened steel shaft could be relaxed if this method of manufacture were to be used.

Although continuous unlimited movement of a shaft in an axial direction plus free rotation around that axis are permitted by the bushings, a jig seldom requires more than 5 mm linear travel of the specimen-holding central shaft and no rotation at all. Localised wear of the shaft produced by limited motion is in practice unnoticeable but, should it become significant in time, an advantage of the bushings is that they allow a simple remedy: the shaft can be rotated through a few degrees to provide a series of new tracks for the steel balls. A critical feature of this type of polishing jig is the straightness, cylindricity and surface perfection of the central shaft. Most of the jigs in use at the moment employ centre-ground shafts which have received no subsequent treatment. The ground surface gives a slightly gritty 'feel' to the sliding motion when the optimum fit has been obtained. Some super finishing of the shaft would almost certainly improve the 'feel', if not the angular stability, of the jig.

The construction in detail is shown in figures 2.58,

59, 60 and 61. A fixed glass or stainless steel conditioning ring (1) is bonded to an adjustable-depth control nose (2) which allows very long work to be accommodated by screwing the nose on the body thread forward and locking it firmly in place with the ring (4). A thin flexible, metal diaphragm (3), with a 2 mm hole in it which allows air to pass, is clamped to the body (5) by its outer diameter using the retaining ring (17). The inner edge of the diaphragm is clamped with a smaller ring to the tilting shaft bearing housing (15) which holds the recirculating ball bushings (16). The tilt control head (6), housing two micrometer screws (7) and (13) at 90° to each other, is fitted at the upper end of the body (5).

The micrometer screws (7–13) actuate the wobble pins (14) and these in turn control the position of the inner housing (8) which is bolted to the tilting bearing housing (15). The return spring (12) is sufficiently strong to deflect the tilting system on its flexible diaphragm (3) at the extreme outward position of both micrometer screws.

2.58 Mk II polishing jig (Material: brass and stainless steel).

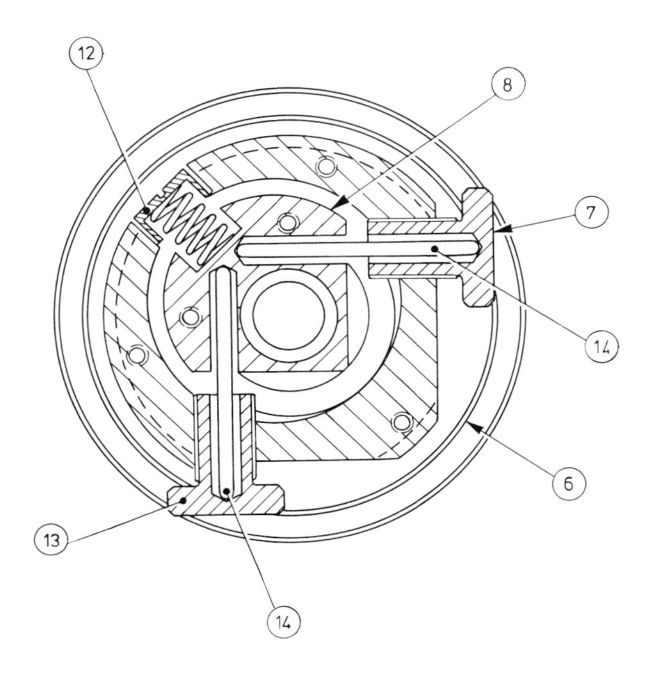

2.59 Tilt mechanism for the Mk II jig.

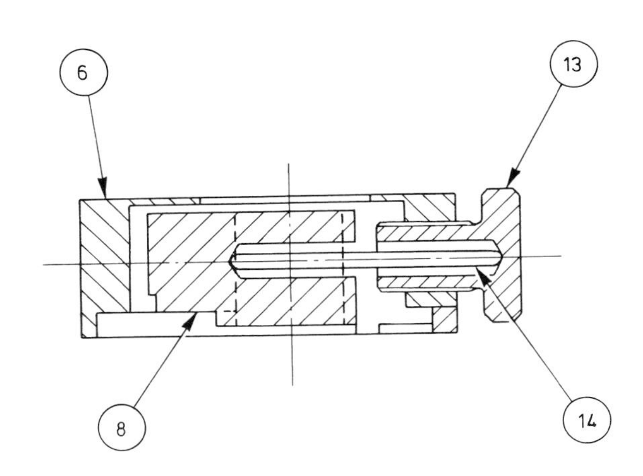

2.60 Tilt actuator for the Mk II jig. Section on axis of tilt control—No. (7) on figure 2.59.

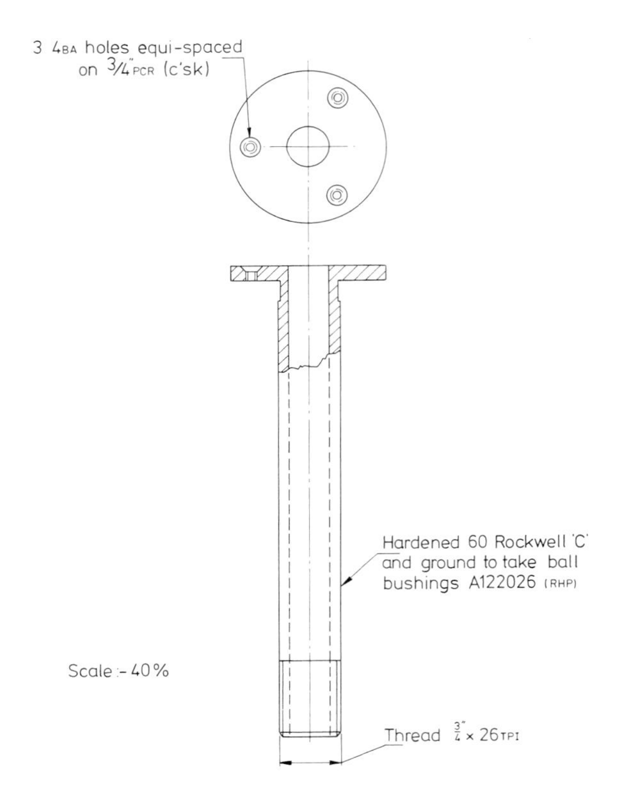

2.61 Precision hollow shaft for the Mk II jig.

The hardened central steel shaft (10) is flanged at its lower end to serve two purposes: the formation of a platform for specimen work plates (19), and as a clamp for the inner edge of the abrasive-excluding seal (18). The outer edge of the seal is clamped to the body (15) by the clamping ring (17) which also clamps the diaphragm. The abrasive-excluding seal serves a secondary function of preventing the central shaft from rotating relative to the body—a vital function in view of the possible tilt-angle of specimen or rod being worked.

The normal polishing load, applied via the central shaft and work plate, is varied from the 'dead-weight' condition (as in the Mark I jig) by the knurled nut (9) and coil spring (11). By screwing down the nut and compressing the spring the weight on the specimen is reduced until ultimately it is lifted clear of the lap or polisher surface. This is a necessary condition to have in some operations. It follows, as before, that the maximum weight on the specimen is the sum of its own weight, the work plate and the shaft and nut. If the area of the specimen is large, the weight per unit area can be below an optimum and in this case additional weights are added at the top of the nut and located by the shaft bore (see figure 2.56).

The Mark II jig has a greatly improved and more convenient angle control system. Its two-micrometer control tilt system, with the actuators at 90°, corresponds to the graticules on auto-collimators. The arrangement gives a surprising degree of rectilinearity. The movements must be, of course, as curved as the radius given by the length of the wobble pins (5 cm), but because a very small angle—usually less than one degree—is visible at any one time, the movements appear both rectilinear and rectangular. Since the actuating point to fulcrum line (push point to pivot line) is approximately three times farther in the Mark II than in the Mark I jig, it is that much more sensitive. In practice, the accuracies achieved by the Mark II jig are limited by the auto-collimators which we use, and we do not expect to read these to better than two seconds of arc. In its standard form, the jig has been designed to give about 100 g cm^{-2} on its conditioning ring and will take specimens up to 3·8 cm in length and of 5 cm diameter. The hollow shaft can accommodate rods up to 1·4 cm with no limitation on length. Its stainless steel construction allows its use in many chemech polishing modes, in particular with silica sol solutions. Figure 2.62 is a photograph of a Mark II jig with a glass-faced conditioning ring.

2.62 Complete Mk II jig.

2.3.4 The Minijig

However precise and adaptable the adjustable-angle jigs may be, they are complicated and fairly expensive to manufacture. Consideration of the push rod, conditioning ring and separator principles shows that a significant amount of accurate work can be carried out very inexpensively by a device incorporating all three modes. This has been termed a Minijig (figures 2.63 and 64) which serves to date its conception quite well. In addition, it is invaluable for free (i.e. unmounted) polishing of slices and is increasingly employed on the work. It can be used on ring-lapping machines and all the types of polishing machines described in §2.2. Basically, the jig comprises a heavy conditioning ring with provision for the location of a weight-loaded push-rod and all other fittings made specially to suit the particular shape of work.

The conditioning-ring body (1) is usually of stainless steel: for use on 15–20 cm diameter laps or polishers it has an outside diameter of 7·5–9 cm. It is often glass-faced, especially for working on pitch-polishers. For many applications (with, say, soft metal matrices or silica sol polishing processes) the ring may be made from solid stainless steel without additional facing material. A sector-shaped piece of stainless steel or Tufnol (2) is bolted to the top face of the ring and provides a loose fitting guide for the push rod (3). A PTFE grommet is a refinement, fitted to the central hole in (2). The push rod is of 3 mm diameter stainless steel, pointed at one end and flanged at the other to provide a platform for the load weights. An assortment of these are required to suit different surface areas of specimens.

For the sake of complete explanation, a specimen is shown within a Tufnol collar (4) which fits the inside of the ring and the outside diameter of the work. In many cases, like the one illustrated, the collar drives a free-fitting cap (5). Again, Tufnol is the preferred material for these caps which usually have two or more conical depressions in which the push rod can be located. One of the dimples must be concentric—that is, it must be above the centre of the specimen (6) and ideally at the centre of the jig. Caps may have relieved internal faces as illustrated, or for thinner work, velvet or napped cloth linings are used to distribute the load evenly.

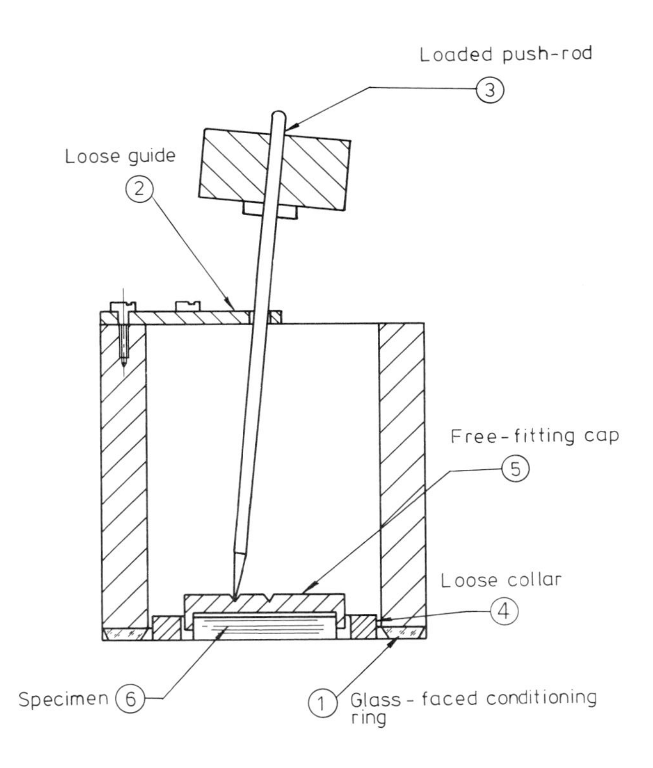

2.63 Diagram of Minijig.

2.64 Minijig in operation.

The conventional cap is not used for very thin single discs but is replaced by a thick collar that registers the specimen, plus a synthetic rubber disc and pressure plate (figure 2.65). Pieces of neoprene sheet have very good flatness and parallelism and are ideal for interposing between pressure surfaces which must be lapped to the required flatness. In our case, this involves running them for a few minutes on a corrected ring-lapping machine to obtain $\lambda/10_{(He)}$ and made just sufficiently specular enough to produce fringes (chapter 3). For the polishing operation the Tufnol collars should be fairly close fitting on both conditioning ring and the work. A curious form of wear takes place on the internal diameters of the collar, particularly during long polishing runs for parallelism correction. It is termed pocketing and refers to the gradual wearing of the Tufnol by the chamfered edge of the work, until there is a slight lifting effect on the latter due to unevennesses in the chamfer and wear-recess (figure 2.66). An unexpected convexity results from a polisher known to be flat. The cure is either to renew the collar or to rebore the existing hole; if too loose a fit results, an 'O' ring may be fitted.

A further application for the Minijig ring, as distinct from the complete jig, is in the polishing of groups of discs (figure 2.67). A Tufnol separator, somewhat thinner than the finished, polished plates, is required. It is bored in a number of positions with holes to suit the specimens, usually with more than five in a ring. The outside diameter of the Tufnol disc itself should be a free fit in a Minijig. A neoprene pressure pad is used, slightly smaller in diameter than the separator, and is

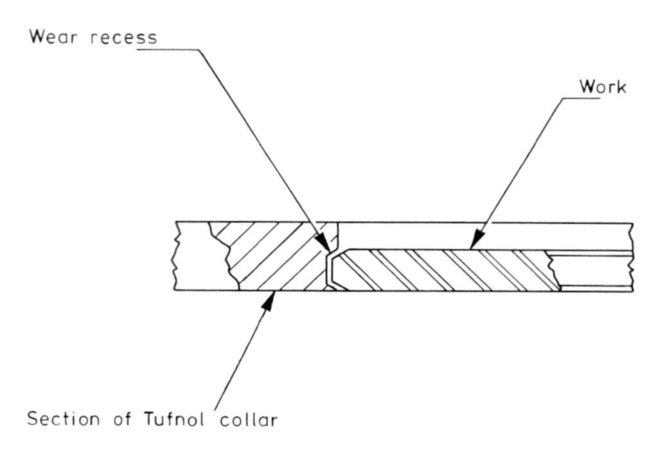

2.66 Conditions resulting in pocketing.

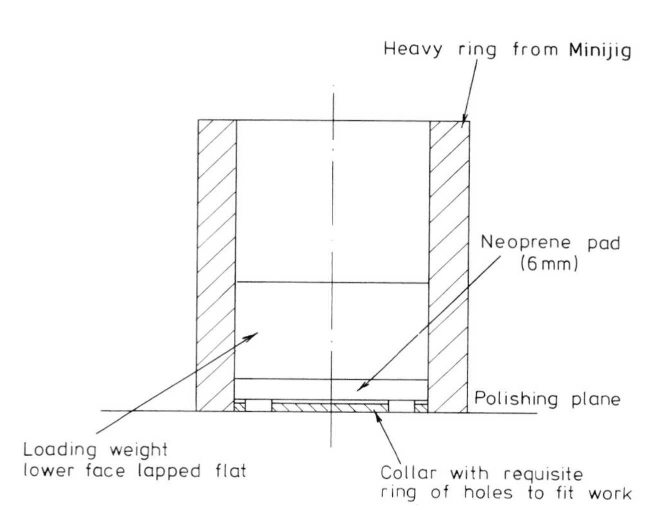

2.67 Multi-polishing arrangement in Minijig.

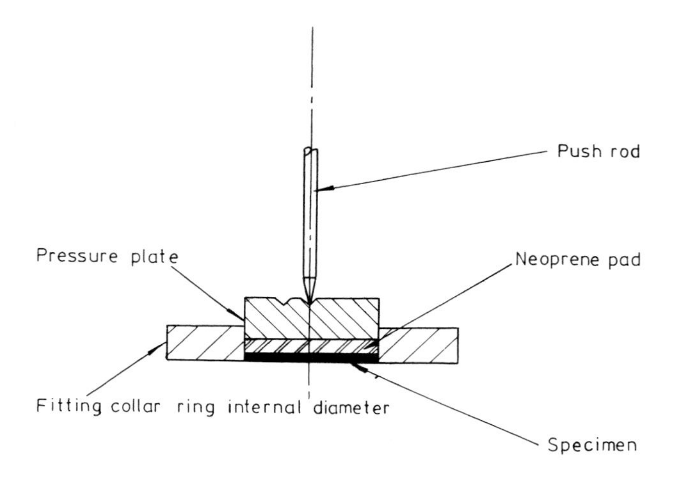

2.65 Tufnol collar and rubber pad load-equaliser for use with the Minijig.

from 3–6 mm in thickness. Finally, a load weight that has been flattened on its lower face—but not necessarily to specular limits—applies the calculated load to the workpieces. The system has been used with great success when producing glass and crystal windows; excellent parallelism is obtained by periodic interchange of the specimens. Sawn wafers can be polished without the need for an intermediate lapping stage: a procedure that depends very much on the quality of the cutting. Both slow-speed annular and reciprocating-blade sawing (chapter 1) have given satisfactory finishes, parallelism and uniformity of thickness for this polishing process. The main reason for operating with weight loading is that it rarely required any parallelism correction. It is slightly more easily carried out with a combined weight/pressure plate. Alternatively, the system favoured by Kasai *et al* (1973) may be applied, which makes use of wedge-shaped blocks to give the required differential loading. Probably the overriding advantage of the general system is that optical components are not waxed down or blocked. When lapping stresses are relieved by polishing, the work is relatively free to change shape and, if turned over from time to time, a state of equalised stress is obtained at last. Thin specimens polished thus no longer 'spring' as they are prone to do when normally deblocked.

References

Bennett G A and Wilson R B 1966 *J. Sci. Instrum.* **43** 669–70

Bond W L 1962 *Rev. Sci. Instrum.* **33** 372–5

Dickinson C S 1968 *J. Phys. E: Sci. Instrum.* **1** 365–6

Draper H 1864 (repub. 1904) *Smithonsian Contributions to Knowledge* **34** 1–55

Fynn G W and Powell W J A 1969 *J. Sci. Instrum.* **2** 756–7

Gilbert K R 1971 *Henry Maudsley, Machine Builder* (London: HMSO)

Johnson S 1848 *Johnson's Dictionary*

Johnston S G 1968 *Machinery* November, 1180–6

Kasai T, Noda J and Ida I 1973 *CIRP Paper* (presented by Kobayashi)

Kumanin K G 1962 *Generation of Optical Surfaces* (London: The Focal Library) 187–209 pp 187–209

Lord Oxmantown 1828 Sir David Brewster's Journal

—— 1840 *Phil. Trans. R. Soc.* **22** 503–27

Rumsey N J 1955 *J. Sci. Instrum.* **9** 338–9

Smith R 1738 *A Comleat System of Opticks* **3** 281–301

Strong J 1938 *Procedures in Experimental Physics* (Englewood Cliffs, NJ: Prentice Hall) pp 69–72

Twyman F 1948 *Prism and Lens Making* (London: Hilger and Watts)

3 Laps and Polishers

'Tis amazing what a polish the world have been brought to' (Thomas Hardy)

It is our intention to refer to a non-specular surface as 'lapped' or 'greyed-off'—that is, produced by a lap and rolling abrasive particles. From this, a polished surface is produced by a polisher and not by a lap and, of course, the term pitch polisher has long been in use in the optical trade. The material which holds the abrasive particles embedded in it is the matrix. Unfortunately, the evolution of polishing matrices other than pitch has established 'lapping' for 'polishing' generally and 'soft metal laps', 'wax laps' and 'cloth laps' are becoming conventional descriptions. Thus laps might well be polishers, though polishers cannot be laps! We can only suggest, with apologies, that the specimen surface should be the arbiter and thus define the term applied to the tool which produces it.

Reference may be made to Appendix II to obtain more information about specific uses for the laps and polishers detailed below.

3.1 Laps

3.1.1 Flat laps

For most medium accuracy work, plane cast-iron laps, (figure 3.1) typified by the Cooke, Troughton and Sims design are adequate as laps and as bases for polishers. Their relieved section and consequent lightness is an obvious advantage. However, for the more accurate greying-off, especially with jigs, we use Harper Meehanite spheroidal graphite cast-iron laps, grade WSH 1 for example, made with a thickness-to-diameter ratio always equal to, or greater than, 1:8.

They are machined and surface ground then either machine or hand-flattened. In the latter technique, the surface-ground faces of the cast-iron blanks are lapped together using BAO 303 abrasive till the grinding pits are removed. In order to check the flatness at intervals, the greyed surface is brightened and an optical flat and monochromatic light source

3.1 Cast-iron plane surfaced lapping plate.

are used to show the fringe pattern. One way of achieving this is to rub two dry plates together and thus burnish them†. A suitable specularity can be obtained by rubbing the lapped face on a sheet of paper which has been stretched over a cast-iron flat and then coated with a few milligrams of 1 μm diamond paste and some Hyprez fluid. This process can, however, result in a turned-down edge. A safer technique—and our standard one—is to use a grooved, fabric bonded Tufnol polisher. This is initially lapped flat, cleaned and coated with 1 μm diamond paste and Hyprez fluid. It is stroked firmly across the metal lap (figure 3.2) and produces a specular surface which is adequate for flatness testing.

There is, of course, the possibility of contamination of the plate by the diamond paste. Though a remote possibility with high-quality spheroidal graphite materials, it does exist and if the lap is to be used with abrasives smaller than 1 μm particle size it should be avoided. Sufficient brightening may be obtained at the sacrifice of speed by using soft, flat pitch polishers with any fine abrasive or Linde A.

The correction of errors shown by the test is made by suitable bias pressures applied to the work. Progress is made, too, by observing the manner in which laps turn when dragged by their edges during lapping: thus a convex lap on a sensibly flat surface, pivots centrally; and a concave lap pivots at its periphery, opposite the drag point (figure 3.3). However, skilled effort rarely can be spared for the hand preparation of optically flat laps and the present practice is usually to prepare them by machining and surface grinding, then flattening them to within one band in four inches on a 'Lapmaster' machine.

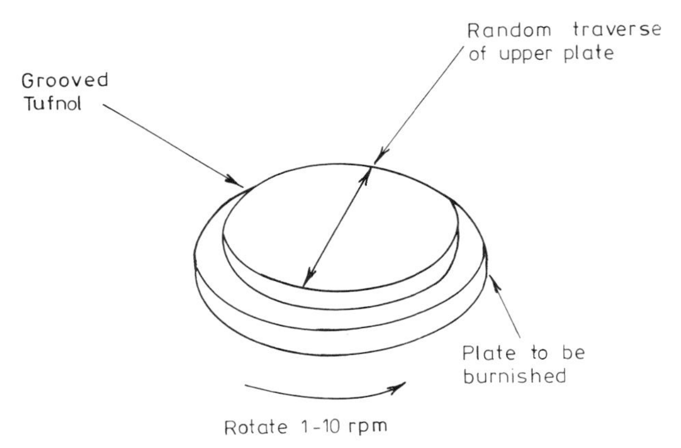

3.2 Burnishing a lap with a Tufnol polisher.

† Although this is a very good method with cast-iron rubbing on cast iron, if any other lap material such as mild steel is used, the following processes are preferable.

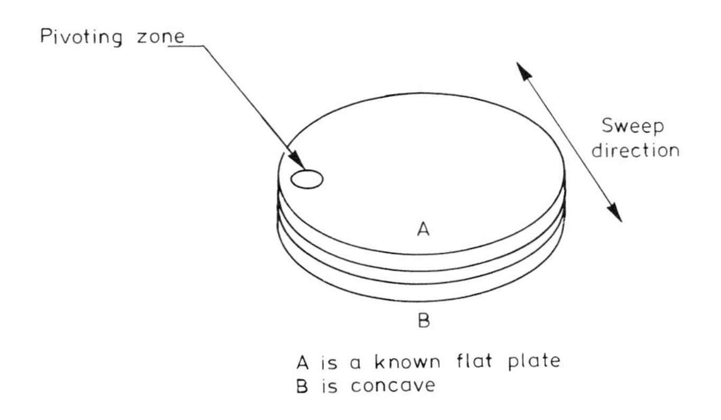

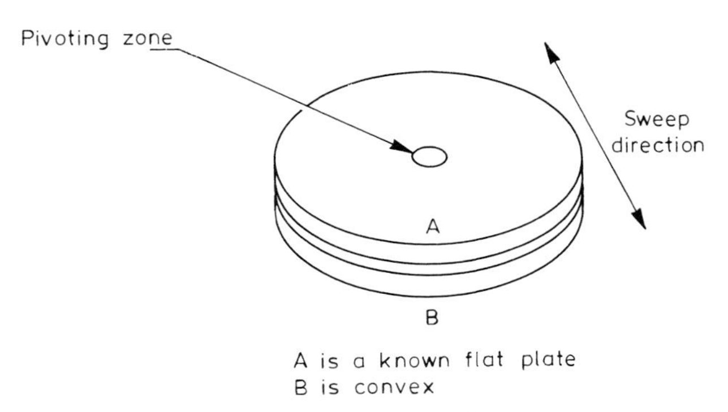

3.3 Behaviour of concave and convex laps.

The 24 in machine produces this from the surface-ground state in less than ten minutes and takes about twice as long when used directly on a very good turned surface. One of the brightening processes already detailed is used to prepare a surface for flatness-checking.

With large, continuous lap surfaces, some form of grooving which retains and releases the abrasive slurry gradually has the advantages of speeding up the lapping process, preventing wringing and extending the life of the lap. Traditionally, surfaces are criss-crossed with a series of slots (figure 3.4). Lord Oxmantown (1840) found that a combination of these reticulations plus annular grooves gave improved results. The logical step from this to spiral-grooving (scrolling; Amberg, 1940; figure 3.5) gives even faster results, both on laps and polishers.

In our experience, a surface interrupted by a spiral groove can lead to about an order of magnitude increasing in lapping or polishing speed. The loss of free abrasive and fluid from the edge of the tool can be minimised by providing a stepped edge, thus leaving the scrolled area as a plateau.

3.4 Cast-iron lapping plate with reticulated surface.

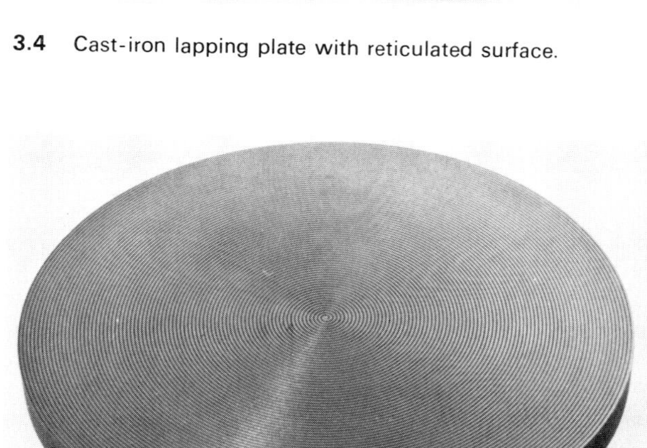

3.5 Cast-iron lapping plate with spiral grooved surface.

3.6 Cast-iron spherical laps (tools).

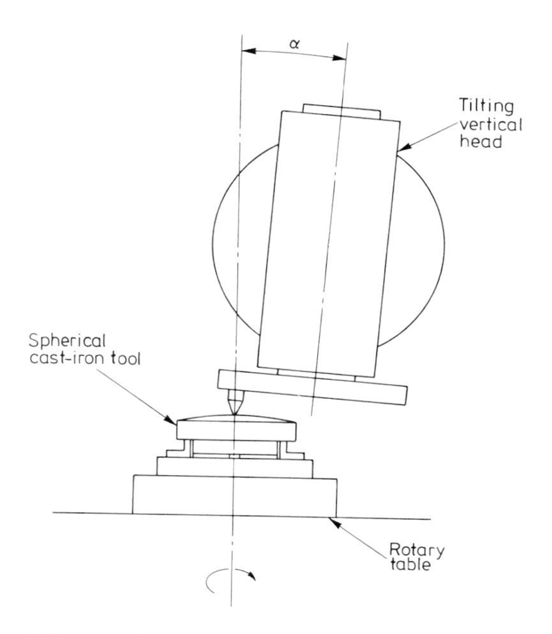

3.7 Milling machine set-up for spherical tools.

3.1.2 Spherical laps

In general, concave/convex laps, or tools as they are called in the optical trade, can be bought off the shelf. However, we often find that the particular radius required for a development project is not available and therefore has to be manufactured (figure 3.6).

The generation of these tools is a simple operation on a milling machine, (figure 3.7). It requires a vertical head which is inclinable and calibrated in degrees. A conventional flycutter is needed—that is, a rotating disc with a single-point tool facing forwards and various holes by which the cutting radius is adjusted. A means of rotating the lap steadily is needed and this is usually a rotary table on a dividing head equipped with a chuck for holding it. The offset angle of the head is calculated from:

$$\sin\alpha = D/2R,$$

where D is the diameter of the locus of the cutter, R is the radius of the tool (lap) required and α is the angle to which the cutter has to be tilted.

Lack of alignment between the axes of the vertical head and the rotary table will cause a 'witness' or step in the generated surface and this must be carefully avoided. Defects such as central pips are removed by spatial adjustment, at 90° to the previous axis. A further refinement in the manufacturing operation is to reduce the closing force exerted by the rotary table chuck on the lap before the final, light cut. This helps to reduce the clamping distortion of the tool, though the evidence of some deflections often can be apparent when lobed patterns appear during lapping.

In the early days of working confocal laser rods, we used a different method for the spherical machining of surfaces: a centre-lathe with a radius pin inserted between a fixed abutment on the head (convex) or tailstock (concave) and an abutment on the moving cross-slide (figure 3.8). It is a convenient method for machining small to medium radii but obviously becomes cumbersome for the larger ones encountered on some laser mirrors. The machined pairs of tools are lapped together with a thin layer of 600 carborundum between them until the entire surfaces appear a dark grey colour. The spent abrasive is repeatedly washed out and replaced with a new, thin layer. The tools may be cleaned, dried and burnished together to indicate the accuracy of conformation in the same way as flat laps are treated.

3.1.3 Lapping

Figure 3.9 indicates the range of average abrasive particle sizes available from several manufacturers†. During lapping, of course, particles suffer fracturing and the slurry of abrasive and lubricant becomes progressively finer in texture and slower in cutting action. For general purpose greying-off, a 20 cm lap flat to $\lambda/2_{(\mathrm{He})}$ is used. A heavy slurry of carborundum (600 grade) and standard Hyprez fluid is kept ready-mixed and is spooned out on the plate, then either spread with a glass disc (a 'crusher') or with the conditioning ring of one of the jigs described in chapter 2. If either the Mark I or Mark II model is used with a specimen mounted on the work-holder, it is advisable to keep the central shaft retracted by screwing down the load control nut. This ensures that the delicate edges of the specimen miss any of the initial bumping during the period when the abrasive is being spread

† The need to treat a stated average size with some caution is shown in figure 3.10 where the graph shows the mean-to-maximum sizes for one of the abrasives that we most commonly use.

3.8 Lathe machining spherical tools.

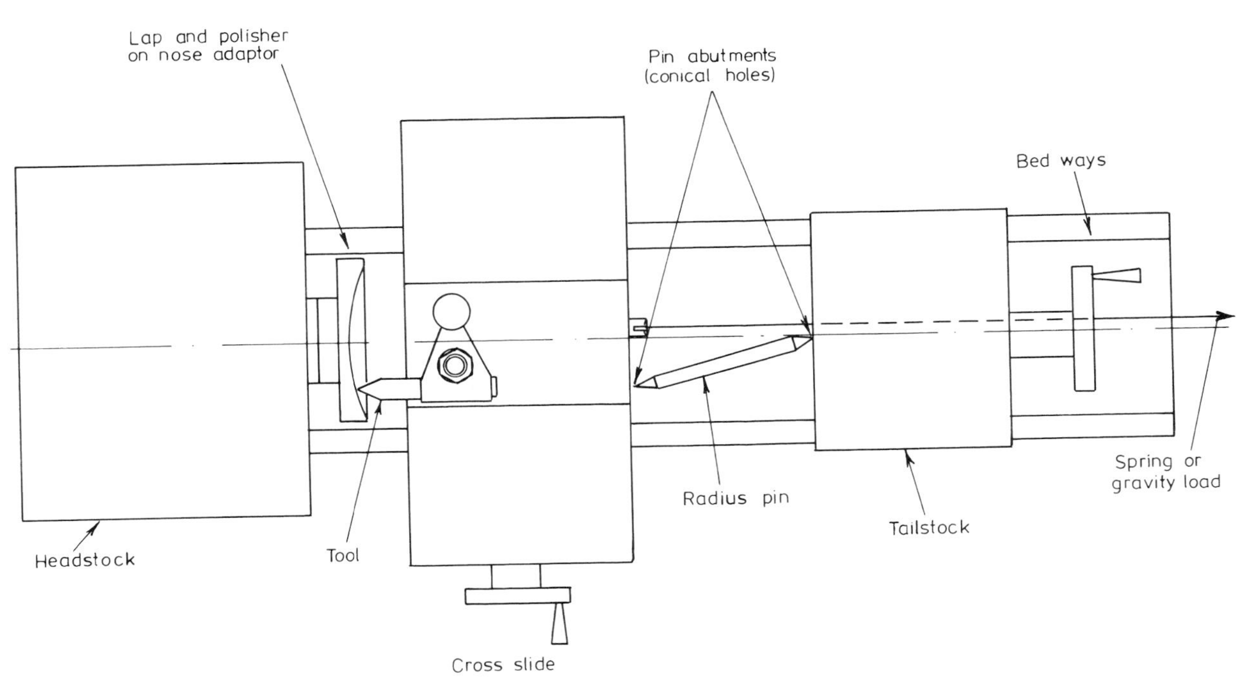

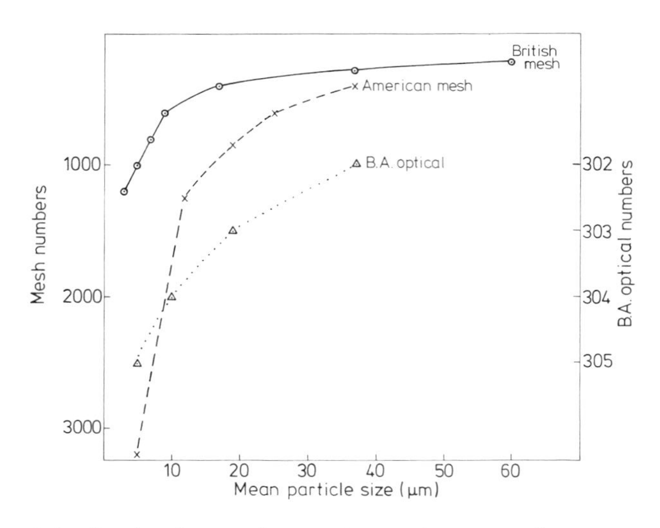

3.9 Graph of several manufacturers' ranges of average abrasive-particle sizes.

and any large agglomerations broken down. When the jig slides evenly over the lap surface, load is progressively applied to the specimen until the entire face is seen to be in contact with the lap. A better 'feel' of this state is obtained initially by hand working since the constantly changing lapping conditions affect the drag of the work and are interpreted accordingly by the experienced operator—and are, moreover, the only way of gaining kinaesthetic expertise. By this we mean that the operator's sensations, via his hands, are projected into the process being carried out and, in time, he feels each subtle change in the process as vividly as if he were being abraded personally. It is a fact that 'feel' gained in this way can be translated directly from hand to machine techniques. It is equally true (for our operators at least) that initiation on, and adherence to, machines for lapping alone, gives 'feel' relatively slowly.

When a different grade of abrasive is used, the lap is washed clean with detergent/water then rinsed and dried. It is advisable to add 5% sodium carbonate to aqueous solutions used in connection with cast-iron laps. This prevents rusting.

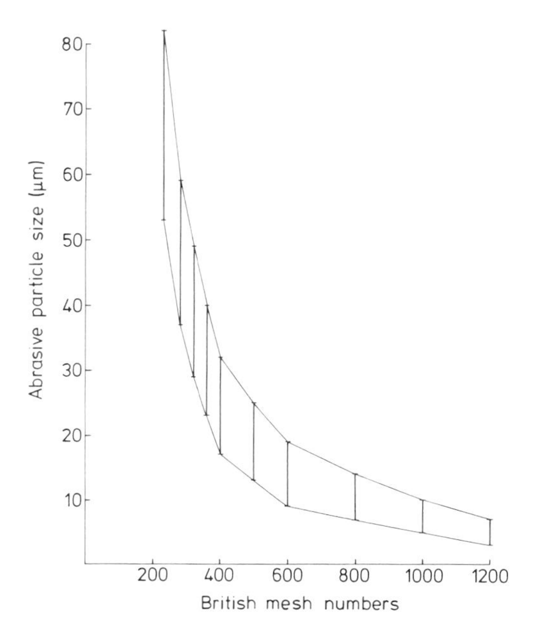

3.10 Spread of mean-to-maximum for a silicon carbide abrasive.

3.2 Polishers

The following descriptions of the preparation of polishers and polishing matrices do not indicate specific uses. These can be found by reference to Appendix II, Summary of Polishing Data.

3.2.1 Cast iron

This material can be used for polishers (figure 3.11), typically for working crystals with a scratch hardness of nine or greater on Mohs's scale. Their preparation is as given in §3.1.1, but since they are used principally with diamond abrasives, once dressed with a given particle size they are labelled accordingly and used only with that size. When one becomes worn and loses its figure, it is resurfaced and used again by completely machining away the old surface by turning-off at least 0·20 mm with one cut on a centre-lathe. This prevents diamonds from being driven more deeply into the surface and causing subsequent contamination of the new and possibly finer-grade polisher.

3.11 Cast-iron polisher.

The method of spreading the WS diamond paste on the polisher is to use a Hyprez syringe with a spreader. It cannot be over emphasised that good housekeeping is all important, and that diamond pastes and contaminated tissues, clothes, card and fingers have to be scrupulously controlled. For a new, scrolled surface a practical initial loading is 0·0015 g for each square centimetre—e.g. 0·25 g of 3-W-45 Hyprez diamond compound is spread on a 15 cm polisher with a 3 cm central recess. Hyprez fluid, oil soluble type, is the lubricant normally employed.

3.2.2 Soft metal

The use of a soft-metal matrix (figure 3.12) in conjunction with diamond abrasives has long been a conventional metal polishing combination (Strong, 1938; Roberts, 1955) and the very good finishes with the minimal edge turndown that speedily result can be obtained, too, on a range of crystals. Thus copper polishers are used when working crystals circa 9 Mohs, indium for those as soft as 2–4 Mohs and tin or solder are generally useful at intermediate values. Flatness better than $\lambda/10_{(He)}$ is readily obtained. Solder, tin and indium are easily cast and, since tin is similar to solder in performance, one or the other is superfluous. Copper is machineable from solid, but is slightly prone to accidental contamination. Currently, we tend to cast solder polishers and electroform indium and copper ones (Fynn and Powell, 1971) on grounds of economy and stability. All our experience increasingly shows that a good electroformed polisher, with about 0·05 mm of pure metal on a sprial-grooved, ground stainless steel master is a most durable and accurate tool.

For all electrometalled polishers, the preparation of the stainless steel masters is the same. The success of the technique depends upon the accuracy of the master, therefore turning, facing and scrolling are as burr-free and precise as possible. The grade of steel used is normally S.80, since it is resistant to chemical attack in acid–copper baths. An example of the preparatory technique is typified by a blank 120 mm diameter, 20 mm thick, blind-tapped on one side to suit 0·5 in BSW. The front surface is faced and has a recess turned at the centre approximately 30 mm diameter and 0·75 mm deep. It is scrolled to give a groove/land ratio of 1:1 at about 6 grooves/cm, 60° included angle. The depth of groove is about 0·5 mm. This machining operation must give the cleanest possible groove, free from any burrs and with no rough, 'chattery' sides since any raggedness has the dual disadvantages of trapping lubricant and of leaving projecting metal. The former causes degreasing difficulties; the latter results in preferential electrodeposition at these places—that is, nodules. Finally, the scrolled master is flattened to better than two bands by 'Lapmastering'.

The preparation of a base for casting is much simpler. A disc of brass of similar dimensions is turned and degreased and is then ready to receive the metal matrix. In the case of the electroformed polishers, almost all the precision is obtained before the coating and it should be free from contamination. It does, moreover, reflect the stability of the base material.

3.12 Soft-metal polisher.

The cast polisher is machined-up after coating and is rather more prone to accidental contamination, fundamentally less precise without extended running-in and less stable and durable. Even so, we continue to use solder in preference to tin, mainly because it requires less specialised preparation, and we have not yet built up a stock of electroformed tin laps which covers a range of diamond abrasive sizes. Of course, the stainless steel master lap is reclaimable by stripping its coating and replating. It is, itself, virtually ageless, apart from accidental damage.

In preparation for electroforming, the master is solvent degreased, usually with tetrachloroethane, then scrubbed vigorously with water and pumice powder. If the machining lubricants prove difficult to remove, the addition of a detergent to the water, and even acetone plus water, is often effective. An electrolytic cleaning process such as cathodic scrubbing (Field and Weill, 1946) may be used for the final operation (figure 3.13). When the surface is scrupulously clean there will be a complete absence of a water-break;—that is, the water wets the surface without forming globules round grease or dirt particles. Figure 3.14 shows the master mounted in a polythene housing A which leaves the front surface and about 3 mm of periphery exposed. There is a clearance hole at the back to allow for attachment to a rotatable cathode shaft B. Three insulating dowels C stand-off a polymethylmethacrylate shield D which is adjacent to the edge of the stainless steel master but separated by about 0·75 mm. This shield is necessary for electro-depositions from sulphuric acid/copper sulphate electrolytes, otherwise the form of the tool becomes noticeably concave even with a coating as thin as 0·05 mm. The throwing power of tin and indium is superior but we tend to use the same shielding process whatever the metal being deposited.

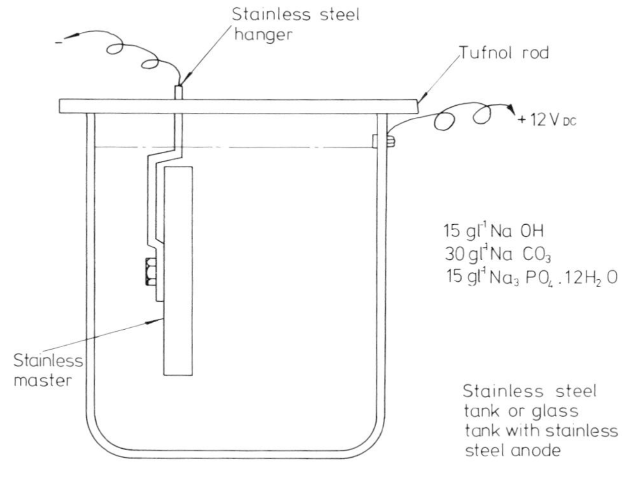

3.13 Equipment and electrolyte for cathodic 'scrubbing'.

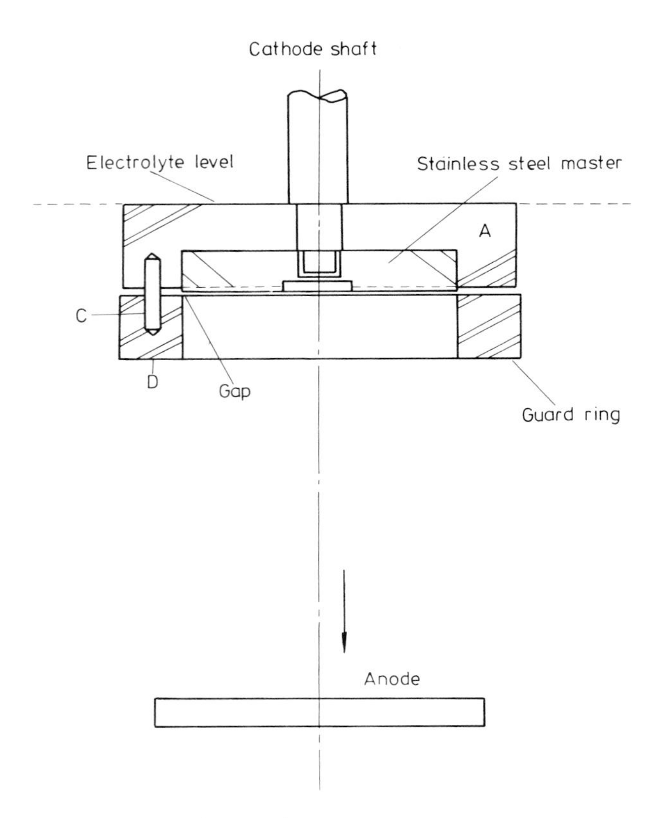

3.14 Diagram of electrodepositing set-up.

Polishers have been made, too, by the commercial Dalic Process which can provide a variety of metals. Though some have required rather lengthier running-in it is a very attractive alternative method and could, no doubt, be improved upon with specially shaped tampons.

3.2.2.1 Copper

At one time, copper-faced polishers were used which had been faced on a good lathe and flattened on a Lapmaster machine. These, as stated, have been superseded by the electrocoppered stainless steel type. One of these is produced by attaching the entire assembly of master, shields and masks, to a rotatable shaft (figure 3.14) and immersing it in an electrolyte comprising $200\,g\,l^{-1}$ $CuSO_4$; 30–$40\,g\,l^{-1}$ H_2SO_4; water. The assembly is at a slight angle (say 10°) to the liquid surface since this assists the spin-off of attached air bubbles. The current density is calculated to give 0·025 mm h^{-1} and deposition is carried out for 2–3 h at about 300 rpm—sufficiently slowly to prevent air bubbles being drawn down under the job. It can be seen from figure 3.14 that the anode is at the bottom of the tank. Further details of acid–copper electroforming technique may be found in Fynn and Powell (1968). At the end of this period, the assembly is removed from the bath and immediately rinsed with water. It should be dismantled under a water film, rinsed again to remove all traces of electrolyte and then dried.

A polisher produced by this method tends to be slightly concave and requires running-in on a machine for a time with the grade of diamond abrasive intended for use on it, and a conditioning-ring loaded to give about 100 g cm^{-2}. However, once pushed into shape it is extremely stable. We have one, in particular, for use with 3 μm diamond which has remained in intermittent use for over three years. If necessary, copper may be stripped in HNO_3:H_2O, 50:50 (vol:vol). Polishers machined from the solid are resurfaced by machining off 0·25 mm, Lapmastering, then etching for a short period in the nitric acid/water solution to free abrasive particles.

3.2.2.2 Tin

Tin is electrodeposited from a commercial electrolyte: Culmo Bright Tin Plating Solution No. 3564A (W Canning and Co Ltd). More information is available from the Tin Research Institute (1964). We use the technique described in §3.2.2.1 but with one side anode of surface area equal to or greater than that of the cathode. The deposition is carried out at a similar rate. It can be easily stripped in a cold solution of 2 g antimony trioxide (SbO_3) dissolved in 100 cm^3 of hydrochloric acid (1·16 sg). A polisher produced by this method needs very little running-in (as detailed for copper) but because the solder matrix pre-dated tin we have relatively few of them and have built up little long-term experience.

3.2.2.3 Solder

These polishers are cast (figure 3.15) from Grade D plumber solder (liquidus 248 °C, solidus 185 °C) on a brass base which is typically 120 mm diameter × 18 mm thick. Pouring should be accompanied by liberal applications of flux—zinc chloride based or resin. An aluminium collar restrains any overspill. The normal thickness cast is about 3 mm and, after cooling and cleaning, the surface is machined flat and scrolled. It can be machined and grooved several times before another casting is necessary. Again,

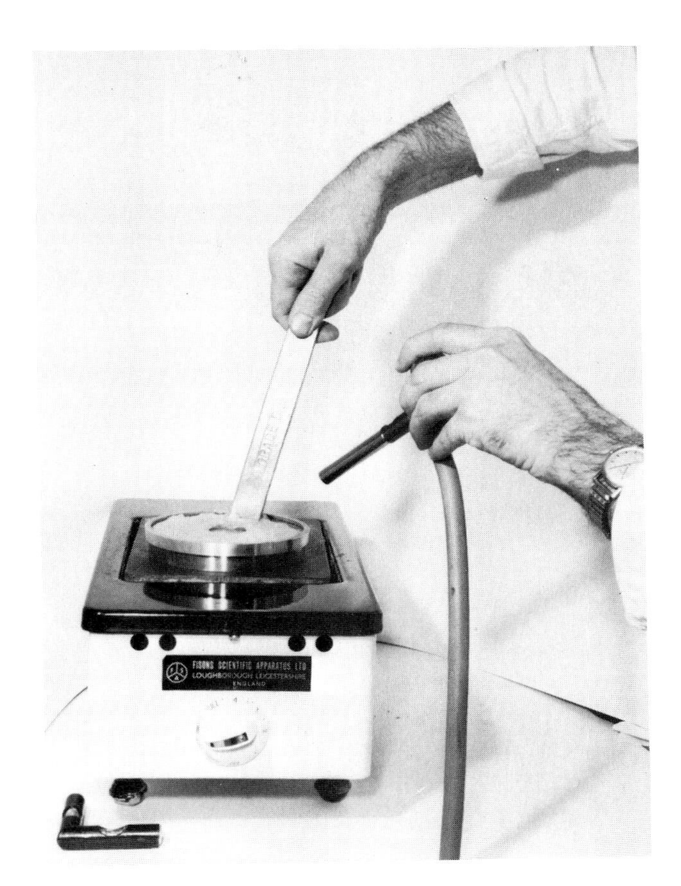

3.15 Casting a solder polisher.

diamond abrasives are normally used, each grade having its own polisher, and the charge of abrasive per unit area is as for cast iron.

3.2.2.4 Indium

We have used a commercial sulphate bath, from W Canning and Co Ltd, for the direct deposition of indium on to a stainless steel base. It has a relatively low cathode efficiency—that is, during electrolysis there is an accumulation of indium ions in solution because more are provided by the anode than can be reduced to metal at the cathode. For this reason it is the practice to alternate indium and insoluble (e.g. carbon) anodes. When the latter type is being used, the removal of metal ions from solution results in an increase in acidity (and a consequent decrease in pH value). Lindford (1941) has shown that good results are generally obtainable when the electrolyte has a pH value in the range 2·0–2·7, and thus the alternation of the anodes can be arranged to balance the bath at about pH 2·5. A value greater than 2·7 is believed to give spongy, non-adherent deposits. A side anode is used, with other plating conditions as for copper and tin. These polishers require very little running-in, but this has to be carried out well lubricated and ideally with the required abrasive. There are several polishers in use produced by the above method, and others very satisfactorily obtained using the Dalic plating process. We have, too, a few—expensive—cast indium laps. The metal has proved itself to be a most valuable 'finisher' for many crystals. It is safe to summarise the advantages of electrometalling as:

(i) the electrodeposit is a faithful extension of the accuracy of the stainless steel master surface;
(ii) it is in an extremely uncontaminated state;
(iii) it can be removed and renewed (or changed) *ad infinitum* without losing the original profile of the master.

3.2.3 Pitch

A great deal of early experience was gained by the use of pitch polishing and, in spite of many break-aways, it often provides the optimum finish when other more easily obtained coatings have failed. The literature is, of course, rich in descriptions of the elegant processes for making pitch polishers (figure 3.16) and the account given below owes much to Twyman (1951) and Strong (1938). The tools that are made range from pure pitch to blends of pitch, wax, resin and wood-flour. However, the starting point is the same in each case: the hardening of pure Swedish wood pitch (H V Skan Ltd) by boiling it for a period of up to 24 h. The degree of hardness traditionally undergoes a qualitative test—such as biting or thumbnail indenting—but we use a quantitative method devised by Twyman (1951). Any attempt to plot a calibration graph of boiling time versus hardness is defeated by the variable solvent content of the raw pitch and progress has to be monitored periodically as follows.

A sample is taken and allowed to cool to 19·5 °C at which temperature it is stabilised for several hours. Some further hardening takes place during this period and it is important that the temperature be kept constant. The sample is placed beneath the indenting apparatus (figure 3.17) which has a 1 kg load acting upon a freely moving plunger which terminates in a spike of 14° included angle with a 0·5 mm flat at the tip. This bears on the surface of the pitch for five minutes and the depth of penetration is noted on a millimetric scale. Some raw pitches indicate 8–9 mm penetration; a hard unmodified pitch in common use yields by 0·5 mm during the test. Although it is generally true that the softer the pitch the more sleek-free the polish that can be obtained, the attendant plastic flow often results in poor flatness of the specimens being polished.

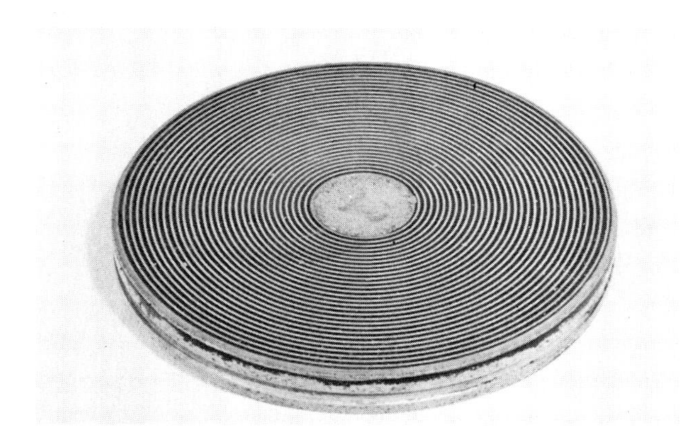

3.16 Spiral-grooved pitch polisher.

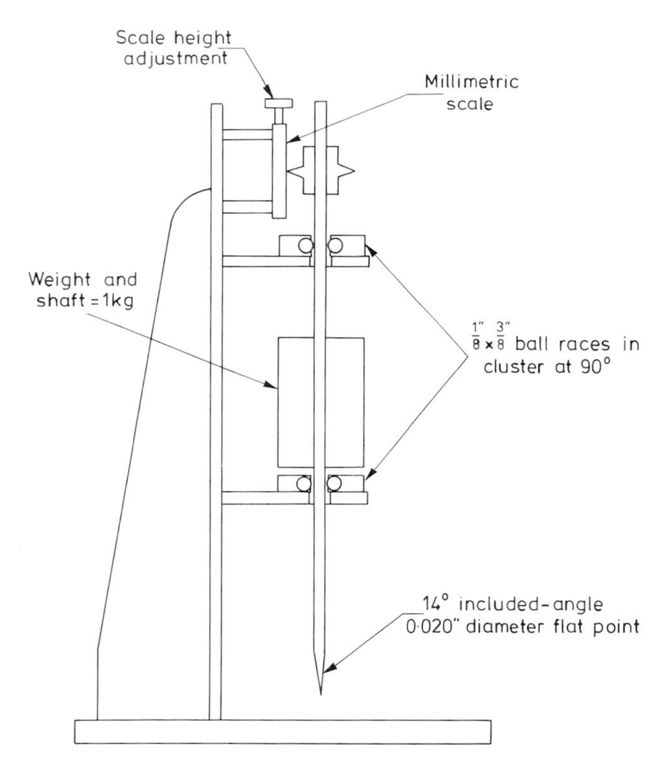

3.17 Pitch hardness testing apparatus.

When flat surfaces together with an abrasive compound and a lubricating film between them, a non-polishing state is gradually achieved because the film becomes greater in thickness than the embedded particle protrusion. Prior to the general acceptance of scrolled matrices, the method of spreading the film has been by reticulating the polisher–surfacer. In fact, since the practice of scrolling calls for good machining facilities, conventional grooving and/or netting is still very much in use in places without these facilities and will be described for the benefit of those who would prefer the technique. The main difference in manufacture is that a rather more heavily loaded pitch is necessary to ease the operation of spiral-grooving, whereas netting can be performed on somewhat simpler mix. Both give equally good results.

3.2.3.1 Flat polishers

For making a pitch polisher which is to be netted a cast-iron base is pre-heated to 70 °C and a mixture of pitch and wood-flour, 1:2 by volume, is poured on it. The hardness of the pitch at this stage is the definitive hardness of the polisher and is spoken of as, for example, a 1 mm pitch. The thickness aimed at is about 3 mm, and with sufficient practice it is possible to adjust the flow so that the pitch is in a state sufficiently viscous to prevent its flowing over the edge. However, this skill can be by-passed by arranging a retaining wall of wet adhesive tape round the periphery of the cast-iron base. The pitch can be hotter with this technique (about 120 °C) and is the more easily levelled. Once the pitch has cooled to a stiff, plastic state the paper can be removed and the turned-up edge trimmed with a knife fed with detergent and water, leaving a wide chamfer. The surface is flattened, either in a heated press or on a hot, flat cast-iron or stainless steel plate with a weight giving about 0·5 kg cm^{-2}. The process should be carried out with detergent solution (Teepol and water) and some of the polishing medium intended for use wetting all surfaces.

Small polishers, typically 4 cm diameter, need no greater reticulation than is given by netting, but larger tools require a grid of large grooves with a finer network superimposed (figure 3.18). The principal grid can be formed by pressing a cool forming tool on the warm pitch as detailed in the last paragraph; then turning through 90° to achieve tessellation. Hand pressure only is involved, but the process has to be repeated several times to obtain adequate depth of grooving. Although there are numerous

3.18 Pitch polisher—reticulated and netted.

ingenious methods of reticulating pitch (e.g. by the use of a heated brass angle in a step-and-repeat jig) we have only a limited experience and would recommend the interested reader to consult either Francis (1890) or Ingalls (1932).

The pitch is allowed to cool and a square of wet cotton net is sandwiched between it and the face of a stainless steel optical flat which has been warmed to about 60 °C. The sandwich is compressed until the pitch squeezes through the holes in the net and flattens out against the flat. The dual operation, of fine reticulating and flattening, can be done in a press as shown in figure 3.19. The polisher is subsequently removed and the net stripped away before cooling makes this impossible. The pitch is slightly roughened by the operation, and is quickly smoothed again for a few seconds against the warm plate before being transferred to a wet, abrasive-coated cold plate and left to attain room temperature.

It is now ready for use with whatever polishing compound and lubricant is required and is, of course, then labelled accordingly. If it has to be stored for more than fifteen minutes or so, it is placed beneath a glass dome to exclude dust and obviate any contact which would in time cause serious deformation. The final smoothing stages are repeated before use.

Those polishers which are intended to be machined flat and scrolled are made typically from a mixture reported by Dickinson (1968). This has the approximate composition (by weight) 30 parts 0·5 mm pitch; 5 parts wood-flour; 1 part beeswax; 1 part

3.19 Netting press.

resin. It is heated to 120 °C and poured on the base plate which has been previously warmed to 70–80 °C. A wall of masking-tape at the periphery retains the liquid mixture. When it has cooled, the polisher is mounted in a lathe (which faces accurately) and is machined as detailed in §3.2..2—that is, with a central recess 30 mm diameter × 0·75 mm deep and a spiral groove about 6 grooves/cm, 60° included angle, but only 0·45 mm deep. This gives a groove-to-land ratio about 1:1. The special machine described in §3.3, provides a better surfacing and scrolling method. Small jobs may require a finer scroll and polishers can be easily custom made by this method. After these operations, the polisher is ready for running-in with the desired abrasive, either with a simple glass or stainless steel faced conditioning ring or a polishing jig. In either case, the pressure is typically 100 g cm^{-2} and the process is continued until monitoring of the ring with an optical flat and monochromatic light source shows the polisher to be sufficiently flat—usually better than $\lambda/10_{(He)}$ is desirable. If the polisher is not intended to be used immediately, it is stored under a dome or crystallising dish and conditioned as and when required. Pitch polishing is usually carried out in the damp-dry state, with a minimum of abrasive and lubricant being used. When a pump system is used to recirculate the polishing slurry, a flow of liquid at about 1 l min^{-1} is used. Both netted and scrolled polishers can be renovated by facing off about 0·5 mm in the special lathe with a steeply raked tool. They are then treated as detailed before.

3.2.3.2 Small spherical polishers

When, for example, a convex polisher is needed for a laser mirror of less than two metres radius, it is a good plan to rough-machine a spare cast-iron plate with the machine settings used for the spherical lap (§3.1.2). A thin layer (~1 mm) of the recommended pitch mixture is cast on the tool in exactly the way described for flat polishers. When it is cool, it is returned to a milling machine (or lathe) for radiusing. This technique results in a great saving of effort, compared with repeated pressing with a hot concave tool. If the polisher is to be used in the ring-polishing situation, a central recess is machined in it. The polisher is finally netted using wet cotton and sufficient pressure is applied with the warmed concave tool to leave a full impression of the net in the pitch. The two are left in contact to cool and the polisher acquires the shape of the cast-iron former. Tendencies for the pair to stick together can be minimised by using liberal coatings of polishing abrasive and soap.

A point worth mentioning about pitch polishers of either flat or spherical form is that, if they should stick due to over-long contact, the much recommended 'sharp rap' as a means of separating them should be treated with circumspection. We find that the particular kind of 'rap' mentioned by the doyens of pitch polishing is a highly skilled performance if it is not to remove the occasional facet! The authors prefer to use hot water as follows: the cold, stuck pair of plates are placed on edge in a sink and a sudden jet of hot water is applied to the back of the forming tool by means of a short pipe. The two slide apart when a small side force is applied.

3.2.4 Wax

There is general agreement that wax polishers (figure 3.20) can provide surfaces on glass and crystals freer from sleeks than would be possible normally with pitch. Moreover, edge definition is largely retained and this is often of paramount importance when allied to surface finish. They are frequently in use and the fact that is is difficult to form and flatten them is circumvented now by the facing and scrolling technique. Their narrow melting range is responsible for the inability to flow to conform to a flat surface. Beeswax has been used but Okerin 100, from Aster

3.20 Wax polisher.

3.21 Scrolling a wax polisher on a lathe.

Petrochemicals Ltd, is preferred. It is heated to about 95 °C and filtered, then cast on a bronze or steel base as detailed for pitch. After cooling, it is faced, recessed centrally and scrolled (figure 3.21) by either of the method previously described in §3.2.3. Alternatively, it can be grooved after it has been faced by drawing a chaser across it, guided by a straight edge. A special machine has been designed (Brooke, British Patent) which sucks away molten wax through a row of heated nozzles as the tool is passed slowly between them. Some of these polishers are still in use but all latter-day manufacture is by the scrolling method.

Flattening can be carried out by running-in the wax surface with a heavily loaded glass or stainless steel conditioning ring at about about 250 g cm^{-2}. Nowadays, we use a specially prepared flattening tool which consists of a 75 mm diameter billet of stainless steel (figure 3.22) with a spiral groove of some 4 grooves/cm machined in it with sharp crests, then 'lapmastered' to produce an optically plane contact line. This device, which has been idiomatically termed a 'stroller' is run on the wax surface until a high polish is seen on the wax.

It can be readily appreciated from the geometry of this device that considerable pressure is required to push a wax polisher into shape. The corollary is that it maintains its shape well—with the proviso that the temperature must not be allowed to rise much above 25 °C. Even though polishing environments are, ideally, temperature controlled, it is appreciated that many workers carry on in ambients that introduce very unfavourable variables.

In use, wax polishers are fed with a plentiful supply of polishing compound and should not be allowed to dry-out to the point where the work exerts increased drag. Wax polishers show a tendency to pick-up under the damp-dry conditions considered ideal for pitch. Because of this inability to withstand much drag even at 20 °C the polishing action is slow, but finishes are usually excellent. They are much less susceptible to accidental contamination from coarser abrasives than pitch is, probably because particles are more readily embedded (Twyman, 1951).

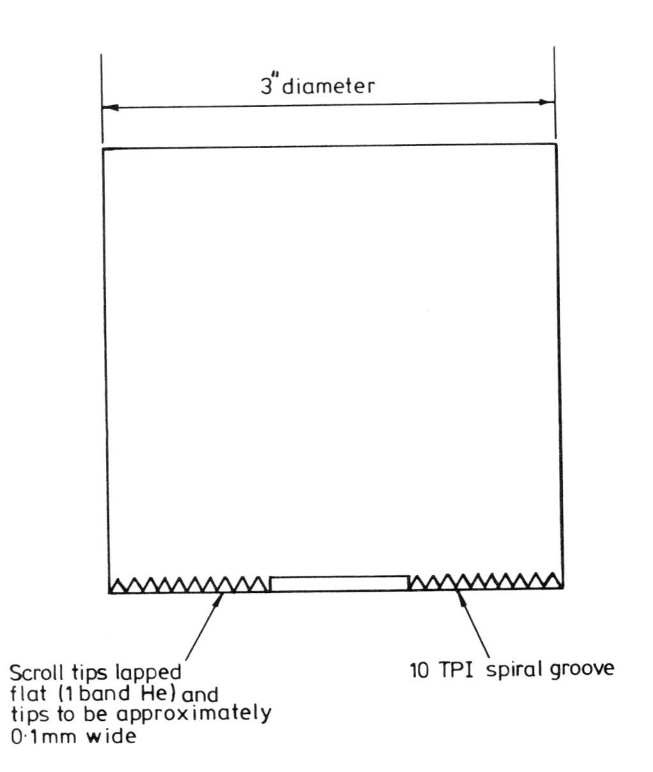

3.22 Diagram of a 'stroller'.

3.2.5 Wood

When the variety of matrices for a polishing compound is considered, wood seems to be either obsolescent or a 'gimmick'. It is fair to state that at one time many of the hard laser rods were finished on pear wood and the material might still find favour with some traditionalists. A cast iron or mild steel disc is faced with a 3 mm thick layer of end grain pear wood (figure 3.23)—that is to say, the growth axis of the timber is normal to the working surface. The wood is cemented to its base with an epoxy resin and this operation is carried out in two stages to preclude the probability of the slice becoming entirely impregnated with the resin. Firstly, the wooden disc is coated on its underside with resin and gelled at 30–40 °C; secondly, it is cemented with the same flexible resin (Araldite AY–HY 111) to the metal base. The polisher is cured at the same temperature, since any attempt to use higher ones for shorter times usually leads to cracking of the wood facing. When it has cooled, it is machined flat, grooved and 'Lapmastered' or hand-lapped flat. Before being used, it is cleaned of all abrasive and given a coating of the appropriate grade of diamond. At no time should water be used since this causes severe swelling or even stripping of the facing. A non-polar fluid such as Lapmaster oil, or oil-soluble Hyprez fluid, is suitable. The combination of diamond paste, oil lubricant and pear wood gives rapid polishing of sapphire—a fact that can be confirmed by jewel-bearing manufacturers.

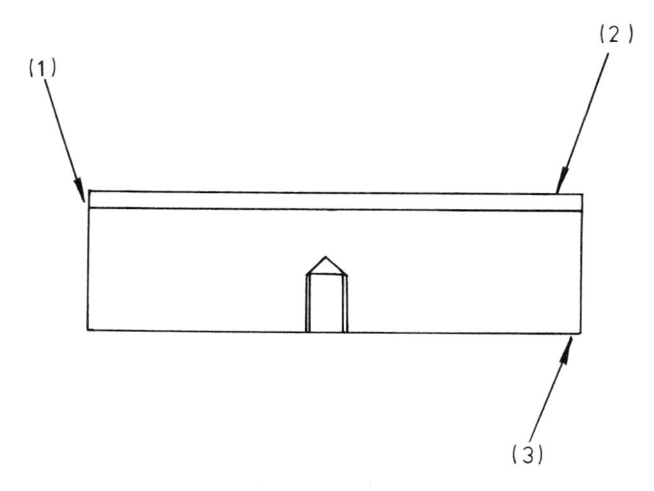

3.23 Wood-faced polisher. (1) Araldite twin-tube bonded at 40 °C for 24 h using a rubber foam disc with weight above. (2) $\frac{1}{8}''$ layer of pear wood, end grain vertical. (3) 3″ diameter × $\frac{3}{4}''$ thick cast iron or mild steel disc. (For $2\frac{1}{2}''$ lapping disc.)

3.2.6 Polyurethane foam

This material, in sheet form, can be used in a substitute for pitch when polishing glass (Mendel, 1967a; figure 3.24). It is capable of conforming to a moderate radius when warmed, but our experience has been limited to plane-polishing. All forms of pressure-sensitive adhesive films sold for mounting pliable facing have turned out to be largely ineffective in the hands of the authors, since the polishing slurry detaches the edge and results in a raised lip round the polisher. A special mounting technique has been devised to prevent this, and other unwanted excursions, from taking place (figure 3.25).

3.24 Polyurethane foam polisher.

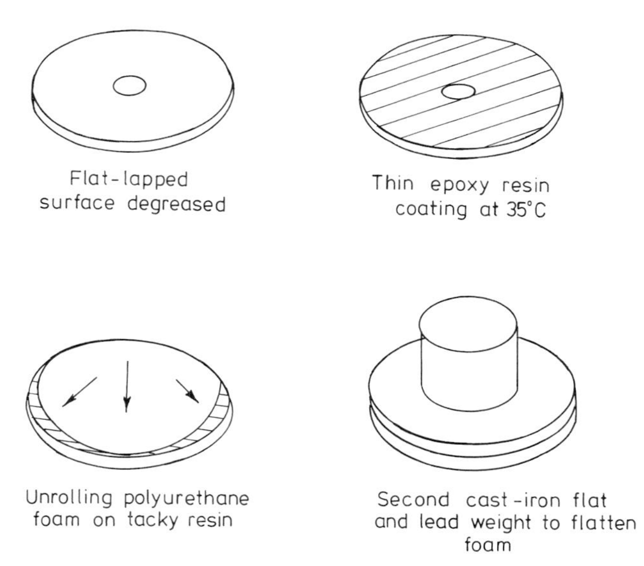

3.25 Stages in mounting flexible facings.

Two stainless steel plates are prepared by machining, surface grinding and lapping to one or two fringes in 150–200 mm. These are warmed to 35 °C and a quantity of epoxy resin (typically 'Araldite two-pack') is smeared on and 'combed' with a finely serrated spreader. The thin, uniform layer thus obtained has space for lateral expansion under pressure. It is most important, moreover, that the layer should be just thick enough to wet the underside of the facing, yet thin enough not to permeate it. Many successful foam polishers have been made by spreading the resin with a filter paper rolled into a ball (good wet-strength grade) and then dabbed to produce a uniform stippled effect. The disc of polyurethane foam, cut to the required diameter, is lowered on the plate whilst ensuring that it does not make contact at the edge before the centre is touching (beware of entrapped air). The second plate is pressed upon it and the whole assembly kept for six hours at 40–60 °C on a hot plate. Polishers prepared in this way give work flatness to better than three bands in 80 mm without further running in.

They are somewhat slow as polishers when used in conjunction with cerium oxide, which may be due to aquaplaning of the work. Whether or not this is the reason, a fact that has been noted is that the rate of polishing is increased when a spiral groove is cut in the foam. Acceptable speeds can be obtained, too, by grooving it at approximately 1 cm intervals. The polisher works particularly well when used in conjunction with silica sols (Mendel, 1976b) for example on silicon, sapphire and many other optical and electro-optical materials. The foam is extremely durable when used under conditioning-ring conditions and has lasted for up to twelve months, at 24 h a day, five days a week. One essential operation is to scrape the surface to break-up the acquired glaze at least once in 12 h. This operation can be carried out with the ground edge of a file. If, for any reason, the equipment should be stopped for an extended period, the polisher surface must be kept wet: allowing it to dry out is tantamount to loading it with coarse abrasive for many materials less scratch-hard than glass.

The end of its working life is usually signalled by the sudden appearance of turndown at the edge of the work, caused by the incipient breakdown of the adhesive bond. The facing may be removed by heating the metal plate to 160 °C at which temperature the resin bond has been sufficiently weakened for the foam to be easily stripped away, leaving the metal plate free.

3.2.7 Cloths

Many users find that cloth polishers (without further itemising the wide variety available) yield adequate, and therefore economically optimum, surfaces. However, we seem to be constantly expected to provide something better than adequate and cloths generally have become classified as quick but fundamentally not quite good enough. They usually give an excellent micro-finish but a poor macro-surface—either in terms of figure, or orange peel, or turndown, or all three. This statement should perhaps be qualified by observing that is is extended polishing operations that encourage these undesirable surface conditions. A relatively quick 'lick' on a cloth, when the principal conditions of flattness and smoothness have been attained on other polishers, can be beneficial. For example, our lowest damage polishing of silicon and gadolinium–gallium–garnet (GGG) have resulted from a final twenty minutes on Microcloth with silica sol slurry. The method for mounting and stripping the cloth is ideally the same as that for polyurethane foam. However, because the life of the cloth under alkaline conditions is barely a few hours at pressures of about 150 g cm^{-2}, and but little extended by lower pressures, self-adhesive coating can be used warily. They have the advantage that they can be mounted easily and strip readily in water at 90 °C.

3.3 Precision Surfacing and Scrolling Machine

One of the major attractions of pitch as a polishing matrix has been the relative ease with which it can be made to conform to a polished master surface. However, as our experience of a great variety of materials increased, the postulated superior abrasive embedding characteristic of wax (Twyman, 1951) often produced the optimum finishes, while many quick and satisfactory results were achieved with soft-metal polishers. Since none of these matrices conformed readily to a master, the rapid attainment of polishers flat to about $\lambda/20_{(He)}$ cm^{-1}, and spiral grooved, became a necessity. A good facing-lathe was used at first and, though the results were satisfactory, some running-in was always needed. Obviously, it requires very little play in a headstock, or excursions of the compound slide, to produce a surface which may be convex, concave, scalloped or a combination of these.

Moreover, though conventional machine tools may be freely used in research institutions, demarcation may limit their use in industry or industrially based research. The answer has been a special machine with accurate but restricted movements and low power consumption. This speedily faces laps and polishers flat to one band in four inches and seems to escape classification as a machine tool.

When its design features were first considered, it was thought easier to produce two rotary motions rather than a rotary and a linear one. This suggested a combination of rotating shafts and spring-loaded thrust bearings. At the same time, the design should avoid overlapping the functions in the controlling parts of the bearing system. Thus, the journal bearings should perform journal functions only, and the thrust races and their attendant heavy spring loading should control only thrust duties. Since rolling components and steel balls are available at very high accuracies, their use in the very precise thrust races was decided upon. The upper and lower race members could be produced to optical limits with ease if they were designed as plane surfaces. Sufficient accuracy from the journal bearings could be obtained with several simple arrangements but standard double-row self-aligning races were chosen because they interfere least with the thrust forces.

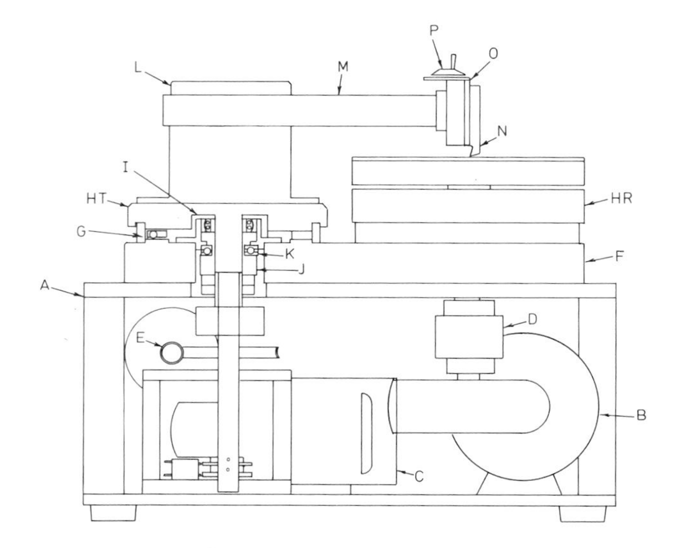

3.26 Diagram of precision surfacing and scrolling machine.

3.3.1 Main components

The machine (figure 3.26) consists of a light frame of plates and spacers A. The lower plate carries two motor–gearbox units: the rotary drive to the polisher-shaft B and tool arm traversing drive C. Both motors are fed from independent power supplies. Rectified 240 V mains are applied to the field windings and the armatures receive the rectified outputs of independent variable auto-transformers (figure 3.27). This arrangement provides continuously variable speed control and negligible speed change with applied load. The rotary-shaft motor–gearbox drives the polisher via a Reynold spider pattern universal joint D. The tool-traverse motor–gearbox operates its shaft through a further stage of worm reduction gearing E and an Oldhams coupling.

Mounted on the top plate of the frame A is a heavy casting F, the upper face of which forms a precise thrust surface for the twin-bearing system. The thrust surfaces are lapped and lightly polished, just enough to see fringes and to an overall tolerance of one fringe in 25 cm ($\lambda/50_{(\mathrm{He})}$ cm^{-1}). This face forms the lower surface of the two co-planar thrust races. Above it are the two sets of double row, caged, precise ball bearings G—6·35 mm diameter, 48 in each set. The upper rotating members of the bearing system are two cast-iron discs H_R and H_T lapped and lightly polished on their faces to the same tolerances as the baseplate F.

The journal loads of the rotating shafts are carried by double row self-aligning ball races I (one on each shaft). Finally, the two systems are held in considerable tension by two spring units J through standard ball thrust races K. The force exerted to keep the precise thrust bearing in contact should be at least equal to a weight of 25 kg on the vertical axes of both the polisher and the traverse arm. Although this obviously causes some plastic deformation of the cast-iron thrust surfaces, the formation of shallow grooves runs-in the bearings and obviates any springiness or transient jitter caused by gearbox vibrations.

The heavy boss above the traverse drive L carries the tool bar M. The clamp for the tool bar is designed to give some radial and tilt adjustment for the cutting tool N. Fine control of cutter depth is provided by a small dovetail slide O with a hand-controlled lead screw P. The traverse may be used manually by switching the operation of the worm E from the gearbox

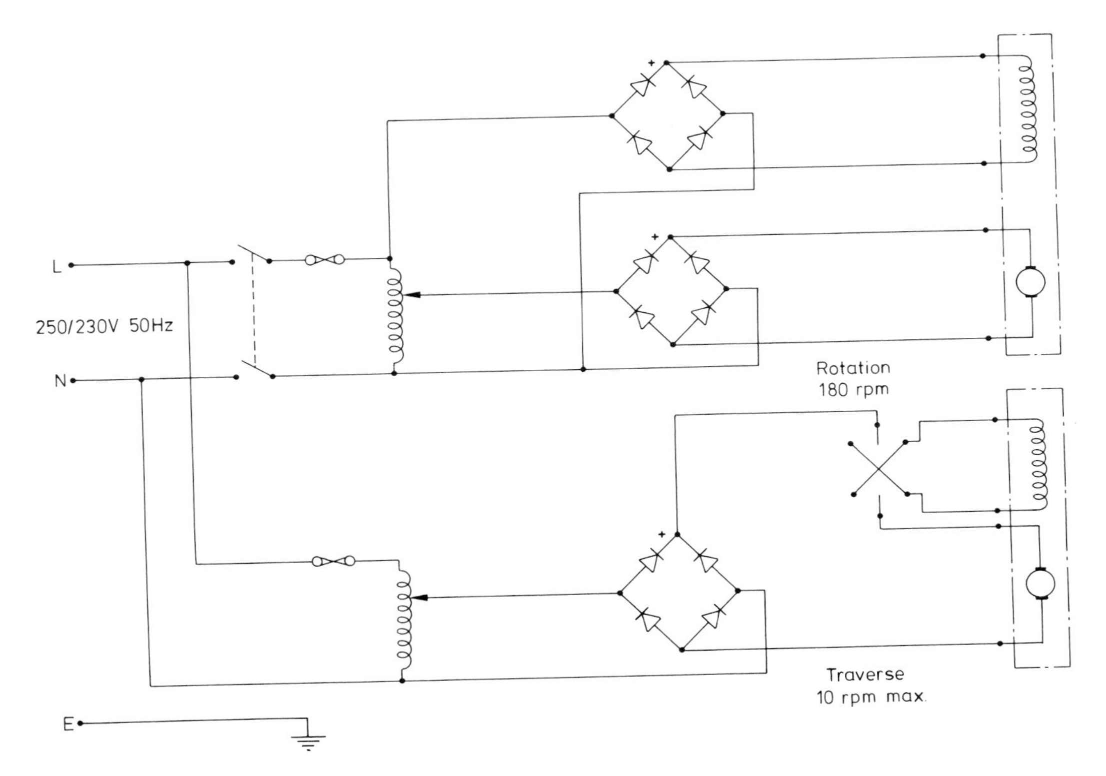

3.27 Circuit diagram of precision surface and scrolling machine.

drive to a panel control handle which can be seen on the left of figure 3.28. Both the manual traverse control and the vertical tool slide may be regarded as convenient refinements. The former could be dispensed with and the latter simplified to a pre-set clamp.

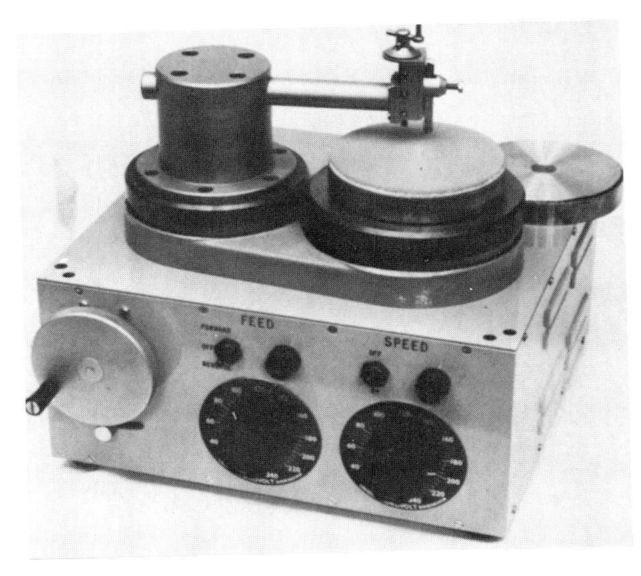

3.28 Machine scrolling a wax polisher.

3.3.2 Cutting tool

The tool used for the facing and grooving operations on most polisher matrices in common use is shaped in all the major angles like one ground for machining aluminium alloys (figure 3.29). The common practice, however, of 'flatting' the cutter tip is unacceptable for polisher maching. Because the extreme tip is the only region which affects cutting in these operations, it must

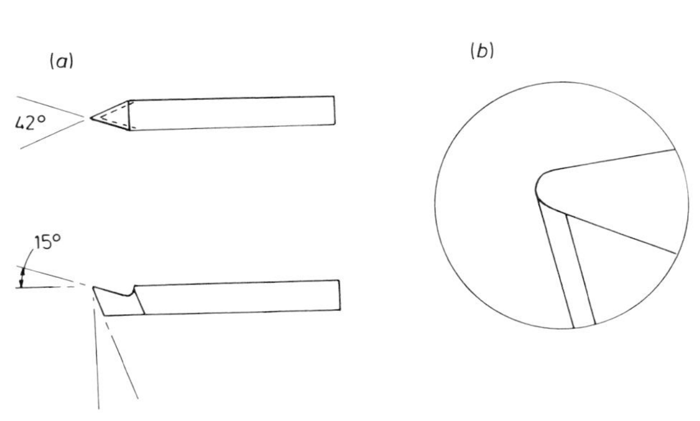

3.29 Cutting tool angles: (a) approximate tool shape; (*b*) enlarged view of the lapped areas of the tool tip.

be as nice as possible in form—that is, the intersection of a lapped part cylinder with the plane surface of the tool top. Thus, the two side faces are joined with a radius of 0.1–0.2 mm, produced with a fine but fast-cutting oilstone. Obviously, any run down or raggedness in this region is in effect a change of tool angle at a point where things are most critical.

3.3.3 Operation

The register faces (i.e. mating areas) on the polisher or lap and the machine are carefully cleaned before being lightly screwed together. The tool is mounted vertically in the post and the polisher rotated at the highest speed—100 on the auto-transformer scale, approximately 180 rpm. Cuts are set and made by means of the depth and traverse controls. The setting on the traverse speed control depends upon the tool tip radius and the surface finish required, but is normally between 25–50. Central recesses are turned-in at this stage by deepening the cut at approximately 30 mm diameter and continuing this cut to the centre of the polisher.

Spiral grooving is carried out by reducing the rotational speed to about 90 rpm (50 on the scale) and increasing the traverse to its maximum. A cut of 0.3–0.5 mm in depth is taken across the active area of the polisher.

Endless variations of groove pitch and surface reticulation are obtainable and some small work has required a correspondingly fine scroll. An effect equivalent to grooving and netting is obtained by superimposing a relatively rapid fine cut over the polisher after scrolling. Polisher matrices that are machined are wax, pitch, solder, indium and polyurethane foams. It is fair to say that our extensive use of wax has been greatly encouraged by the ease with which a contaminated or glazed surface may be precisely regenerated in a matter of minutes with this machine.

References

Amberg K 1940 *Metal Progress* **43** 147

Brooke S C J *Brit. Pat. 1040476*

Dickinson C S 1968 *J. Phys. E: Sci Instrum.* **1** 365–6

Field S and Weill A D 1946 *Electroplating* (London: Pitman) pp 192–6

Francis E A 1890 *Amateur Work Illustrated* (London: Ward, Lock and Bowden Ltd) 50–4

Fynn G W and Powell W J A 1967 *The Engineer* **224** (5839) pp 834–7

—— 1971 *Ind. Diamond Rev.* 270–4

Lindford H B 1941 *Trans. Electrochem. Soc.* **79** 443–53

Lord Oxmantown 1840 *Phil. Trans. R. Soc.* **22** 515–7

Mendel E 1967a *SCP and Solid State Tech.* p 32

—— 1967b *SCP and Solid State Tech.* pp 36–7

Metadalic Ltd *The Dalic Process* (Booklet)

Roberts E W 1955 *Trans. Brit. Ceram. Soc.* **54** 120

Strong J 1938 *Procedure in Experimental Physics* (Englewood Cliffs, N J: Prentice Hall) p 33

Tin Research Institute 1964 *Publication No. 92* (6th ed)

Twyman F 1951 *Prism and Lens Making* (2nd ed) (London: Hilger and Watts) pp 30–3

—— 1951 *Prism and Lens Making* (London: Hilger and Watts Ltd) pp 90–1

4 Specimen Mounting, Demounting, Cleaning and Packaging

Day after day, day after day
We stuck (S T Coleridge)

4.1 Mounting

Normally, a specimen must be affixed to a holder so that both are held unambiguously during the various lapping and polishing processes. Although this suggests cement of some kind, it is instructive to list the possible methods of obtaining a solid or semi-solid relationship between two or more components. Those which have proved to be the most used are:

(i) clamping mechanisms, for example, bolts, screws, dogs, vices;
(ii) freezing a molten metal, for example, Wood's metal, soft soldering;
(iii) magnetic and electromagnetic chucks;
(iv) air pressure, for example, vacuum holders, wringing;
(v) pressure-sensitive adhesives;
(vi) solvent-losing adhesives;
(vii) embedding or encapsulation in a compound which undergoes irreversible chemical change, for example, thermosetting resins, calcined gypsum;
(viii) cementing with polymerisable and/or cross-linking resins, for example, cyanoacrylates, polyesters, epoxides;
(ix) freezing a molten non-metallic solid, for example, waxes and water.

It is easy to dismiss the first three as scarcely applicable to polishing techniques. All the mechanical fixing methods cause a comparatively large amount of distortion but when the workpiece is sufficiently large and the wanted surface remote from the clamping constraints, satisfactory results can be obtained. Solidified molten metals could be used with thermally robust, wettable specimens. Heat distortion can be minimised by the use of low melting point alloys such as the bismuth/antimony/tin/cadmium/indium/lead mixtures. There is at least one notable example of magnetic fixing for lapping very precise components: the NPL (National Physical Laboratory) method of producing slip gauges (Rolt, 1929).

The fourth method has been applied successfully by some workers, especially by the development of vacuum blocks and chucks, but we experienced cracking of delicate, thin semiconducting compounds in our early work and did not persevere with the technique.

Pressure-sensitive adhesives—that is those which, like sticky-tape, adhere under the influence of applied load—have a limited value. They do not provide strong bonding on rough surfaces and are often readily attacked by polishing fluids and solvents. Moreover, they creep gradually under load. Because of their fundamental elasticity they are resistant to impact. Stefan (1874) derived an expression for a liquid acting to unite two discs (in essence as a pressure-sensitive adhesive) which de Bruyne (1957) applied to a highly viscous fluid between two surfaces in air:

$$F = \frac{0{\cdot}75\pi\eta R^4}{tD^2},$$

where: F is the normal force applied; η, the viscosity of the fluid; R, the radius of the discs; t, the time of separation and D, the initial separation.

Thus the force holding the bodies together is directly proportional to the viscosity of the adhesive and inversely proportional to the square of their separation. As the gap increases, therefore, the strength of the bond falls and as the viscosity increases, the bond-strength increases. These two factors are competitors: in practice we end up with thick films that have low mobility and thus do not flow into surface iregularities, They work best between polished surfaces. Pressure-sensitive adhesive tapes have their uses, especially for some light-duty sawing operations (chapter 1). A consistently satisfactory application to a lapping and polishing process that we can recall has been in a single-surface operation—that is, where only one side of a robust, thick specimen requires a flat specular surface.

The provision of a quick polish for crystallographic orientation checking of a slice taken from a boule-end is often achieved by holding the specimen in a block with double sided, pressure-sensitive adhesive tape. Apart from the disadvantages enumerated above, it should be noted that it is difficult to demount a fragile, large-area specimen stuck down in this way. A satisfactory method is to warm the block to about 60 °C, when the adhesive becomes more mobile and the application of a slight, continuous side pressure often serves to effect removal.

When a solid is either dispersed in or liquified by a solvent, it can set by drying out. At least one of the adherends must be porous for this to be possible, thus solvent-losing adhesives find little application in lapping and polishing operations. Even so, we have encountered devotees of 'Durofix', where no heating is necessary, but their tolerances tend to be generous.

Embedding a specimen in a thermosetting resin is a technique often used in the preparation of metallurgical specimens (Samuels, 1971) in order to protect the edges and extend the surface area of the work. In this sense thermosetting is used to indicate a resin or plastic which cures when heated to form a new substance. The reaction is irreversible. The designations 'thermosetting' and 'thermoplastic' are slightly woolly : originally they were applied to condensation and polymerisation reactions respectively in plastics. However, an uncured epoxide is termed thermoplastic and the cured substance a thermoset. Although the reaction is irreversible, different formulations of the resins and hardeners can be polymerised at temperatures ranging from 5–150 °C. They are not invariably thermosetting in the same sense that, for example, the phenol-formaldehyde resins are. 'Thermoplastic' is conveniently applied to those substances which, like paraffin wax, can undergo repeated heating and cooling cycles without changes occurring in their properties. Demounting is not a problem with an embedded specimen since it is in effect a composite block and there is usually no need in practice to separate the components at the completion of the polishing operation. Plaster of Paris (formed by an exothermic reaction which occurs when calcined gypsum is hydrated) is used for blocking prisms and lenses in the optical trade (e.g. Horne, 1972). Because the work is effectively wrung to the bottom of a flat mould, only the edges are held by the plaster. Extremely accurate work is obtained but the method cannot be applied to thin specimens.

We have occasionally resorted to embedding techniques but the great majority of mounting operations are achieved with thin films of re-frozen non-metallic solids: thermoplastic resins and waxes. On superficial examination, there is an inclination to use the polymerisable epoxides and cyanocrylates, especially the former which are exceptionally strong, chemically inert and produce very little strain as they cure. In the uncured resin, the termination of the bonds in polar hydroxyl or ether groups give ready adsorption onto the surfaces of the adherends, while low shrinkage during cure does very little to disturb the bond. Hence strong adhesion is obtained and, with a very small amount of strain within the joint, high cohesive strength as well. The problem is not one of sticking but of unsticking. Epoxides are gradually attacked by such chlorinated hydrocarbons as dichloromethane, especially when they are newly cured, since cross-linking in a cold-cure resin continues for a considerable period. However, a thin film beneath a specimen of moderately large area is removed very slowly, even under gentle reflux conditions. An alternative method is to heat the bond to its carbonising point but that is all too often a process which destroys both resin and work! With all these possibilities and limitations in mind, the mounting techniques which will be described hereafter are based almost wholly on a range of thermoplastic waxes and resins.

Generally, waxes can be esters of monohydric alcohols of higher homologues, or mixtures of higher-order alkanes; often crystalline, translucent and mouldable. Resins occur naturally as secretions from various plants, trees and insects; are amorphous, fusible, more or less brittle and often transparent. Common usage has confused the nomenclature since, for example, sealing wax was originally beeswax, but is now based on the non-vegetable resin, shellac. We have included our waxes and resins under the general heading of cements—where a cement is an adhesive that forms a solid union between two (or more) bodies.

4.1.1 Thermoplastic cements

What is an ideal cement? Probably one that is (i) easy to apply; (ii) strain relieving; (iii) wets a surface readily at a temperature not much above blood-heat; (iv) has high adhesive and cohesive strength; (v) low viscosity (except where gap-fillers are required in special instances); (vi) is compatible with polishing slurries; and (vii) leaves either no residues or ones that are readily dissolved with a non-toxic, cheap solvent—preferably water. Ice fulfils most of these requirements; the ones it contravenes are (iii) and (vi): the former indirectly since though it wets a surface most readily it solidifies at a difficult temperature for operation and the latter because most polishing fluids—except oils—would solidify at the bond temperature!

Consider table 4.1 which lists some of the relevant data for the cements we use most commonly.

One thing is immediately apparent: even the weakest cement provides a bond which, in shear alone, can withstand the normal lapping and polishing pressure of 100 g cm^{-2}. This has been confirmed for thin specimens. However, as the work becomes thicker, the forces imparted become increasingly similar to peel (figure 4.1) and, due to the rotation of the work, the situation is dynamically unfavourable. In re-circulating-slurry polishing, the friction between specimen and polisher can be disregarded and the force applied at the cement-line is effectively the polishing pressure. Where damp–dry conditions are employed however, the friction between work and polisher may be high and the force at the bond correspondingly greater, even though the apparent

Table 4.1

Cement	Bond strength† (g cm^{-2})		Thermal properties (°C)		Fluid attack (at 35°C)				Remarks
	Shear	Tension	Flow	Liquid	H_2SO_4 10% HNO_3 10%	NaOH 10% KOH 10%	Solvents‡ (See table 4.2)	Oils	
Paraplas	4500	2700	57·5	58	No	Slight	Trichloroethane, Toluene, Xylene	Yes	
Dental wax	7200	9000	55	56	No	Slight	Trichloroethane, Toluene, Xylene	Yes	
Beeswax/resin	20 000	12 000	56	62	Slight	Yes	Trichloroethane Acetone Dichloromethane	No	
Tan wax	> 54 000	> 54 000	60	(125)	Slight	Yes	Methylated spirits Propan-2-ol Acetone	No	Barium sulphate filler
Glycol phthalate	43 000	> 54 000	60	100	No	Yes	Ethyl acetate Methylated spirits Acetone	No	
Pitch	54 000	4500	60	130	No	Yes	Acetone Dichloromethane Ethyl Acetate	Yes	
Red wax			Plastic at 20	55	No	Yes	Dichloromethane Acetone Trichloroethane	Yes	

†Cement thickness 20 μm; 1 × 1 cm polished glass specimen 0·5 mm thick on lapped (600 carborundum) stainless steel block.
‡Solvents are quoted in order of efficacy. Only three are given in each case, though, for example, Beeswax/resin is attacked to some extent by all the solvents listed in table 4.2.

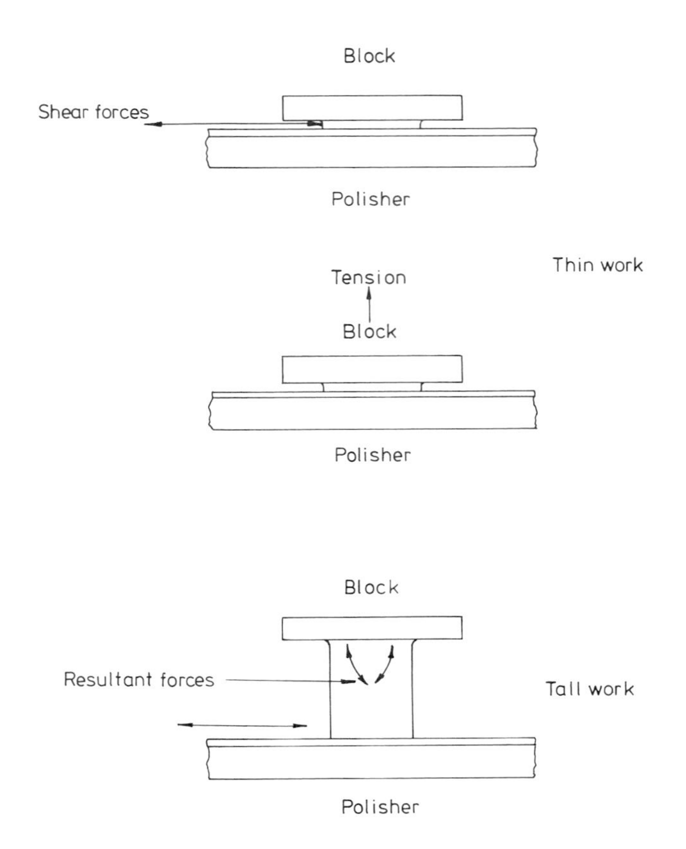

4.1 Forces at cement interface during lapping and polishing processes.

applied pressures. Conversely, one containing an inert filler, though not easy to arrange in a thin and uniform manner, resists shrinkage and increases the bond strength. Experiment alone can indicate which thickness of specimen at a given surface finish and flatness will remain safely stuck in place during a specific polishing process. The columns listing fluid attack are partly a guide to solvents which we use most commonly for demounting and cleaning (§4.2) and partly a pointer to the lapping vehicles and etch-polishing slurries that can cause trouble. Flowing and melting temperatures are somewhat arbitrary, since the cements are not always consistent. These cements—as we have listed them—often have to cope with materials which are affected by thermal shock. Therefore a low melting point, ideally associated with strain-relieving properties, is an advantage. Of course, that a cement should flow at a low temperature is not the only consideration: it must wet readily when it melts and be low in viscosity. Consider figure 4.2: a typical graph of viscosity versus temperature has the curve as shown. Where a specimen is robust enough and the cement stable at elevated temperatures, a thinner film is obtained by heating the adhesive well above its wetting point. (The wetting point may be taken as the temperature at which the forces of adhesion to an adherend equal the cohesive intermolecular forces in the cement.) The upper limit is the temperature at which it either vaporises or becomes solid.

polishing load might be normal. Increased pressure is attended by increased temperature and these factors combine to cause movement of the specimen and even complete breakdown of adhesion. Users are often wary of paraffin waxes because of bond-failures but these can usually be traced to one or more of the following factors: poor wetting initially; temperature rise in the machine and slurry; high frictional drag between polisher and work; bonding uneven surfaces with a non-gap filling cement.

Note (i) to table 4.1 describes the surfaces of the adherends tested. In every case the bond failed at the polished glass/cement interface. Since hot-melt adhesion is a combination of physical (keying) and chemical (adsorption) bonding, the cohesive strength of thin films seems to be greater than their adhesive strength. The expected behaviour of a thick, unfilled cement is that the strain introduced during setting weakens a join and encourages rupture at lower

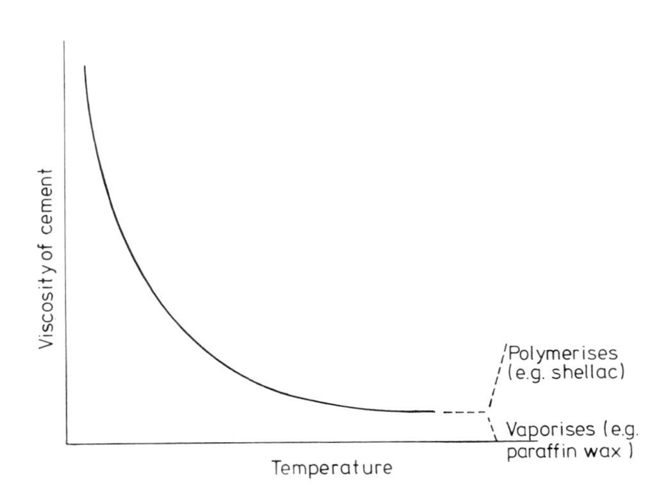

4.2 Typical graph of viscosity against temperature.

4.1.2 Sources and/or manufacture of the cements tabulated

Paraffin waxes are often translucent crystalline or microcrystalline substances. They consist of a mixture of straight and branched-chain alkanes with greater than fifteen carbon atoms per molecule and can be obtained by de-waxing light lubricating oil stocks. Paraplas is obtained from British Drug Houses Ltd and Dental wax from M S Cottrell and Co., London. Though apparently very similar in thermal behaviour, the former wets a surface more readily but is weaker both in tension and in shear. We believe this to be due to the presence of a small quantity of dimethyl sulphoxide in the wax. The faint dye in our dental cement does not have sufficient colour density to add anything to the visual evidence of successful coating seen through a transparent specimen: both waxes clear on melting and becomes cloudy on solidifying

Beeswax/rosin (colophony) has been a standard laboratory cement in various proportions (sometimes with additives) since 1827 (Faraday, 1827). Beeswax, secreted by bees, consists mainly of an ester of palmitic acid plus some cerotic acid. It melts at 60–65 °C. Rosin is the sticky residue which is left when tree resin is heated in order to distil-off turpentine. Its melting point is 100–140 °C. When the two are compounded in the proportion of $3\frac{1}{2}$ parts of rosin: 1 part beeswax, as recommended by Twyman (1957), the melting point falls to about 63 °C. We obtain both components from British Drug Houses—sold as beeswax and colophony—melt and stir them together, then pour the mixture in strips on to a sheet of mylar film spread over a cold·cast-iron surface table. The sticks thus produced are broken into convenient lengths and stored vertically in glass containers which are kept in a cool place.

Tan wax is an example of a filled, shellac-based cement. Shellac is a purified insect resin, produced on twigs of trees in India by the lac beetle. It is a complex mixture of acids and esters and has two quite distinct states: a softer thermoplastic condition (the desirable one) which is quite pliable at 80 °C and a harder, polymerised form that finally becomes irreversibly solid with prolonged heating at above 100 °C. At the first state, shellac is readily dissolved, for example, by alcohols, but at the second it is denser, very resistant to solvent attack and, if accidentally caused, not easily remedied. Overheating with the application of brutal side pressure can free a specimen—if it can withstand such treatment. Alternatively, a long soak in caustic soda solution can be effective but again the specimen may possibly suffer. However, in both cases, it is Hobson's choice, since work permanently bonded to a block can find few applications apart from acting as a permanent warning. We obtain Tan wax in sticks from F Lee and Co. (Coventry) Ltd. The filler is barium sulphate.

It cannot be said to wet a surface as, for example, the paraffin waxes do. The presence of the filler results in cement films which, at their most liquid, probably deserve the term pastes. By the application of heavy loads during setting, the join may be thinned down to about 25 μm but it is the gap-filling, high shear and tension strengths of the cement which are its most attractive features. A thick specimen can be basted with Tan wax applied in drops from a stick heated in a flame; the resulting fillets around the area of the joint give exceptional strength and many of the special operations listed in Chapter 5 rely upon the adhesive properties of this cement.

We have studied glycol phthalate in some detail because it is strain-relieving to a considerable extent, provides very thin films when fluid and is extremely tenacious. Moreover, it is readily removed without leaving residues. Kienle and Hovey (1930) and Kienle and Ferguson (1920) investigated the alkyds thoroughly (*al*cohol and aci*ds*)and reported that the esterification of stoichiometric quantities of ethane diol (ethylene glycol) and phthalic anhydride produces a long-chain polymer which does not become a gel (infusible resin) when heated:

$$CH_2OH . CH_2OH + COC_6H_4COO \xrightarrow[\text{in 1 h}]{250\,^\circ C}$$

$$HOCH_2CH_2O(COC_6H_4COOCH_2CH_2O)_n$$

$$COC_6H_4COOH + H_2O \qquad (95\%\ \text{esterified}).$$

At 250 °C the condensation reaction goes virtually to completion in about three hours. However, at 200 °C the rate of the reaction is slower and it is possible to produce resins which vary in flow-point for different 'cookings'. Figure 4.3 shows esterification versus time at 200 °C. The resins have been tested for penetration much as pitch is tested and on the same indenting apparatus (chapter 3). All the softer mixes healed with time—that is, the dimple produced by the loaded steel point disappeared when the resin was left to stand in a dark cupboard at room temperature. The esterification process continued insensibly too, since on retesting, there had been some hardening. The data is displayed graphically in figure 4.4. We have a resin which is fusible, soluble and can be

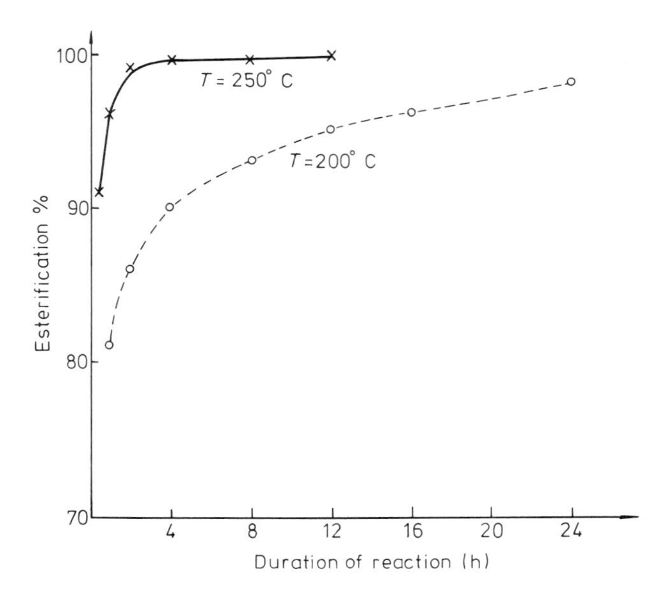

4.3 Esterification of glycol phthalate against time at 200 °C.

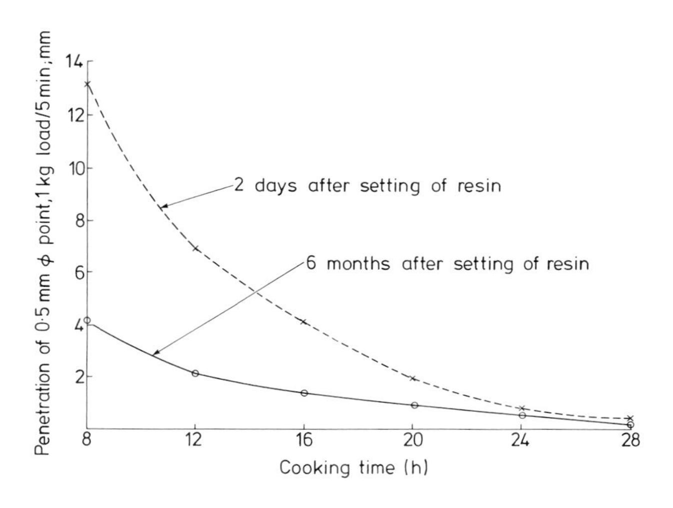

4.4 Increase in hardness of esterified resin over an extended period.

tailored to suit specific materials: where strain-relieving is important it can be arranged accordingly. Some snags are that glycol phthalate is not gap-filling and flows freely only at a fairly high temperature (~100 °C, depending upon the completeness of esterification).

Our standard method of manufacture is to take equi-molar proportions of Analar grade reactants, mix them thoroughly together and heat them at about 135 °C until they clear. Since phthalic anhydride sublimes readily, the reaction vessel must be covered to minimise the formation of long, fine needles in the air. These are extremely irritating to the inhaler. If the chemicals themselves are not contaminated and the glassware is clean it has proved unnecessary to filter the mixture but it must be stirred at this stage. It should be recovered, returned to the oven and cooked at 200 °C for the length of time required: say 16 h for a soft resin and 24 h for a brittle one. Typical quantities that we use are: 186 g ethane diol + 444 g phthalic anhydride, which yield about 550 g glycol phthalate. The only by-product of the reaction appears to be water. At a temperature of 400 °C the resin vaporises completely.

Pitch is a cheap and time-honoured cement in many optical polishing processes. We use 1 mm Swedish pitch (so-called) which is obtainable from H V Skan Ltd. Although it is chemically the same as tar, the term pitch is here applied to resinous wood tars, derived from softwoods by distillation. These differ from hardwood tars because traces of the pleasant-smelling mixture of terpenes (turpentine) are still present in them. They are more generally known as Stockholm or Archangel tar and comprise a complex mixture of esters, alcohols, ketones, fatty acids, phenols and waxes.

As a mounting medium, pitch has several special applications. In the process of lens making, cylindrical pieces of it—known as mallets in the UK and as buttons in the USA—are used for cementing the grey lenses to a block. Two requirements are thus satisfied: the alignment of the lens-faces with the correct radius cast-iron tool and, when cool, the thick pitch can still flow enough to accommodate slight changes in the shape of the work. Its dense pigmentation provides a non-reflecting background which results in increased contrast when fringes are viewed, and it makes enhanced photomicrographic records of polished surfaces possible. The latter advantages are true, too, with other darkly tinted cements. The practice of using colouring as an aid to detection of uneven adhesive films between specimen and block is not often applicable with pitch, since a thick, dark, strain-relieving layer is uninformative. Perhaps a major snag is that the residues are extremely tenacious and many electro-optic materials cannot be scrupulously cleaned after they have been mounted with pitch.

Soft red waxes of the SIRA type (Walden, 1936) are useful in sawing operations for mounting crystals that are prone to cleave from thermal shock. They have, too, some applications in lapping and polishing for the same reason. No great angular stability should be expected however and joints should always be as thin and as large in area as can be arranged. It is an easy matter to check and correct the angle at intervals on an adjustable-angle jig. Since the mixture is of beeswax, rosin, oil and red pigment, it is readily attacked by most solvents and by oils. We obtain SIRA wax from British Drug Houses Ltd.

4.1.3 General techniques

Probably the greatest disadvantage in the use of thermoplastic cements for work mounting is the strain that they introduce between the specimen and the block. Any differences in coefficients of expansion are potential strain-inducers. A non-uniform cement thickness, too, affects the work because of eneven shrinkage on cooling and setting. Even a uniform film of cement inevitably solidifies from the periphery of the join and imposes a bending moment on the adherends. When these factors have been optimised by arranging for good matches in the thermal expansions of work and block (there are some happy coincidences such as calcium fluoride and aluminium) and for a very even adhesive film, further troubles can still arise. They stem from the use of the more rigid cements with the consequent inability of the work to 'move' as, say, the lapping strains are exchanged for the lesser ones of a polished surface. This is manifested by changes in the shape of the work after its removal from the block. The problem is a fundamental one and a great deal of ingenuity has been applied to minimising strain: the component must be fixed to a block, rigidly enough to prevent its moving during lapping and polishing operations while allowing it sufficient freedom to change shape with varying stresses.

Fortunately, in a lot of electro-optic work the final flatness is of secondary importance. However, we have met the problem in the shape of lightweight single-sided $0{\cdot}9 \times 1{\cdot}3$ cm rectangular mirrors about 0·05 cm thick, flat to $\lambda/2_{(\mathrm{He})}$. These were made from fused quartz, mounted in batches of twenty on stainless steel plates with 1 mm pitch as cement. The thickness of the adhesive layer was about 0·01 cm, which would be regarded as very thick by waxing standards and, conversely, very thin if it were a pitch mallet. Nevertheless, its function as a flexible mounting medium was satisfactory, provided that it was warmed by 10–20 °C twice or three times at intervals during the polishing cycle. The observed shape changes after each warming and cooling were polished out to the point where little or no departure from a true plane could be seen on removal of the work.

There are several different methods of applying cement for the purpose of affixing specimens to blocks. Some workers favour coating the entire surface of a heated block with cement then placing the work, which must be at the same temperature, on the adhesive. The assembly is taken to a cold plate, such as a large cast-iron surface table and the work immediately covered with a large sheet of polyester (Melinex) film followed by a sponge-rubber disc of the same diameter as the block. A flat weight, giving 100 g cm^{-2}, is positioned on the disc. Figures 4.5–8 depict these steps. The polyester film acts as a renewable release agent because, although it sticks to some cements, the fact that it can be bent through an obtuse angle allows it to be peeled off.

Painting the block with cement in this way tends to enclose pockets of gas beneath the specimen, even though many voids find their way out during the pressing operation. Our own technique is akin to that of cementing optical components together: one spot of cement for each slice or plate is dropped on the previously cleaned block and the work allowed to settle on this (figures 4.9 and 4.10). When the specimen

4.5 Wax-covered heated block.

4.6 Work in place on melted wax.

4.9 Wax spots on block.

4.7 Transfer of block and work to cold plate.

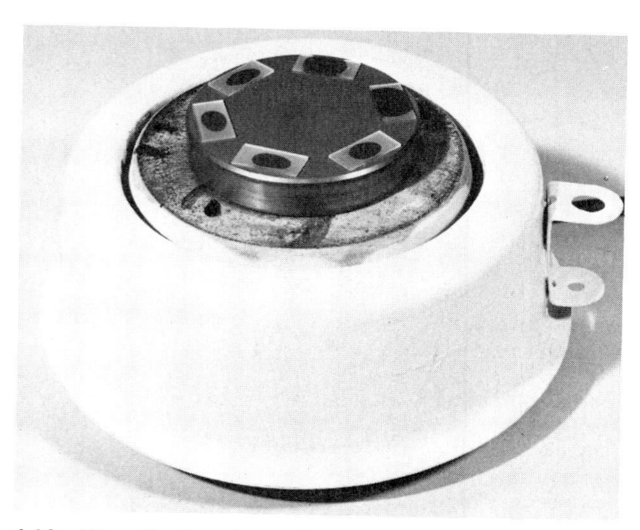

4.10 Wax flowing beneath specimens from the centre outwards.

4.8 Complete assembly of block, work, polyester film, sponge rubber pad and weight.

is rectangular the drip of cement is lengthened proportionately into a stripe. No air is entrapped and provided that the amount of adhesive is gauged correctly, the surplus will produce a small fillet at the edge of the work. Cleaning off superfluous cement with solvents, or chiselling it away from the finished block is thus minimised.

An additional satisfactory way of applying adhesive is to assemble the work—say crystal quartz surface wave plates—in position on a stainless steel block, place a grain of cement at the edge of each plate followed by the Melinex sheet, foam rubber pad and loading weight. The assembly is slid onto a hotplate, warmed gently to 100 °C and allowed to cool equally slowly. During the cycle the cement penetrates under the specimens by capillary flow. At the same time, it produces an even fillet around the edges of the work. Capillary techniques are only possible with cements that become properly fluid, typically paraffin waxes, beeswax/rosin and glycol phthalate. Shellac, stiffened with inert fillers, is not recommended. Figure 4.11 shows capillary flow with the pad and weight removed for the purpose of illustration.

The techniques so far detailed are related to work which has to be polished on one side only, making the protection of the remaining lapped (or grey) side unnecessary. Methods vary considerably for the protection of polished surfaces which are being cemented to a block. Conventionally, either lacquering or the insertion of waxed paper between the polished face and the mounting plate is employed. The polishing process in use determines the technique to some extent. Waxed paper—or, better still waxed impregnated optical tissue—is adequate for short polishing runs such as obtaining a specular surface with a machine using pitch and cerium oxide/water slurry. A process like this is usually completed before any swelling of the edge of the paper tissue occurs. However, after 24 h of alkaline silica sol polishing most of the cement fillet at the edges of a specimen will have been eroded away, exposing the paper. Despite its wax impregnation, it swells and bends the edges of the specimen upwards (figure 4.12). Polyester film forms a good separator for work which has to be subjected to extended attack by alkaline slurries.

4.11 Capillary flow of wax from one edge inwards: specimens unweighted for the purpose of illustration.

Cleaning off surplus cement is very important where polishing follows lapping. Most polishing processes, in practice, erode any exposed adhesive on the blocks even though no actual contact takes place between the polisher and the surplus cement. The chief danger lies in the possible release of carborundum grains embedded in the cement during the lapping processes: the release of relatively large silicon carbide or alumina particles into polishing media causes havoc. If sufficient care has been taken during the early stages of mounting there should be very little surplus cement on the face of the block and what there is can usually be removed with a solvent. However, where larger quantities have strayed onto the surface, a wooden or Tufnol chisel should be used to pare the cement away—not screwdrivers and nickel spatulas, which can chip the edges of even well-chamfered work all too easily. Similarly, wooden or Tufnol probes should be used to manipulate work

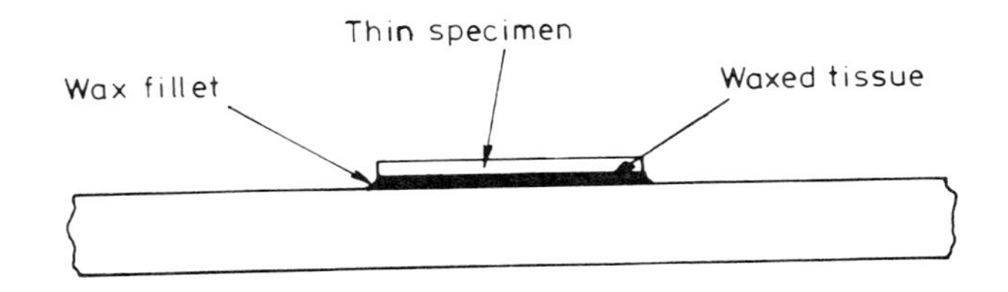

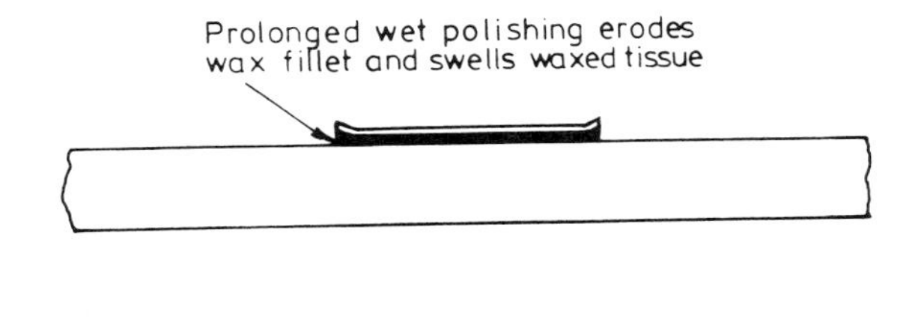

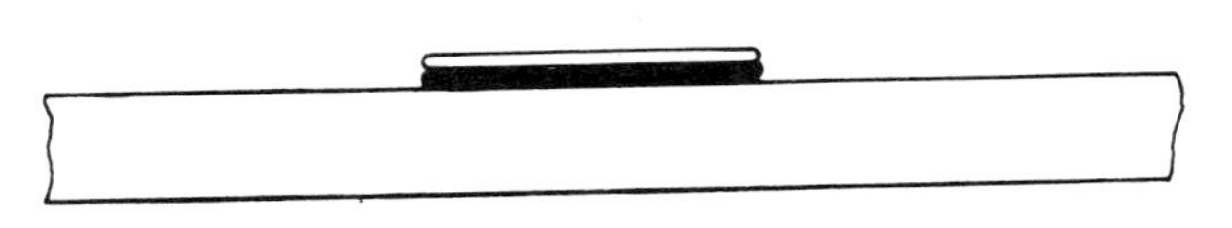

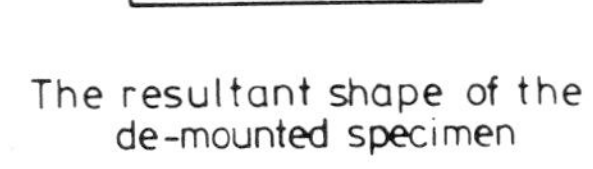

4.12 Effects of edge swelling of tissue on specimen form.

into position on the block prior to pressing and cooling. If there are several slices they should, of course, be evenly distributed in the interests of good, uniform flatness of polished work.

At the pressing stage, some workers have difficulty with specimens sliding out of position when load is applied. This can be prevented by using identical diameters of both block and load weight; the former is arranged against two magnetic blocks and the latter lowered in place against these constraints (figure 4.13). Production or semi-production would encourage the development of a more refined jig, possibly of tubular form, made to slip over the work block and just high enough to register the load block. An alternative way of preventing work from sliding on a film of liquid wax is to use a fairly large piece of polyester film and anchor it down to the cold plate plate with magnetic bases in three places. The rubber pad and weights can then be applied without the specimens moving. A very simple method of applying pressure, both to squeeze out excess cement and to apply load during cooling is to use a rubber teat (figure 4.14).

The surfaces of work plates have been assumed to be optically flat. Although a grey (lapped) face is invariably used, our workplates seldom depart from a true plane by more than $\lambda/4_{(\mathrm{He})}$ in 9 cm diameter—probably much better than is required. However, it is easily obtainable from the 60 cm ring-lapping machine.

4.13 Magnetic bases used to prevent lateral movement of weight and consequent specimen-slip on wax film.

4.14 Application of light pressure to a single specimen using a rubber teat.

Lesser standards can be adopted: for example, high-grade turned or fly-cut surfaces which are free from ribs and ditches. Any significant surface irregularity will hold excessive quantities of cement which shrink on cooling, pulling thin workpieces out of shape. After it has been well lapped and polished, such work replicates the machined block-face demounted and freed from the cementing stresses. Fine surface grinding ($\sim$4 microinches centre-line-average or better) probably provides the best alternative surface to an optically prepared one, especially when the same care is taken as in the techniques for manufacturing lapping plates (§2.1). Some surface-ground blocks obtained commercially have been flat enough to produce fringes under an optical flat: their surfaces are more than adequate for routine mounting of slices. Slightly satin finishes from grinding, as with lapping, provide a good keying surface and hence a strong physical bond for cements.

So far, mounting techniques have related to stainless steel blocks. It should be stressed that it is the more general practice in the optical industry—and the occasional practice in our environment—to use glass for the purpose. Soda-glasses need a much modified heating and cooling treatment, since a cold glass disc, larger than 4·0 cm diameter × 0·6 cm thick, almost always cracks when slid onto a hotplate ($\sim$100 °C). There are expensive alternatives of borosilicate and fused quartz that are unaffected by this degree of thermal shock. The glass discs (usually cut from float-glass sheet) are conveniently buffered by interposing a filter paper between hotplate and disc. Blocks and workplates up to 9 cm diameter and about 1·2 cm thick can be satisfactorily treated by this procedure. It is equally important to use a paper disc when the glass is cooled on a heatsink.

A simple way of monitoring the temperature of a hotplate is to drill a brass cylinder about 5 cm diameter × 5 cm in length so that a mercury thermometer can be fitted in it (figure 4.15). The depth of the hole approximately equals the immersion depth (~5 cm) of the thermometer and the block is self-standing too. Most simmerstat controls operate down to 100 °C. Hotplates without them can be fed from a variable auto-transformer of the variac type. The arrangment allows control at temperatures lower than those achievable with conventional simmerstats and is particularly useful with paraffin waxes and beeswax/rosin. It is worth reiterating that shellac-based cements must not be overheated; because their wetting properties are poor the temptation is to give them prolonged heating. The bond gradually polymerises at temperatures above 90 °C and quickly polymerises above 150 °C.

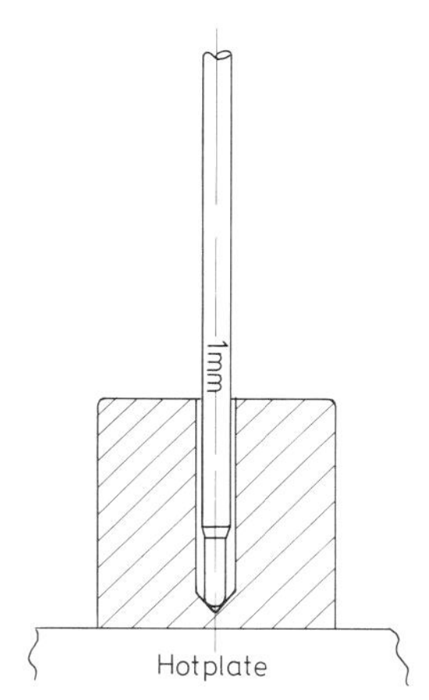

4.15 Brass block for standing thermometer on an electric hotplate.

4.1.4 Jig techniques

The techniques described for blocks apply equally well for mounting adjustable-angle jig-work. Special care has to be taken with thicker specimens when a surface, already polished, is mounted over a hole in the workplate in order to expose a section for optical alignment purposes. (chapter 6). The work must be thick enough to be held over the hole without being noticeably deflected: ideally it should be cubic. The problem with all work mounted in this way is to prevent damage to the cemented polished surface—as described before—and to restrict the cement from creeping over the exposed face. If a waxed tissue is to be interposed, as it can be for all but very long, wet runs, it should be impregnated with a minimal amount of cement such that the finished mounted join is free from voids. When polyester film is used the join has to be a capillary one, most usually with paraffin wax which produces a thin, clear bond. Of course, the specimen must not slide about during the operation: firstly because any movement might damage the polished face; secondly because the wanted exposed area may become smeared with cement and thus destroy its reflectance. If this happens, there is a temptation to clean the face by poking a tissue down the hole, but in our experience it seldom results in a clean face while circular scratches have resulted from detritus gathered from the bore. In order to prevent work from sliding about whilst cementing it, the plate is sometimes machined with a recess which accommodates the diagonal (or the diameter) of the specimen. (figure 4.16). To be effective, the recess should be as shallow as possible yet still register the work. Two shallow pins protruding from the workplate form a very effective alternative solution. They have the advantage that the plate can be lapped flat after the drilled holes for the pins have been made. These do not have to be a press fit in their holes but are more easily made a slack fit and subsequently cemented in place.

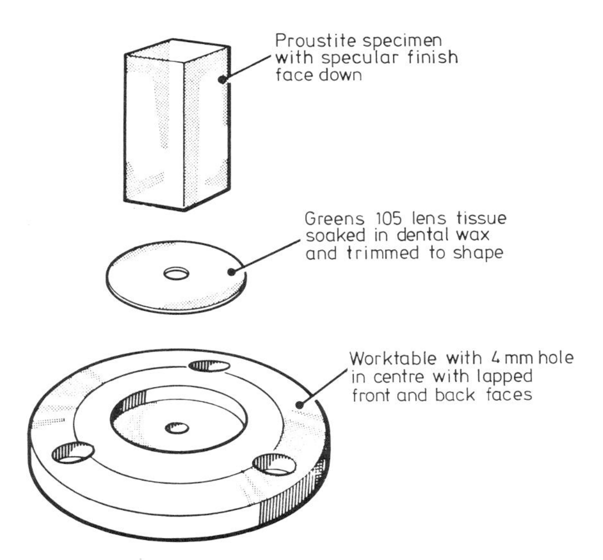

4.16 Recessed jig-workplate: insertion of tissue to minimise damage to a soft specimen.

4.1.5 Uniform cement layer mounting

The production of uniform cement films of known thicknesses presents difficulties and a complete solution would be of great value in polishing angularly related surfaces. Plaster blocking offers a partial solution since the process relies on mating precisely lapped faces and it can produce accuracies of one minute of arc in very skilled hands. It is however limited to thick specimens while a great amount of electro-optic work is too thin for it to be applicable.

Various techniques for producing uniform cement films have been developed with some success. Block surfaces can be interrupted to provide reservoirs for the cement: for example, by lapping a substrate with very coarse carborundum (~ 120 grit) which produces severe pitting. A subsequent light polish forms an array of small plateaux with well-rounded edges and provides a discontinuous optical surface. Good images appear under an auto-collimator and clear fringe patterns show with an optical flat. Thin wax films (~ 5 μm) can be obtained by carefully applying this technique.

Another method which involves thicker adhesive films and is useful on more substantial specimens employs a miniature three-ball plane: the work stands on three equally high protrusions and the interspace is flooded with cement (figure 4.17). The substrate is made by dipping a wire in liquid epoxy resin and stippling a series of dots on a spare glass slide whilst viewing the action through a low-power microscope. After several dots have been made, their diameters seem to vary very little. At this stage a previously prepared disc of glass (or metal) has three dots placed on it, arranged at 120° on a pitch-circle-diameter suited to the diameter of the work. The resin is allowed to cure at room temperature or slightly above it. Too great an application of heat, though serving to accelerate the cure, causes spreading of the dots. However, epoxies benefit from a post-cure at an elevated temperature after the gel stage has been reached and this may be conveniently ascertained by testing the dots on the spare glass slide. They can be given the same post-cure, too, and tested for complete cure at intervals.

Assessment of uniform height of the three dots is made by standing a small optical flat on them and observing the interference fringes between the flat and the substrate. Under these special conditions where appreciable gaps are present between the components, parallel monochromatic light should be used: an interferoscope as opposed to a diffuse source, that is. The latter arrangement will produce fringes of concave appearance between two flat surfaces and though it can be used it should be remembered that, for the parallel state, the fringe system needs to be concentric with the three dots.

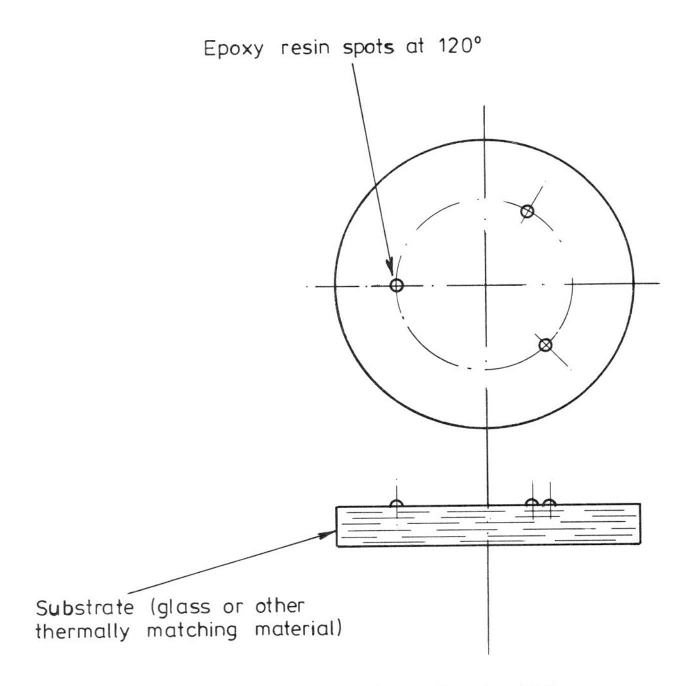

4.17 Miniature three-ball plane for uniformly thick cement layer.

If the dot-height is unequal, small corrections can be made by lightly polishing the high dot or dots until the fringe pattern fluffs out. In this arrangement, where up to 100 μm cementing thickness may exist, strain-relieving pitch helps to reduce bending of specimen and substrate as the cement solidifies. The film can be very uniform in thickness—variations of less than 0·25 μm are easily achieved.

4.1.6 Porous substrates

In spite of its good results and satisfyingly kinematic design, the three-dot substrate imposes a fairly tedious and quite demanding technique on the operator. The antithesis of the thick, even film is the zero, or near-zero adhesive layer. It can be achieved by the application of high pressure to such robust materials as ruby but many specimens cannot be loaded heavily with safety while they are supported on a fluid cement film owing to their fragility and irregular surface area. The method now described uses a near-zero cement layer ($\sim 0{\cdot}125\,\mu m$) on the surface of an impregnated porous substrate which has been prepared plane parallel and is analogous to an optical flat (Fynn *et al*, 1971). Since we originally devised the process for polishing thin specimens of silicon the details given relate to this but it has been used subsequently on a wide variety of metals, semiconducting compounds and electro-optic materials.

Up to the present time the porous substrates have been made from a medium grade (180–220 grit) 'India' oilstones from Norton Abrasives Ltd. The particle range seems to be fairly critical since a much finer grade/cement combination results in a filtering effect which leaves a non-congealing film. A coarser grit will not acquire a good specular surface—a condition which is important for jig and auto-collimator techniques. Other sintered materials such as glass and bronze may offer alternatives but it must be remembered always that thin polishing often involved etch/abrasive (chemech) polishing at the final stage and that the substrate must resist chemical attack by the etchant.

The oilstones are trepanned to 3 cm diameter (see chapter 5), any oil or grease in them removed by heating and solvent-degreasing, then lapped or surface ground until they are 0·6 cm thick. These dimensions are related to the adjustable-angle jigs on which the polishing operations are most conveniently controlled and carried out but in principle there is no bar on larger substrates. The final parallelism required can be obtained by hand polishing on brass, copper or solder laps with $3\,\mu m$ diamond abrasive and Hyprez fluid. Frequent checks are made by revolving the substrate slowly on a kinematic (three-ball) plane (chapter 6) using a sensitive comparator to monitor variations in thickness. Accuracies to two seconds of arc are obtained in this way ($\sim 0{\cdot}3\,\mu m$ in 3 cm). The polish on one or both sides must be sufficiently specular to show fringes when viewed by monochromatic light through an optical flat and the standard of flatness required is better than $\lambda/10_{(He)}$ in 3 cm. An improvement in the specularity of the surface is obtained by giving it a short duration polish with a silica sol–alkaline solution process (Mendel, 1967). When auto-collimator techniques are being used to check parallelism of the work, both sides of the substrate have to be flat and parallel; if, however, specimen parallelism is obtained by interference techniques on the front surface, one flat and polished surface only is required.

Just prior to cement impregnation, the substrate is cleaned by ultrasonic means or by firing at 400 °C. The impregnation technique which is used for glycol phthalate will be described. The substrate is heated to 100 °C and the cement spread on the surface (figure 4.18). Even at this temperature the esterified resin is quite viscous and sinks slowly into the oil-stone. Some areas appear matt and must be re-covered with cement until the surface appears liquid for 2–3 min. The substrate is then placed on a cold surface, when the sudden increase in viscosity effectively stops further impregnation. While the wax is still plastic, all surplus is sheared off the surface with a sharp straight edge, such as a new razor blade (figure 4.19). The first stroke removes most of the cement and subsequent strokes made successively at right angles shave the cement level with the polished porous surface. Latterly, we have found that many of the lapping and polishing operations which were considered to require a cement with high shear strength can be equally well carried out with paraffin waxes. Porous substrates are wax-impregnated and shaved in the same way but at a lower temperature, of course.

The specimen is mounted on the porous substrate by sandwiching a piece of optical tissue between the faces then withdrawing the tissue, which brings most of the interfacial detritus with it (figure 4.20). Once the specimen is in contact with the substrate care must be taken not to slide it and in order to avoid movement during cementing-down the substrate is

4.18 Impregnating a porous substrate with cement.

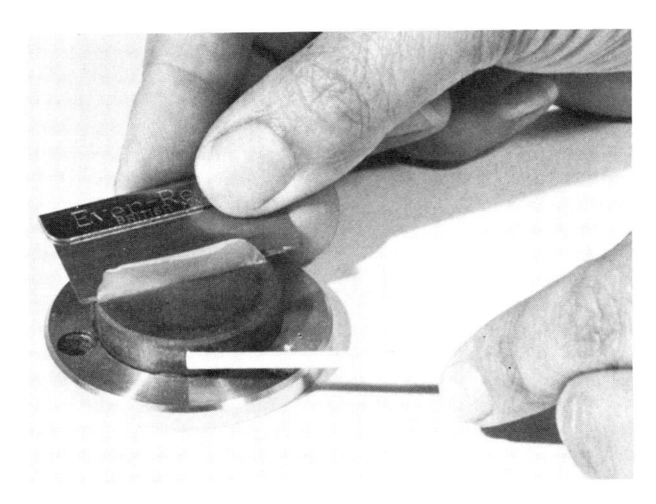

4.19 Shaving-off superfluous cement from the upper surface of a porous substrate.

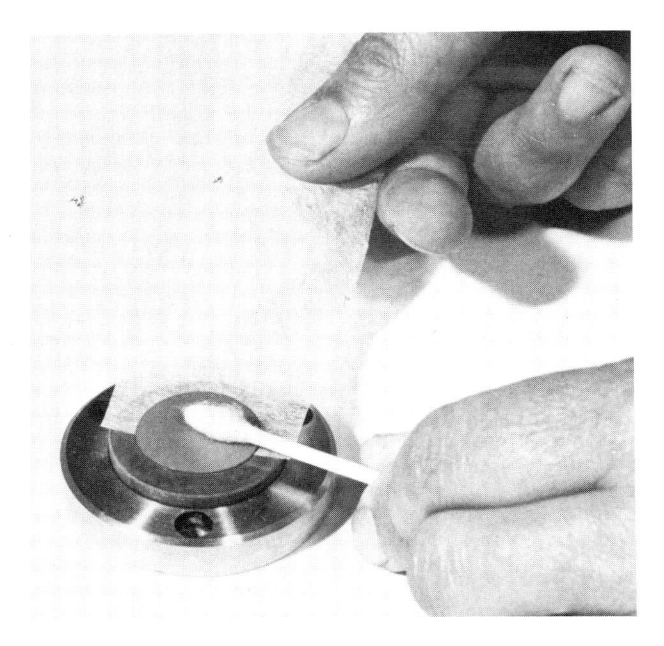

4.20 Mounting a specimen on a porous substrate.

placed in a jig (figure 4.21). Light pressure is applied to the specimen with a small clamp using a polyester-film disc, rubber pad, rigid plate and steel ball. The ball reduces the transmission of side forces during clamping. Cementing is carried out by heating this assembly in an oven or on a hotplate to the melting point of the cement and then allowing it to cool slowly. A very convenient hotplate is shown in figure 4.18. The fork design allows the substrate to be lifted away easily and then cooled on a heat sink. The jig is finally dismantled and the polyester film peeled off the specimen and substrate.

The one remaining mounting operation is to cement the outer edge of the porous substrate in its turn into a brass or stainless steel jig workplate which has been bored to receive it, as in figure 4.16. The operation is a little critical if the same cement is used since the upper surface must be kept well below 90 °C. Heating the workplate to 120 °C and coating the recess with a fillet of glycol phthalate, then pressing the substrate quickly in place before heat sinking, works satisfactorily. Alternatively the entire assembly of specimen, substrate and workplate may be mounted together in the same type of jig. A semi-permanent bonding of plate and substrate by an epoxy resin can be used but subsequent heating and cooling cycles and attack by solvents used at the cleaning stage eventually weaken the bond.

The specimen is mounted in an adjustable-angle jig and aligned for thickness reduction and polishing to the final plane parallel state. In our experience, no allowance need be made for cement thickness which we have measured by subtraction and consistently appears to be about 0·125 μm. Uniform results in specimen thickness $\pm$0·5 μm in 2·5 cm are readily obtained and the tolerance is largely attributable to the standard deviation in porous substrate parallelism ($\sim$0·3 μm) plus auto-collimation setting limitations. It is pointed out in chapter 6 that an operator cannot be expected to read the

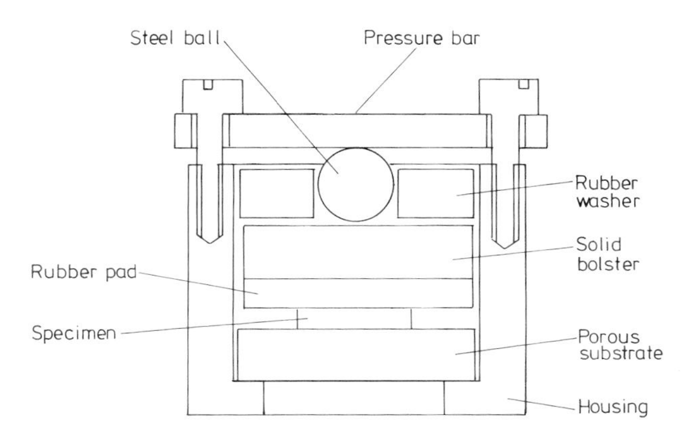

4.21 Mounting jig to apply light pressure to a specimen on a porous substrate during cementing.

simpler types of auto-collimator or Angle Dekkor to much better than five seconds of arc. Electro-mechanical measurement methods, when applicable, can improve the parallelism if needed. The process has been used for polishing parallel optical flats greater than 0·1 cm thick: for example four 1 cm diameter × 0·15 cm thick sapphire windows have been worked simultaneously to about three seconds of arc.

4.1.7 Related-angle techniques

One of the more difficult cementing problems occurs when an end face has to be produced at right angles to a cylindrical surface (classically on a laser rod) or when a face has to be worked at 90° to a pair of existing faces. When the accuracy required is only a few minutes of arc (in practice, down to one minute we find) end faces may be produced by cementing the work in a special plate which holds a vee-block previously machined at 90° to the polished face of the workplate (figure 4.22). This is a transfer arrangement and is highly dependent on the initial squareness of the vee-block relative to its end face. Although more precise registration and better accuracy could be obtained by having a long vee, cementing the entire length of a rod to a block could bend the work. It is not recommended for more than the first end, say, of a flimsy rod, in order that good parallelism may be obtained; the second end is better cemented into a suitable hand-drilled plate (figure 4.23). In this way the upper 90% of the work is air-spaced and unstrained, at least while the assembly is positioned vertically—the normal case for both polishing and testing operations.

4.22 Miniature vee-block in a jig.

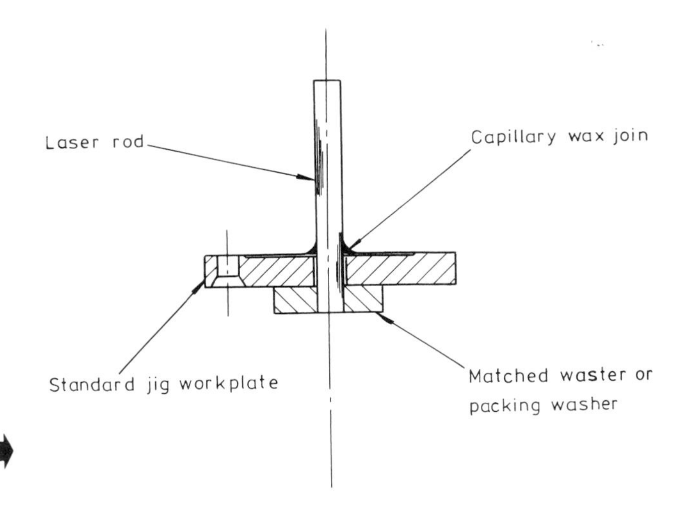

4.23 Free rod, one end constrained in a standard workplate (removed from an adjustable-angle jig).

4.2 Demounting and Cleaning

After any polishing process which has involved the use of mineral oil as a lubricant, it is advisable to remove the oil before warming the block. Hydrocarbons can become very firmly 'fixed' on a lapped or polished surface by being heated. Since paraffin waxes are themselves softened by oils, it is presupposed that they will not be used in any lengthy polishing process. Table 4.2 (Solvents) includes those which are commonly used for degreasing—both on grounds of safety and efficacy. All the cements given in table 4.1 are normally removable by one or more of the solvents listed. For an extended selection,

Table 4.2

Solvent	Formula	Specific gravity (at 20°C)	Boiling range (°C)	Flammability(i)	Flash point(ii) (°C)	Vapour pressure(iii) at 25°C	Toxicity TLV(iv) (ppm)	Solvent power(v) (KB value)	Remarks
Dichloromethane (methylene chloride)	CH_2Cl_2	1·33	38–41	Non-flammable	—	400	350	136	
Trichloroethane	$Cl_3{:}CH_3$	1·33	72–74	Non-flammable	—	120	350	124	Many stabilised products: e.g. Genclene, Inhibisol
Trichlorotrifluoroethane	$Cl_2F.ClF_2$	1·57	48	Non-flammable	—	350	1000	31	Many stabilised products: e.g. Arklone P, Fluorisol
Toluene	$C_6H_5CH_3$	0·87	110–111	Highly flammable	+4·4	20	200	100	
Xylenes	$C_6H_4(CH_3)_2$	0·86	138–144	Highly flammable	+17·2	7	100	94	
Acetone	$(CH_3)_2CO$	0·79	56–56·5	Highly flammable	−16·7	200	1000	(130)	
Butanone (methyl ethyl ketone)	$CH_3COC_2H_5$	0·80	79–80	Highly flammable	−7·2	100	200	(125)	
Ethanol	C_2H_5OH	0·81	78·5	Highly flammable	+12·7	50	1000	(125)	Principal ingredient of methylated spirits
Propan-2-ol (iso-propyl alcohol)	$(CH_3)_2CHOH$	0·78	81·5–82·5	Highly flammable	+21	40	400	(125)	
Ethyl acetate	$C_2H_5COOCH_3$	0·90	77	Highly flammable	−4·4	73	400	(120)	

(i)Flammability: solvents with flash points below 22·8 °C are designated 'highly flammable'. They are subject to the petroleum spirits regulations regarding use and storage. 'Flammable' solvents have flash points between 22·8 °C and 65·6 °C; the halogenated solvents have no recorded flash points and are thus 'non-flammable'.

(ii)Flash points are those determined by closed cup Kensky–Martens apparatus, published by BDH.

(iii)The vapour pressure of a solvent is the pressure which the vapour exerts on the liquid from which it forms. The higher the vapour pressure, the more volatile the solvent.

(iv)The threshold limit value (TLV) is the concentration of a species in air which it is currently considered safe to breathe for periods of eight hours per day indefinitely without suffering minor discomfort. These figures are constantly being revised and should not be taken as indisputable. The values are quoted in parts per million (ppm) but are often given as mg M^{-3}.

(v)The Kauri–Butanol value (KB) is a measure of relative solvent power. Strictly, it is not applicable to ketones, alcohols and esters and their relative values are therefore given in brackets.

the reader is recommended to consult Durrans (1971).

It should not be forgotten that a fine aerosol spray of degreasing solvent from a pressurised can may be at a temperature as low as −10 °C at the nozzle. The risk of cracking from thermal shock is high. It is preferable to use bulk solvent with light brushing; or a short-duration vapour degrease in an apparatus like that shown in figure 4.24.

Very robust specimens may be demounted rapidly by impact. Most of the cements withstand quite heavy steady pressures but have relatively low resistances to mechanical shock. Optical workers favour a sharp blow from a hardwood mallet whilst at the same time holding the work in such a way that any noticeable movement of the component after the bond breaks is restricted. The technique is not used for freeing thin slices but can be applied to glass components and some crystals of average proportions.

Demounting single-sided work is a simple procedure since no great care need be taken over the rear surface. The block is heated and wooden sticks (e.g. medical

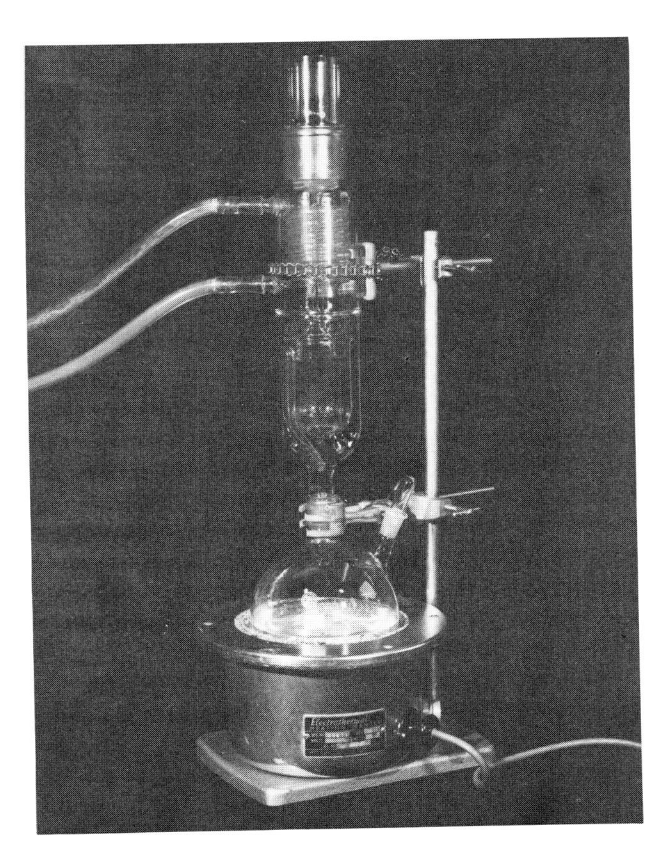

4.24 Vapour degreasing apparatus.

swab sticks) used to push-off the specimens when the cement becomes fluid. The work may be guided on to thin cardboard strips then transferred to a cleaning bath, or it may be slid directly into the solvent-dish which contains tissue paper to cushion the descending specimens. With work polished on both sides, some measure of protection can be given to the rear surface by adding fresh cement to the heated block and floating the work off on to thin card or filter paper. When all the specimens have been transferred to this temporary holder, it can be lowered into a dish of the appropriate warm solvent. Filter paper soaks immediately and very quickly releases the work but the additional rigidity of card, though slower in release, is advisable with thin, inflexible slices. Alternatively, the work may be allowed to cool before being immersed in cold solvent. The removal operation can be accelerated by warming and brushing. It is, however, only a preliminary cleaning process.

After the specimens have been removed in the warm solvent, their transfer for more rigorous cleaning, inspection and packaging depends upon their fragility. Robust work is easily and safely handled with steel or plastic tweezers (figure 4.25) but the more delicate slices (and dice) can be transferred by the brush technique. A sable-hair brush, sufficiently large (or small) to support the slice, is held at a low angle in the solvent and manoeuvred beneath the work. An additional brush is often useful in persuading the work to stay in position on or within the bristles and supports it at the critical moment when it breaks the surface. Small dice are very easily handled in this way and the relatively gentle pressures involved make it a particularly good technique for dealing with thin slices of delicate materials. Figure 4.26 shows the method in use on lead germanate.

Very small dice (125 μm square × 75 μm thick) can be separated and cleaned with a modified dropper pipette (figure 4.27). When the nozzle is in the solvent, the bulb is given a slight squeeze, then a specimen touched and sucked into the tube as the pressure is released. It can be carried to the next bath and expelled gently.

The cleaning process itself varies with the demands made by the surface properties of a device: for example, a polished optical window does not require the cycle undergone by, say, a photoconductive detector. A

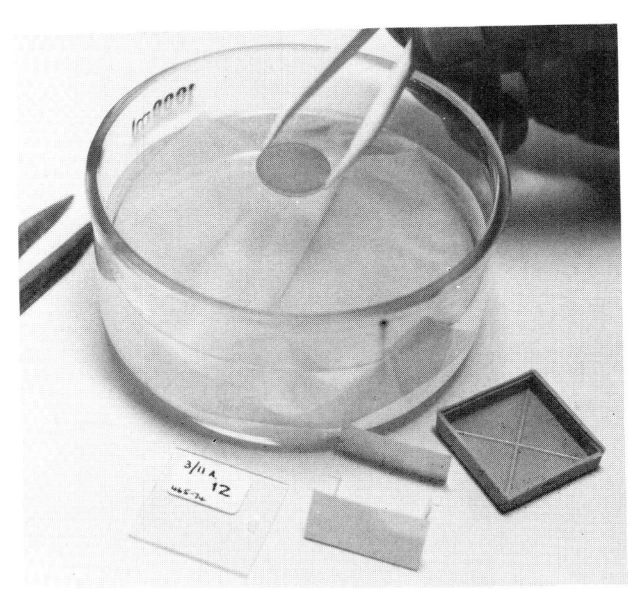

4.25 Manipulating a robust specimen with plastic (or steel) tweezers.

a solution for cleaning analytical glassware which is very effective for rigorous, short-duration cleaning: 30/5/5/60 nitric/hydrofluoric/Teepol/water by percentage.

Ultrasonic solvent cleaning may be applied to robust specimens where a filled cement such as Tan wax has been used. It does not, as a general rule, improve the solvent action to any marked degree, but it loosens insoluble particles very effectively. Some of the light metals, for example aluminium, are slightly attacked when ultrasonically vibrated in a chlorinated solvent such as trichloroethane. The reaction involves a small amount of gassing and is both deleterious to the metal and disconcerting to the operator.

The method of transport from bath to bath may be, under favourable circumstances, in a special holder. Polished slices can be held in a segmented basket (figure 4.28) like the ones described for sawn slices

4.26 Two-brush manipulation of a fragile specimen.

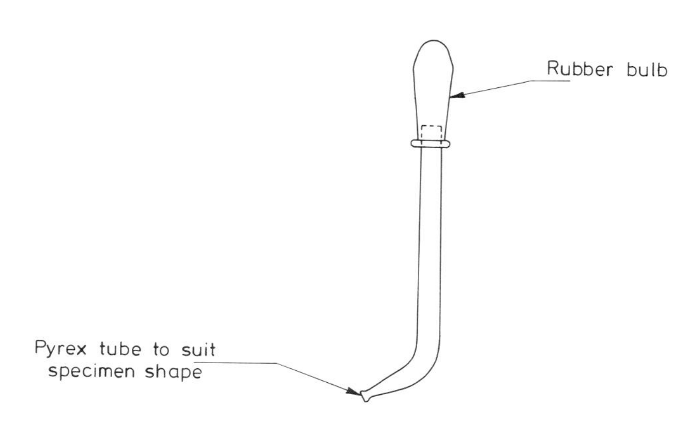

4.27 Modified dropper pipette.

typical process for the latter which removes the greater part of all normal contamination is: two or three washes in a clean warm solvent to remove the mounting cement, each taking about five minutes; immersion (or vapour cleaning) with propan-2-ol, again for the same period; subsequent transfer through two or more baths of de-ionised water; final blow drying with dry nitrogen. An effective glass-cleaning cycle is from the initial solvent using detergent and water (with ultrasonic agitation or brushing) to propan-2-ol. Gorsuch (1970) recommends

4.28 Segmented basket for holding a number of slices for cleaning.

(chapter 1). Small perforated plastic, wire gauze or fabricated glass baskets are often used to accommodate single specimens as in figure 4.29. Inevitably, the refinement of any technique brings its own impedimenta but there is very little that cannot be carried out by an experienced operator using tweezers or brushes, glass containers like petrie or crystallising dishes, tissues and a hotplate.

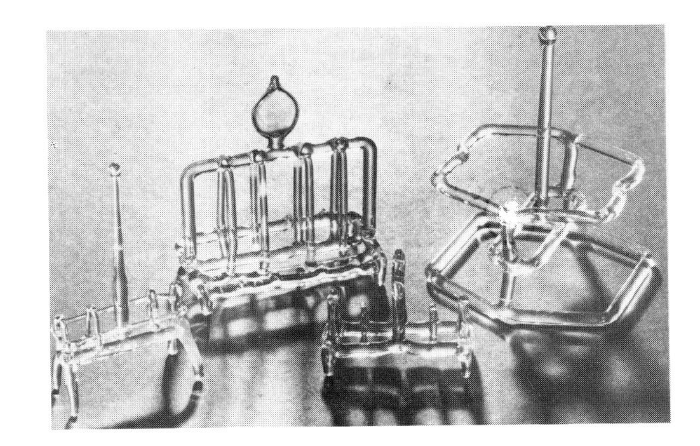

4.29 Various small cleaning baskets.

4.3 Packaging

An appraisal of the problem of transporting specimens without their being damaged leads first of all to the question of their destination. Any work which is interdepartmentally required has the obvious solution of being sent superficially cleaned but still mounted on a block. The onus is on the user to demount it and return the block. Although this can hardly be termed packaging, it does indicate a safe and more general method for carriage over any distance: cementing specimens to glass or metal slides which can be fixed in slotted boxes or held in a container with double-sided pressure-sensitive adhesive tape. The work is normally immersed in a paraffin wax, since it is relatively easy to remove. A specific application of the technique is shown in figure 4.30 where numbers of small dice have to be sent by post. They are waxed to impregnated tissue paper.

Polished slices—mirrors, windows and the like—destined for one customer can be packaged in a multi-compartment box, as shown in figure 4.31 and more often used for sawn specimens. These have the advantage of preserving the identification of material (say slice position in a boule) and of preventing sliding about during carriage. The foamed plastic padding is not, however, suitable for very delicate work since it can scratch and even crack it.

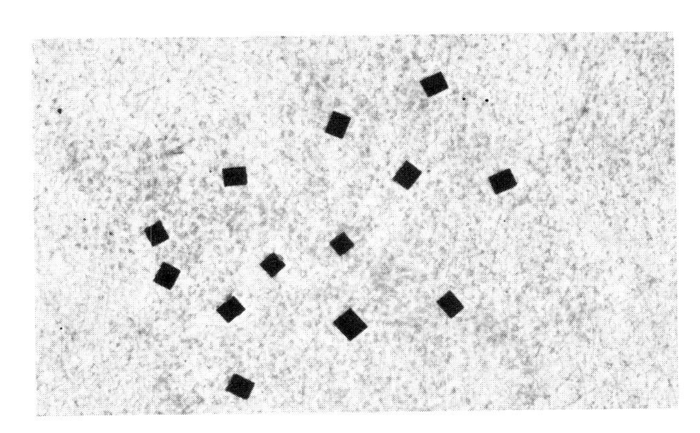

4.30 Waxed, small dice on tissue for safe transport.

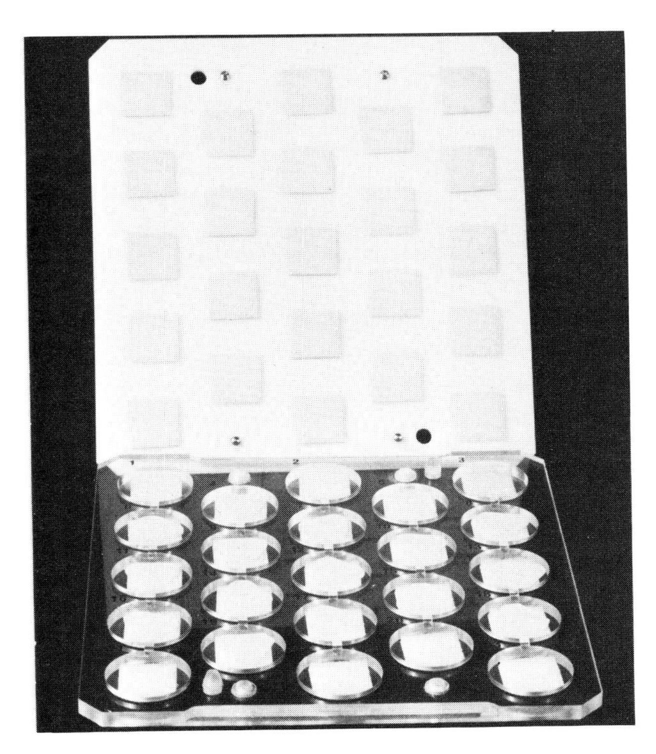

Multi-compartment box.

Single specimens can be placed on rubber or polyurethane foam pads fitted in plastic boxes then sandwiched lightly in place with an additional pad. Identification data is written on a card, again cut to the size of the box and positioned beneath the cover—especially when it is a sliding lid. Very soft materials are best transported in solidified wax, but where this is impracticable (as with cubes of proustite) they are wrapped in washed chamois leather and carefully boxed.

Glass components require far less cosseting but even so the tendency to wrap them in hard tissue paper, which may contain abrasive particles, is avoided. Lens tissue is the preferred medium and their boxing is correspondingly less critical.

In most instances, the user is the arbiter. If he requires a component delivered clean and ready for use it cannot be sent encapsulated in wax, however safe the technique. On the whole the optical industry and related users of its wares prefer the work ready for use, whereas in the electronics field there may be many additional etching processes and cleaning steps to be performed and a film of wax presents no problems.

A final word on the subject of handling, whether for mounting, demounting or packaging. The temptation to handle many materials with bare hands must not be indulged without careful thought, since some electro-optic crystals and all of the associated solvents are hazardous in some way. For example, gallium arsenide, proustite and cadmium telluride are worked to fine limits: they and their waste are potentially toxic. Two informative books on general laboratory chemicals are those by Gaston (1970) and Gray (1966). Contamination is very much a double-edged weapon: to the component as well as to the handler. Handling work at all stages, indirectly or with tweezers and brushes, has its tactile snags. In some ways the blurring of kinaesthetic sensation is akin to eating a sweet with the paper on and there is little doubt that components which can be safely touched are often best handled. But beware of solvents, injurious materials and cements on the skin and avoid contaminating the work at all times.

References

de Bruyne N A 1957 *Nature* **180** 262–6

Durrans T H 1971 *Solvents* (8th ed) (London: Chapman Hall)

Faraday M 1827 *Chemical Manipulation* (London: John Murray) (repub. Applied Science Pub. Ltd 1974)

Fynn G W, Powell W J A, Blackman M and Jenner D 1971 *J. Phys. E: Sci. Instrum.* **4** 391–2

Gaston P J 1970 *The Care, Handling and Disposal of Dangerous Chemicals* (Institute of Science and Technology)

Gorsuch T T 1970 *The Destruction of Organic Matter* (1st ed) (Oxford: Pergamon) p 12

Gray C H 1966 *Laboratory Handbook of Toxic Agents* (London: Royal Institute of Chemistry)

Horne D F 1972 *Optical Production Technology* (London: Adam Hilger) pp 223–5

Kienle R H and Ferguson G S 1920 *Ind. Eng. Chem.* **21** 349–52

Kienle R H and Hovey A G 1930 *J. Amer. Chem. Soc.* **52** 3636–45

Mendel E 1967 *SCP and Solid State Technology* August, 36–7

Rolt F H 1929 *Gauges and Fine Measurement 1* (London: MacMillan) pp 84–5

Samuels L E 1971 *Metallographic Polishing by Mechanical Methods* (2nd ed) (London: Pitman) pp 4–22

Stefan J 1874 *Sitzb. Akad. Wiss. Wien.* (*Math. naturw. Kl*) **70** 713

Twyman F 1957 *Prism and Lens Making* (London: Hilger and Watts) pp 113–7

Walden L 1936 *J. Sci. Instrum.* **13** 345

5 Special Techniques

'*When found, make a note of.*' (Charles Dickens)

This chapter does not describe processes which demand sophisticated skills or equipment, but covers some of the shaping operations associated with electronic components. It deals mainly with methods by which specimens may be, variously, trepanned; made accurately cylindrical and chamfered. A description is given, too, of one method of making small diameter, well defined holes for measurement purposes. Graticule making methods are not dealt with, although a simple scribing technique is given in chapter 6.

5.1 Scribe-and-Break Glass Cutting

Since many sawing operations require an expendable glass mounting strip, the dimensions of which seldom need to be precise, the scribe-and-break method is a quick inexpensive way of producing regularly shaped pieces. Very little additional skill is needed in producing washers, too—for example, those used for facing some polishing jigs' conditioning rings.

5.1.1 Straight line fracture

It would be hard to find a technique which is as easy to some yet as mystifying to others as simple glass cutting. Apart from glaziers diamonds almost anything harder than glass (such as wheel-cutters, sapphire or tungsten carbide chips suitably sharp) can be used to produce a scratch-line with which to initiate a fracture. A successful procedure, using a wheel cutter on glass of thicknesses 0·06 cm to 1·2 cm and up to a metre or more in length, is:

(1) place the clean (degreased) glass on a paper tissue, resting on a flat surface;
(2) arrange a rule or straight-edge on two pieces of plasticine, (or silicone putty) placed at the extremities of the cut;
(3) adjust the rule to the correct position, making allowance for the cutter width, and press it hard into the plasticine to prevent its slipping on the glass;
(4) draw the wheel cutter over the surface firmly enough to obtain a fine, unsplintered line without retracing it;
(5) remove the rule and plasticine;
(6) lift the glass and insert 0·5 cm of a 0·3–0·4 cm diameter wooden rod under one edge nearer to the worker;
(7) press on either side of the cut and the glass will part along the line. Figures 5.1 to 5.3 show the process. Beware of glass splinters, especially those which become embedded in the plasticine!

Newly manufactured glass is said to be the most easy to cut but much of our supply is 0·6 cm plate between ten and twenty years old which breaks cleanly. On thicker sheets, opening the scribed line can be made easier by tapping the under-face of the plate, directly beneath the line at one end, with a spanner (figure 5.4). The blows have to be aimed accurately in line with the mark and some experimenting is needed in determining the weight of tap required. The difficulty of

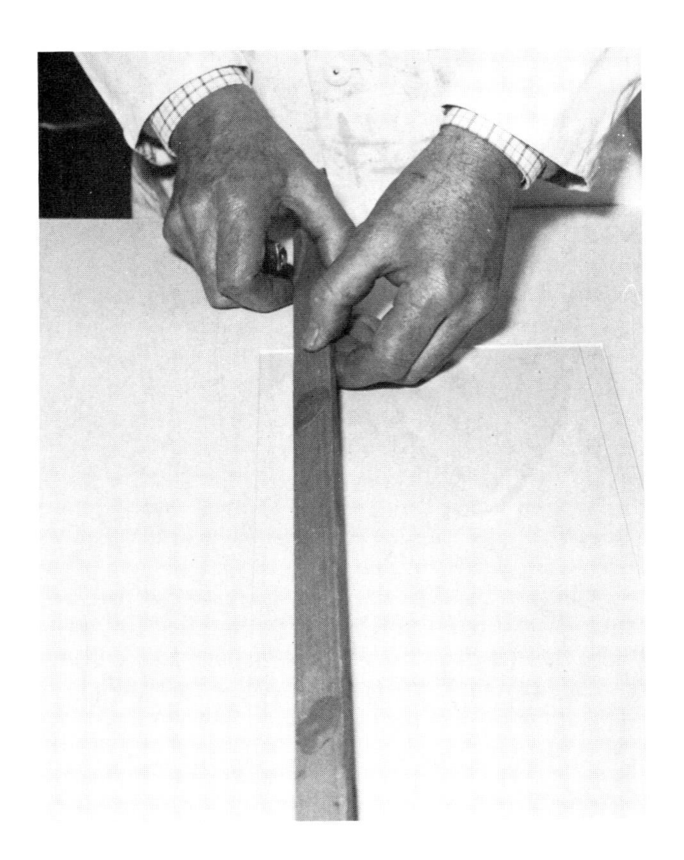

5.1 Initial stage in producing a straight-line fracture of glass.

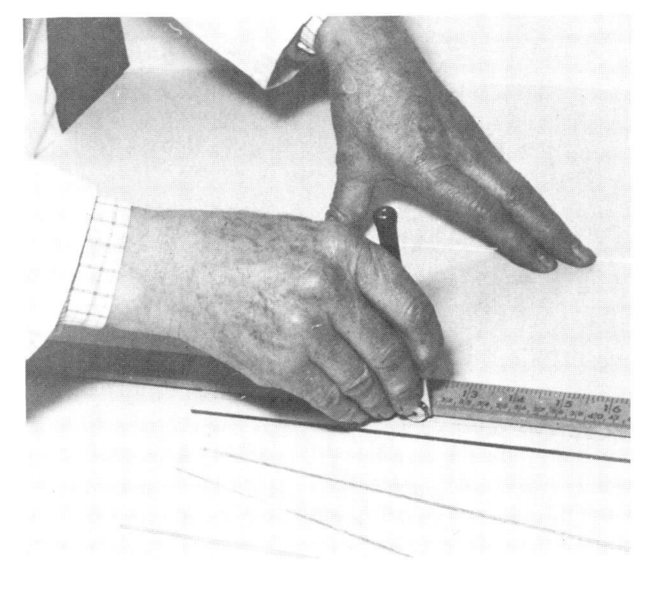

5.2 Application of the wheel cutter.

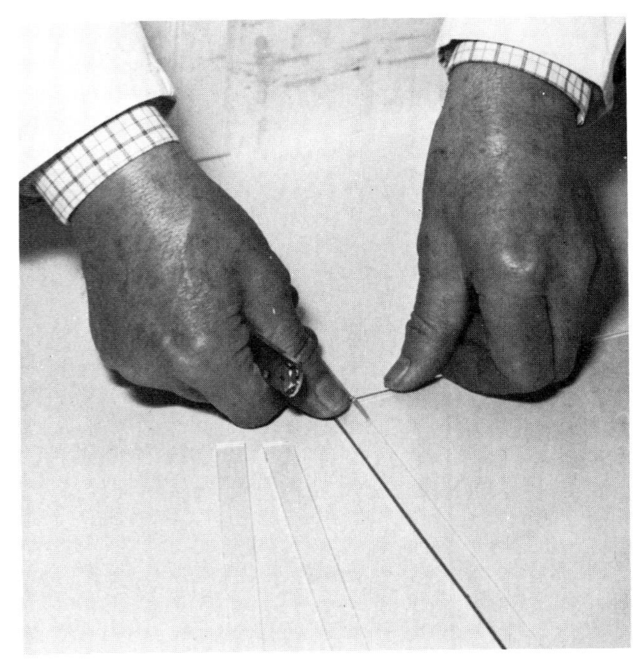

5.3 Opening the scribed line.

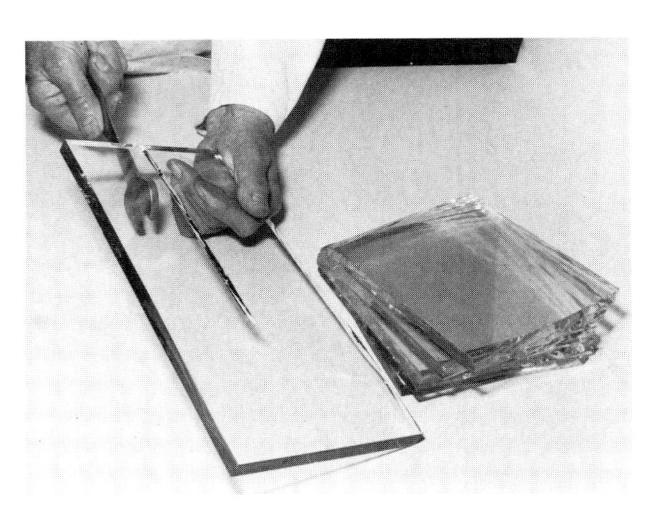

5.4 Opening the fracture by tapping beneath the cut.

simultaneously combining accuracy of aim with sharpness of blow can be sidestepped by using a chisel with a rounded edge, firstly positioned and afterwards struck. This scribe and chisel method is often used for producing mirror blanks from thick plate glass.

5.1.2 Circular fracture

Hole-cutting in glass by scribe-and-break methods can defeat those who are otherwise expert in manipulating it; yet salesmen demonstrating wheel glass cutters at shows and exhibitions scribe an approximate circle and a few roughly diametral lines then deftly tap around the circle until the central sectors fall out cleanly. The main explanation for any lack of success in emulating their techniques is that technicians are often supplied with old glass whereas the salesmen use only very new, relatively thin panes.

There is no problem with hole or ring cutting if one is prepared to use some elementary form of jig. The technique described by Beale (1880) is used to cut 0·6 cm plate, up to twenty years old, although it was originally devised for cutting rings that spaced microscope cover plates from slides holding thick specimens. It uses a thick brass collar whose internal and external diameters match the wanted ring size. Two circles are scribed on the glass with the wheel-type glass knife registering against a collar the dimensions of which allow for the width of the cutting mechanism. The brass ring is cemented in place on the side opposite the scribed circles and as closely as possible concentric with them. The cement used should provide a strong bond: shellac-based Tan wax is satisfactory; the one recommended by Beale was Chatterton's compound which is shock resistant when slightly warm. In either case the thermoplastic resin must be allowed to cool thoroughly then a blow with a hammer is made to the centre of the glass circle whereupon it shatters, principally in radial lines (figure 5.5). Experience has shown that no fracturing occurs beyond the wax junction. Beale recommended that circles be made on both sides of the glass, but a single-sided scribing has proved to be sufficient.

An alternative technique for scribing washers is to wax the brass collar in position and then grip it lightly in a lathe-chuck. One or two scribed lines can be made by rotating the work slowly whilst holding the wheel cutter against a temporary guide held in the tool post of the machine. Precise cuts are made by positioning the cutter-wheel opposite the inside and outside diameters respectively of the brass ring.

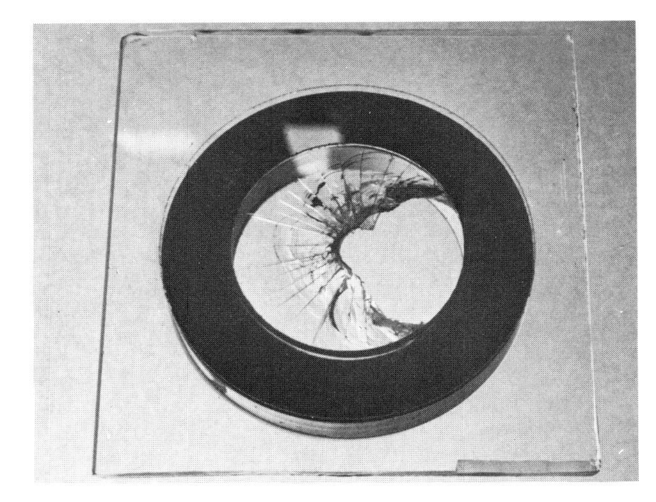

5.5 Circular fractures in thick glass plate.

5.6 Deep chamfers on jig conditioning-ring glass facings.

Any glass components cut by the scribe-and-break techniques should have their sharp edges chamfered. Conditioning-ring facings in particular are heavily chamfered with the bevel extending through the ring (figure 5.6). Section 5.4 deals with the methods of chamfering.

5.2 Trepanning Methods

Holes, discs and washers may be cut in electronic materials by trepanning when machines and appropriately sized tools are available. Fixed-abrasive trepanning—usually with diamonds held in a metal matrix—has generally the merit of speed in operation and durability of tools, but initial outlay is fairly expensive while some materials cut more cleanly with a loose abrasive such as silicon carbide. For example, drilling small holes ($\sim$1 mm diameter) for electrodes through borosilicate glass, with diamond impregnated tools or 'free' compounds, very easily initiates cracks in the glass which cause vacuum leaks. Both free and fixed abrasive trepanning methods will therefore be described.

5.2.1 Free-abrasive trepanning: 'through' holes

Very simple equipment suffices for this operation when the tool trepans completely through the work: a pillar drilling machine and biscuit cutters of the required size, (figure 5.7). A trepanning tool is usually made from brass tubing with one end holding a metal stub turned down to fit the chuck of the drilling machine. Internal/external diameters are machined to suit the required finished components sizes: smaller by about 0·2 mm for a hole; or, for a disc, larger by about 0·2 mm on its diameter. The specific allowance depends upon the abrasive-particle size with other factors such as the material being worked and depth of cut influencing the overall accuracy.

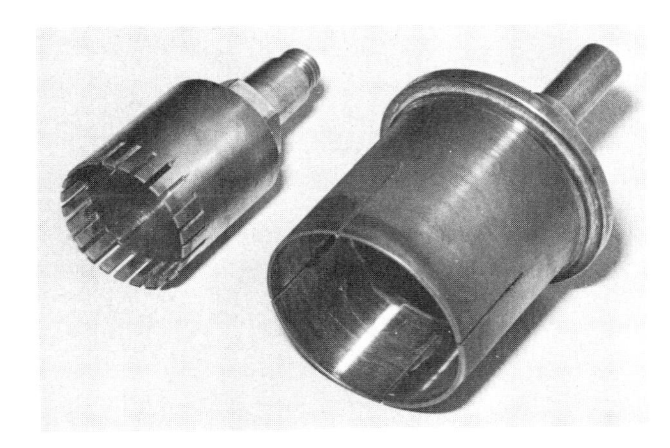

5.7 Typical biscuit cutters.

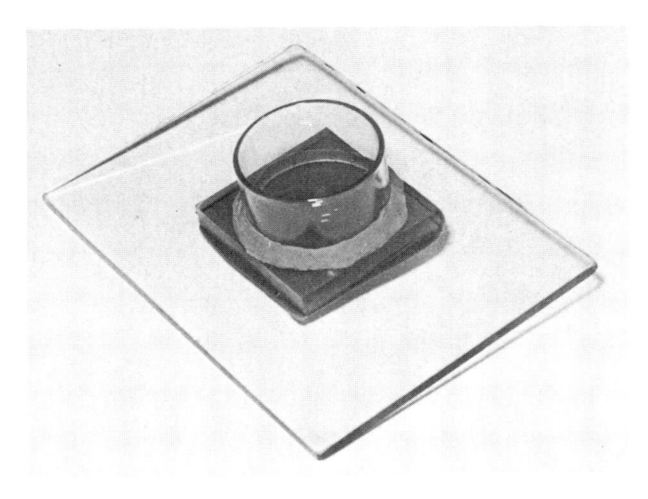

5.8 Plasticine moat for retention of abrasive slurry.

An important requirement in free-abrasive cutting is that the slurry should have access to the cutter face even in the deepest holes. Thus deep saw cuts are made in the face of the tool, preferably a little deeper than the thickness of the work. The mode of applying the abrasive slurry to this region is equally important. Brushing it on is both messy and ineffective. An adequate supply can be arranged by building a moat, on the surface which is being trepanned through, for holding a reservoir of slurry. At its simplest, this can be formed by a ring of plasticine pressed firmly on the work (figure 5.8). A neater moat can be made by cementing a short length of glass or metal tube to the face of the work at the same time as it is being waxed to a baseplate. The great advantage of the latter is that it can be washed clean of abrasive for inspection during, and at the end of, the cut. In this way, no ball of carborundum-loaded plasticine can accumulate. Moreover, oil-borne abrasives slowly attack plasticine and erode it away.

For the best definition of upper and lower edges of the hole, even though it is usually enough to use a lower plate only, glass plates should be cemented to both faces. The thinnest possible cement film is needed since a thick layer with its different abrasion characteristics allows the cutter to drop through before contacting the backing plate, thus chipping the work-edges.

The abrasive particle size used depends on the material being cut: for instance, a 15 cm glass disc 2·5 cm thick requires 80 or 120 carborundum at a typical rotational speed of 100 rpm. At the opposite extreme, 1·5 cm diameter cores are cut from 3 cm thick lead germanate crystals with 600 carborundum again at 100 rpm. Both operations take approximately fifteen minutes to complete. Some form of automatic reciprocating mechanism fitted to the trepanning machine is an advantage if deep holes or repetitive work should be needed. Special pneumatic units are commercially available; however, we prefer a cord and crank arrangement operated from a variable speed motor–gearbox unit (figure 5.9).

For drilling very small holes in crystal or ceramic materials a tubular tool should be used in preference to a solid cylindrical one. Metal capillary tubing (nickel, steel or molybdenum for example) is suitable and it should be mounted in such a way that there is a clear passage through its bore, accessible from the rear, for the removal of cores broken-off prematurely during trepanning. Straight lengths of tungsten wire are used to push or knock these cores out since they do not bend readily and they transmit tapping forces well. In our experience, solid drills are extremely slow in action and even a very shallow blind hole improves their performance. There comes a time with very small diameter tools when the interruption of the cutting edge is impracticable. However, with tools of about 1 mm in diameter, a single, thin diametral saw-cut is worthwhile.

It is sometimes better when trepanning small holes (in ceramics, for example) to use diamond abrasive with capillary drills, though this is not strictly a free-abrasive technique. The method of charging the tool is to place a blob of diamond compound (25–45 μm particle size) on the end face of a hardened steel roller, lower the drill into the blob and tap lightly with the down-feed of the machine. In use, only lubricants like light oils (traditionally olive oil) are necessary with an occasional recharge of diamond compound.

Glass rings, for facing metal conditioning rings on jigs, are trepanned with a twin-diameter tool. This is made from two suitable diameters of brass, copper or steel tubing cemented into machined recesses in a backing plate which has a suitable shank for holding it in the drill-chuck (figure 5.10). The cutting edges of the tubes can be chamfered to produce a bevelled ring from a given thickness of glass.

We have trained many technicians to use trepanning machinery and, generally, few difficulties arise. Initially, some are convinced that high-feed pressures are synonymous with rapid progress, but a reasonable quantitative guide is about 100 g cm^{-2} of cutting edge.

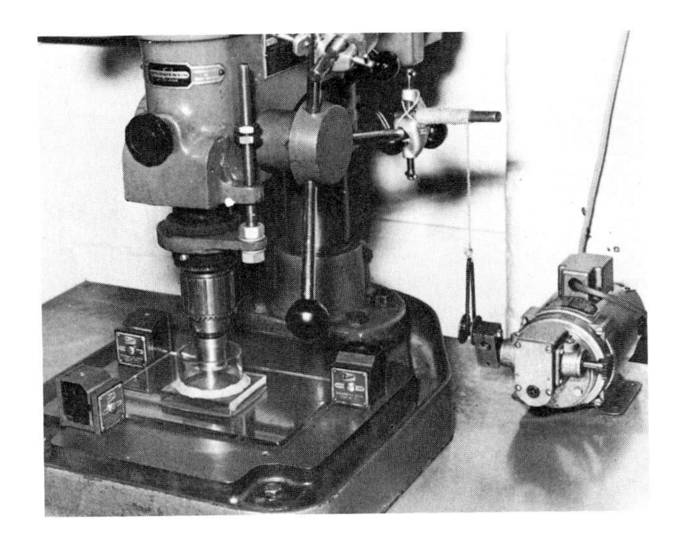

5.9 Mechanism for reciprocating drill downfeed.

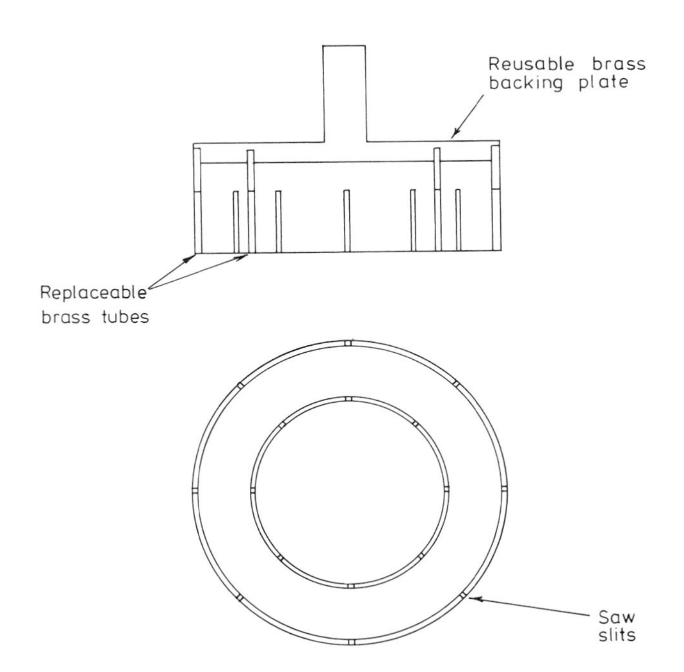

5.10 Concentric twin cutter.

Qualitatively, the sound of the abrasive rolling between the tool and the specimen is informative: the beginner should listen to a light cut being made and increase the pressure over a period of a few seconds. It will be noticed that the average pitch of the sound rises, terminating in a squeak. The moment when the pressure must be reduced and then the cutter's forward motion relieved in order to allow a fresh supply of abrasive to enter the cutting region is just before the squeak is expected. Progress is maintained in this way. Whilst silicon carbide is the abrasive in general use for substances up to and including garnets in hardness, boron carbide cuts faster in both single-crystal and polycrystalline aluminium oxide.

5.2.2 Free-abrasive drilling: blind holes

The manufacture of devices such as silicon hollow cathodes requires the lapping of small blind holes—that is, flat-bottomed cylindrical holes which do not continue right through the material. There is a temptation to use a solid tool with its diameter below that wanted and a face square with the cylindrical axis, but this presents one of the more difficult of lapping problems, that of generating a flat surface at the centre of rotation of the rod. A small area there is in a relatively static state, whereas the requirement for successful uniform lapping is close harmony in speed and pressure. Inevitably, a dimple is worn in the tool and a corresponding central pip protrudes at the bottom of the hole.

The fault can be corrected by machining the tool-face flat again, then continuing the process as necessary until the pip has been lapped away, and many hollow cathodes of 1·0 cm diameter × 3·5 cm in depth have been made successfully by this technique. Where the hole has to be uniform in diameter, a new tool (or tools) is needed since wear occurs of diameter as well as of length. Tools are pieces of steel rod, usually silver steel: not because of any special property of this carbon steel, but because it is obtainable in a closely spaced range of diameters. The choice of diameter for a given hole dimension must be made to allow for the particle size of the abrasive grade in use. Carborundum 220 is the preferred general purpose grade, resulting in adequate cutting progress with surface finish. Precise dimensions for this coarse grit are not easily obtained and vary, moreover, with diameter; but a working allowance is about 150 μm. Carborundum 600 is used with second and subsequent tools and for them about 50 μm (0·002 in) is allowed. Any radial slackness in the machine bearings has its effect, too, and for the best results for a specific size, with a standardised material and on a known machine, trial runs have to be carried out and corrections made from the data obtained.

A more elegant solution to the problem of low motion at the centre of a rotating flat tool is to use one which is a few millimetres smaller than the hole diameter required and make it orbit as well as rotate. The requisite motion can be obtained by arranging two sets of ball-bearings, one within the other, the outer members of the inner races mounted eccentrically relative to the inner members of the outer races. This is done by using a cylindrical component to connect the races together: the bore for the inner, smaller races, is made eccentric by a small amount which corresponds to the difference between the tool and required hole diameters. A double pulley is fixed on the motor-shaft, driving two separate pulleys on

5.11 Blind-hole drilling arrangement.

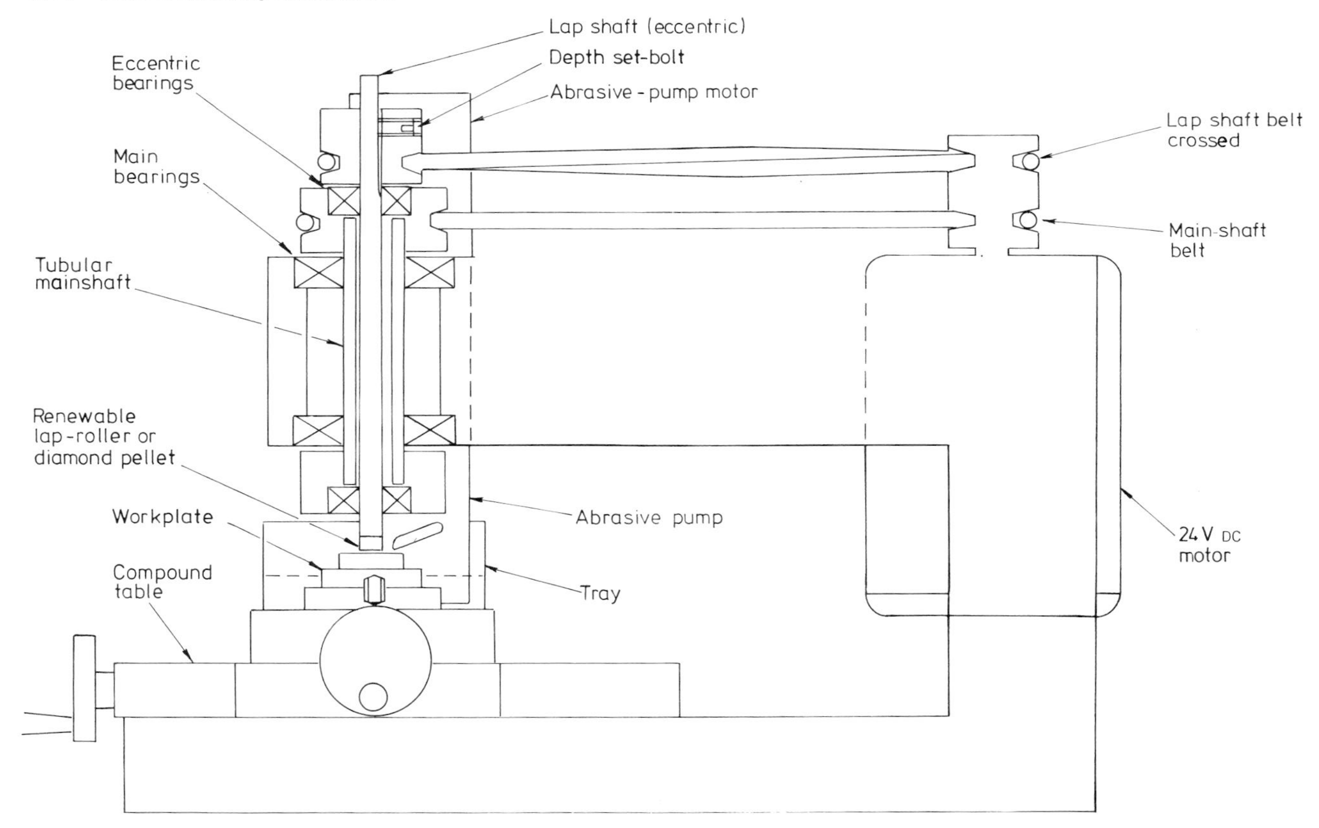

the orbit-and-rotate head: a larger one on the outer diameter of the eccentric housing and the other in the inner lapshaft. One of the two small driving belts is crossed so that its pulley contra-rotates. An experimental machine is shown in figure 5.11. A compound table, fitted below the orbit-and-rotate head enables work to be positioned relative to the lapping tool.

The constant supply of abrasive slurry which is essential to the process is pumped from a tank bolted to the compound slide and surrounding the work. A flat mild-steel plate, 12 mm thick with a threaded stud protruding from it, is bolted through the tank to the slide. Work-holding platforms can be screwed to this stud. At the upper end of the lapshaft a flat is machined which allows burr-free locking with the depth control set-bolt. One method of setting a specified depth is to slide a feeler gauge (or similar spacer) under the end of the tool while pressing on both lapshaft and collar, then lock the collar to the shaft. For holes other than very shallow ones, two depth settings are necessary since with tools of common materials (such as hardened steel rollers or machined cast iron) appreciable wear occurs which results in holes less deep than expected. Materials like tungsten carbide or sapphire could reduce lap wear to negligible amounts.

5.12 Water-feed drill chuck system.

5.2.3 Fixed-abrasive (diamond) trepanning

There are several commercially available trepanning tools which incorporate diamond particles in a metal matrix as the abrading mechanism. We have had most experience with those of the 'Habit' system, especially thin walled, tubular tools with diamonds bonded by electrodeposited nickel to the cutting face. An essential feature in their operation is water, fed coaxially to the tool. A standard drilling machine can be fitted with a rotating joint on the drill chuck which remains stationary as the chuck revolves, and allows water to pass through to the cutting region. Mains water is connected to the joint via a flexible pipe with a gauge in series to monitor pressure in the cutting zone (figure 5.12). It acts, too, as a sensitive indication of broken cores which may become jammed in the bore of a trepanning tool: any sharp rise in the water-feed pressure and its failure to fall when the cutter is withdrawn is a signal to stop the machine and remove the fragment with a stiff tungsten wire.

Most of the harder materials can be trepanned or cored with bonded-diamond, water-fed tools. Soft materials with a tendency to cleave can develop fractures while being diamond-abraded. It is usually safer to shape them by brass-tube and free-abrasive techniques. Strained crystals (especially soft ones) should be avoided. They may disintegrate at the moment of contact with diamond tools.

Squares of 6 mm thick glass, 7 cm and 15 cm side, form convenient, expendable mounting platforms to which work can be cemented for trepanning operations. The glass itself can be heated on an electric hotplate—the larger its volume, the more slowly it should be heated. Turning it over after every 20 °C or so rise in temperature helps to avoid breakages. At about 95 °C, a few drops of a shellac-based wax (such as Tan wax, chapter 4), melted in a gas flame, are dripped centrally on to the platform and the pre-warmed workpiece placed over the molten wax. A sheet of polyester film is placed over the work, then a foam rubber pad and finally a load weight while the assembly is allowed to cool.

It is considered to be hazardous to hold work by hand whilst diamond trepanning and for our staff it is a rule that all work must be solidly prevented from moving on the baseplate of the machine, for example, by clamping it down. An effective compromise between safety and convenience is to surround the glass-mounting platform which holds the work with several magnetic bases (figure 5.12). The arrangement at least prevents the work from rotating and allows the operator's hands to be kept clear. Magnetic blocks are equally satisfactory with biscuit cutters since they retain work-to-cutter relationship during the necessarily continuous reciprocation of the machine spindle.

Since the cutting pressures necessary with diamonds are often much greater than those with carborundum, cementing backing plates to the work is more critical. The thinnest film possible of a hard wax is advisable in order to minimise chipping at the breakthrough point. Although diamond trepanning is regarded as a high-speed technique, in common with our experience with diamond saws running at low speed there seems to be no disadvantage in rotating trepanning tools correspondingly more slowly. Moreover, the range of crystals cut with minimal damage is probably increased.

The lower limit of feed pressure is critical with diamond trepanning; at less than a certain pressure the tool does not cut. This is shown by a flow of clear cooling water emerging from the cutting region. Effective abrading is evidenced by a flow of water which contains powdered material—with most optical crystals and glasses the fluid is white and with semiconductors, black. Prolonged running of a diamond tool with a cutting pressure too light to maintain progress through the work eventually produces a glazed edge. Thereafter, a dangerously heavy pressure is needed to restart the cut. It is preferable to run the water-cooled tool at a slow rotational speed into a carborundum stick and thus restore its cutting qualities.

It is now a fairly well established practice to produce cylindrical rods by Habit diamond trepanning. Figure 5.13 shows a pillar drilling machine which has been converted for the purpose by fitting a water-feed chuck to the No. 2 morse taper in the machine. A three-point steady is mounted on the pillar and the screws arranged to contact the plain-tubular part of the tool above the diamond deposit. The steadying of the tool should be carried out whilst it is rotating, when small eccentricities can be seen and, at least partly, corrected. A special cylindrical wax-filled crystal holder in which the orientation face is adjusted normal to the machines axis is clamped to the baseplate. Set bolts are provided with which to preserve the position of the crystal while the hot wax is being poured in. Figure 5.14 shows a selection of these long trepanning tools. When holes, or cores, are needed at angles other than normal to the surface of the work; or where components such as

5.13 Drilling machine converted to trepan long holes.

Newtonian secondary flats must be trepanned from plate glass mounted at 45° to the worktable, packing is needed for both for entry and exit of the tool (figure 5.15). Plate glass is all that is necessary for the exit side, but for the entry it is desirable to present a surface normal to the drill's axis. Damaged or obsolete prisms are useful for the latter purpose. For trepanning oblique secondary flats at 45°, the standard diamond tubular cutters will have only enough depth to produce components of about 25 mm in depth. Specially deep cutters can be obtained, but it is cheaper to use brass tube and carborundum methods for small numbers of components.

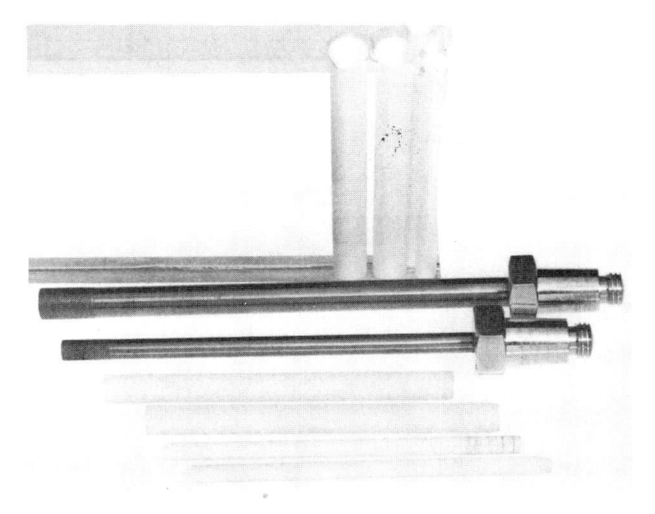

5.14 Long trepanning tools.

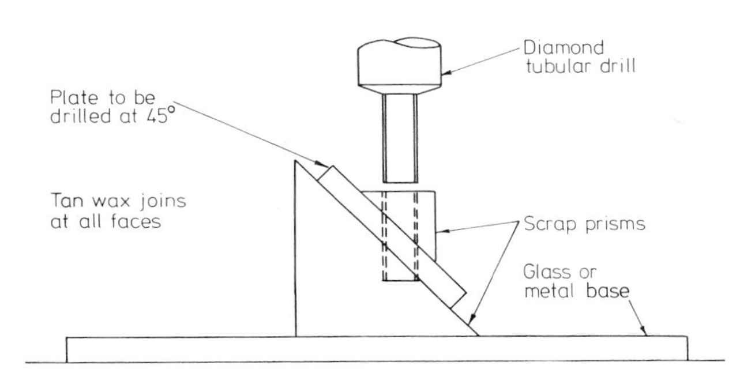

5.15 Packing sandwich for coring off-square specimens.

5.16 Surface-milling on drilling machine.

5.2.4 Surface-milling with fixed-abrasive tools

We have carried out machining operations on the Habit drilling machine that are completely outside those intended: an example is its use as an end-milling machine (figure 5.16). Here the work (usually glass or ceramic) is mounted on a compound table and the trepanning tool brought down to a position fixed by a depth-stop normally provided on the machine. Slots can be made in the workpiece or large volumes of material milled away to produce depressed areas or even components thinner overall.

Spherical surfaces, too, may be manufactured. The work is mounted on a rotary table which is in turn clamped to a tilting angle-plate. The mode of setting and operation is the same as that for spherical lap manufacture described in chapter 3, except that the tilt is applied to the axis of the work instead of that of the tool. Surfaces milled in this way cannot be expected to have the fine surface-finish produced by lens-grinding machines. However, acceptable finishes are obtained after prolonged lapping with BAO 303 abrasive powder. A cast-iron tool may be used for the operation, or a second glass spherical face produced with the same machining conditions but of opposite sign.

5.3 Shaping Cylinders

The trepanning method described in § 5.2.3 can be used to produce cylinders from many of the more robust, strain-free materials; but the range of diameters and care needed to prevent small, brittle cores from breaking, limits its general use. Moreover, the resulting cylinders are not normally accurate enough in form and benefit from further shaping. Centre and centreless grinding, together with hand lapping, are the accepted techniques for the full range of materials and sizes.

5.3.1 Centre-grinding

For a period of some years in the early days of solid state lasers, centre-grinding was our only mechanical method of producing cylindrical work. The equipment was (and still is) a small lathe with automatic feed, a homemade toolpost grinding head and a pumped, recirculated supply of coolant (figure 5.17). The grinding wheels, either diamond metal-bonded or resinoid types, were 75 mm diameter, 3 mm wide at the rim and driven by a 0·1 hp 24 V DC motor.

Since it would be difficult to mount work between centres by the method for metals of drilling centre-holes at the ends, Tufnol bosses carrying extension female centres are cemented on the crystal in a jig which registers them and applies a light spring-pressure (see figure 7.16). The adhesive used is an epoxy resin with a small amount of flexibility to cater for different coefficients of thermal expansion (e.g. Ciba AY/HY 111). The whole assembly shown in the figure can be transferred to an oven and the adhesive cured at 60°C for three hours. Rapid-set epoxies, now available, may be more convenient for the process. A standard hard centre or, preferably, a light-duty revolving centre, is fitted to the tailstock, with one machined from Tufnol in the headstock. This combination removes the need for any arrangement to give positive work drive when the crystal is mounted on the lathe; since the increased friction between Tufnol and Tufnol in light compression as opposed to Tufnol and steel ensures rotation.

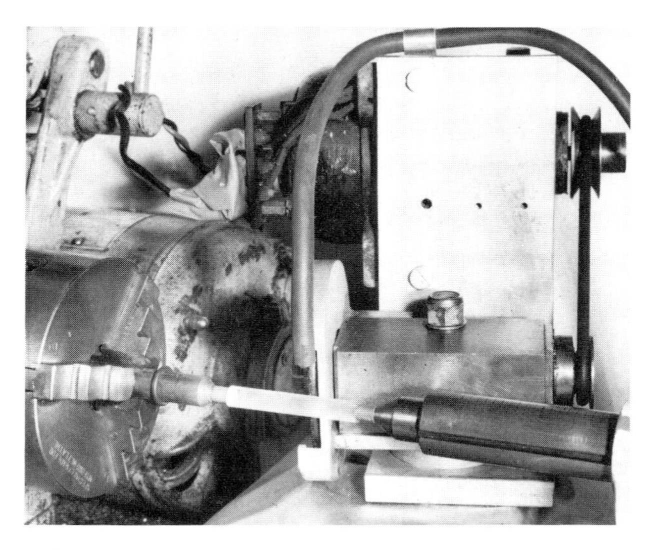

5.17 Centre-grinding on small lathe.

A copious supply of lubricant is needed to minimise temperature rise in the crystal and to retain any toxic dusts. The liquid is pumped from the tray of the lathe by a centrifugal pump of the same design as those described in chapter 1. The flood used makes it advisable to arrange a throw-off ring part-way along the parallel portion of the Tufnol live centre thus preventing the fluid from working its way to the chuck. Even so, extra splash guards are required for a small lathe being used as a grinder since it is not designed to withstand abrasive slurries in the slideways. A plastic sheet has been inserted between the grinding-head and the cross-slide face and draped over the ways. This helps to preserve the mating surfaces, but as an extra precaution the slide lubricators are frequently replenished with oil. When the centre-grinding equipment was first installed, we thought we were sacrificing a relatively inexpensive lathe, but time has shown that the precautions taken have preserved it from undue wear.

In operation, the work and grinding wheel contra-rotate at 200 and 10 000 rpm respectively. The depth of cut possible depends upon the robustness of the material being ground and the longitudinal feed rate. For example, silicon and glasses safely withstand roughing radial cuts of about 125 μm (0·005 in) deep at a feed rate approximately 40 threads/cm (100 TPI). A final cut of about 25 μm (0·001 in) deep is advisable. Ruby, however, is ground, initially at a depth of cut of 25 μm and finished at about 10 μm in order to preserve the wheel; whereas the softer crystals receive the same careful treatment as ruby but for their own preservation. Thus progress must be slow on many soft crystals, especially if they are square in form. In the interests of speed it is well worth removing corners before grinding the work cylindrical, preferably by the techniques described for half-lapping, § 5.3.3. Finally, the Tufnol centres can be removed after immersion in a solvent such as dichloromethane for the epoxide. Alternatively others can be destructively sawn and lapped away.

This method has been used to grind many materials and experience shows it to be safe with delicate crystals: less breakages have occurred than with any other mechanical technique. Of the larger sizes, rods of 2 cm diameter, 23 cm long have been produced from ruby boules; and in the smaller sizes, yttrium lithium fluoride cylinders of 0·2 cm diameter, 2.5 cm long.

5.3.2 Centreless grinding

A centreless grinding machine is probably the most convenient for producing fairly precise diameters, but the work must be nearly cylindrical initially. Thus the method is most easily used for finishing rods which have been trepanned (§5.2) or those which are sensibly cylindrical as grown or pulled. For the majority of boules and all crystals sawn from blocks of material either the Tufnol extensions noted in §5.3.1 or a moulding process is necessary to give the work an effectively circular shape. A series of PTFE moulds are made to accommodate the expected range of square or irregular dimensions. A mould is chosen which fits the crystal's corners as closely as possible and is then about one-quarter filled with an epoxy resin such as Ciba MY753 mixed 8:1 by weight with hardener HY951. The work is lowered into the encapsulating resin and any surplus allowed to flood over the top of the PTFE (figure 5.18). Any excess resin is wiped away and the moulding oven-cured at 100 °C for twenty minutes then extracted whilst still slightly warm. The curing conditions for delicate materials are 60 °C for three hours and the heating and cooling cycle is arranged to minimise thermal shock.

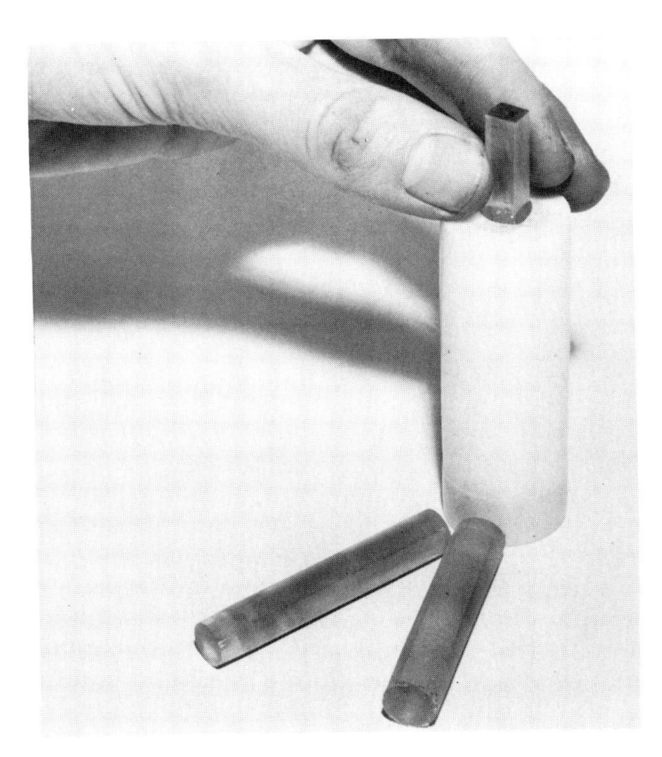

5.18 Resin encapsulation of bars to obtain cylindrical forms.

5.19 Bunter centreless grinder.

We have many years experience with a Bunter centreless grinder (figure 5.19) which uses wheels up to 6 cm wide. One of its most useful features is a variable-speed drive which can be used to reduce vibration on very delicate work. Although the machine is intended for small components (about 7 mm diameter), robust crystal materials up to 2·5 cm diameter have been ground on it, and some of the softer materials have been worked successfully at 3·2 cm diameter. These large cylinders are not thought to overload the machine since they are usually soft and friable.

A tungsten carbide knife which supports the rod has been fitted and it wears very slowly when in continuous contact with abrasive work. Its seems to be a very low-friction material which does not impose unwanted drag on cylinders that may be reluctant to rotate freely. In general, the centre of the work is operated below the centreline joining the rubber driving wheel and the grinding wheel: the height of the knife is suitably adjusted by means of a vertical slideway which can effect circularity of the ground work.

Two types of wheel (figure 5.20) are in regular use on the machine: electrodeposited diamond on steel for the harder materials like alumina and garnets; carborundum for the whole range of glasses and all the soft electronic materials. These carborundum wheels have the great advantage that they can be diamond-trimmed, not only restoring their accuracy but improving their free-cutting qualities, too. The interchangeability of its grinding-wheel shaft assemblies is

5.20 Various types of wheel used on the centreless grinding machine.

one of the Bunter's most valuable features. Each wheel has its own shaft with a machined fitting diameter at the centre. Threaded collars clamp the wheels in place, invariably as a permanent assembly. If paper washers are not supplied with carborundum wheels, they must be cut from thin card and fitted before clamping-up; no washers are needed for steel-cored, diamond-faced wheels. The sections of the shaft on each side of the wheel carry individual bearing-sleeves which are adjustable with respect to the shaft diameter. Insertion of the assembly in the machine is carried out by removing the top halves of the housings by unscrewing four knurled nuts (figure 5.19). The bearing sleeves are then seated in the lower halves of the split housing, taking care to engage the key and keyway in the housing and bearing sleeves respectively. The top-halves are replaced and tightened down with the knurled nuts.

New carborundum wheels (and those accidentally damaged) require trueing, before being used, by bolting-on a diamond trimming-unit to the special slideway on the right-hand side of the machine. A cut is made by setting the trueing diamond at a suitable angle, locking it in position and traversing the wheel face which is rotating at low speed. Water-lubrication is advisable during the operation in order to prevent the dust from becoming airborne. A similar mechanism is bolted to the right-hand side of the Bunter for trueing the surface of the rubber driving roller, but both the grinding wheel and the knife mountings have to be removed in order to make room for this trueing head. Lubricant-feed is not necessary for the rubber grinding operation. The depth of cut can be varied by the work-feed mechanism, setting the micrometer end-stop to control the roller position and using the workfeed micrometer to press the end-stop against the machine body.

A pumped, recirculating coolant system operates from the main storage tank in the base of the machine with a weir arrangement acting as a filter. Any water-soluble coolant is suitable for most of the crystals worked, but those which hydrolyse either to their detriment or, as in the case of gallium phosphide, to produce a toxic gas, need a non-aqueous solution. If sprays which constitute a health risk are evolved during grinding an extraction hood and plastic curtain can be fitted around the machine.

From the moment the grinding wheel makes contact the work must rotate otherwise a flat forms which has to be removed by other means before the work can be expected to rotate freely. Work must either remain within the width of the wheel (that is, it is shorter than 6 cm and is plunge-ground) or must be fed through with very small cuts being taken. Too large a cut would rub against the leading edge of the grinding wheel which, if diamond, usually has no abrasive facing. The work should have as large a length-to-diameter ratio as possible since some crystals, particularly those with uneven hardnesses, tend to drift badly in axis when the ratio is about 1:1. If the finished crystal has to be short, either extensions can be waxed to its ends or the epoxy moulding made long enough to be suitable. In either case, the additional material is ground away simultaneously without abrading the wheels and helps to preserve the specimen as a right cylinder.

We find it better not to use the pre-set diameter stop in the usual way on crystal work, but to obtain the required feed by unwinding the stop slowly while maintaining cutting pressure with the feed screw. In this way, crystals which vary in hardness are kept more closely circular during the stock removal stage. Departures from circularity are a constant concern with centreless grinding and it is very much a matter of optimising the height of the wheel, drive roller and work. Errors are rarely seen by measuring across a diameter with a micrometer because the figure is invariably constant (usually tri-lobar). Some qualitative guide may be obtained by rotating the rod in the fingers—it is surprising how small an amount of lobing can be felt. Another informative technique for indicating lobing is to rotate the rod in a vee-block with the stylus of a DTI registering on the diameter. If precise circularity is very important, the work should have plots made from it on a sensitive profilometer such as the Talyrond. The data obtained can be used to assess improvements (or otherwise) resulting from adjustments of the various parameters. Ultimately, half-lapping may have to be used (§ 5.3.3).

As with many grinding and lapping processes the

edges at the cylinder's ends are weak against the tensional forces imposed by grinding. If the crystal has little length to spare for chipping to be lapped away subsequently, some means of protection is needed. Suitable methods are: complete encapsulation in an epoxide; casting on ends; cementing on glass, crystal or Tufnol discs with a hard wax. Most resins grind down without clogging the stone, as long as they are fully cured. End protection has an additional advantage where the work is being fed through between grinding and driving wheels, since a wooden rod can be used to assist the cylinder in its traverse. The inevitable marking which results does not matter then.

Centreless grinding has proved satisfactory on a wide range of electronic materials: electrometallised diamond wheels have dealt with the more robust substances such as sapphire, yttrium aluminium garnet and quartz; the carborundum wheels with softer materials like calcium fluoride, calcium tungstate, cerium fluoride, lanthanum fluoride, soft glasses, germanium and silicon. Over many years of operation, the breakage rate has averaged 1 in 50 cylinders ground; whereas centre-grinding—especially of strained crystals—has had a mean breakage of about 1:70.

5.3.3 Half-lapping

Manual techniques are probably the safer to use for making cylinders of the most delicate materials and they are not unduly tedious. The following method may be applied to crystals as hard as quartz, though it is those with a Mohs's value close to that of yttrium lithium fluoride which abrade rapidly with 600 carborundum. A finished cylinder of about 3 mm diameter is obtained from a square bar 3.5 mm side—one usually sawn from the boule with a peripheral low-speed saw. Thus a slab is cut, 3·5 mm thick, then a series of square bars sawn at 3·5 mm intervals plus the kerf. The most precise cutting greatly assists subsequent hand-lapping.

Next, the bar is made octagonal by holding it centrally and lapping a second set of sides on the corners of the square section, continuing until a regular figure is obtained. A slurry of carborundum, 600 grade, on a ungrooved cast-iron plate cuts fast enough for all but the harder crystals. In the absence of a cast-iron plate, greyed glass makes a satisfactory lap. During the lapping of successive small flats, inspection of their depth and width with a magnifier is advisable. The material removed can be calculated from gauge measurements (with a micrometer or a DTI, for example) but the sensitivity of the eye to variations in width on a side is very critical, readily to better than 0·1 mm.

Should holding a square rod in the fingers prove difficult or tiring, a small jig can ease the process (figure 5.21). A right-angled groove, into which the bar is cemented is milled across a metal plate. Three hard feet may be affixed to the jig to serve as depth controls. These can take the form of lockable screwed stubs, adjustable to give control of the flats produced on any given rod. Where several similar cylinders have to be made, a number of grooves should be milled, each to receive a square-sectioned bar.

The eight corners are rounded-off with semi-circular laps (half-laps) made from any convenient material such as cast iron, steel, glass and brass, either from tube or drilled rod. The internal diameter should be a small clearance on the octagon. These half-laps can be either used in pairs with 600 carborundum or, more conveniently, singly in the following way. A rubber sheet is placed on a flat surface, spread with a layer of abrasive and water then the bar rolled in the slurry by means of a half-lap. The pressure applied has to be adjusted to ensure smooth and continuous rotation. This action quickly reduces the corners of the work and, depending upon the skill of the operator, it becomes fairly circular and constant in diameter. A jig which sizes the rod and achieves circularity can be used now, in the form of a short length of metal, drilled with two diameters concentrically (figure 5.22). For the entry of the rod, a hole that is a sliding fit is needed and the sizing section is made by a drill 0·1 mm bigger than the rod's final diameter. It is pushed slowly through the jig whilst being rotated slowly with a suitable abrasive slurry to produce the wanted size and surface finish.

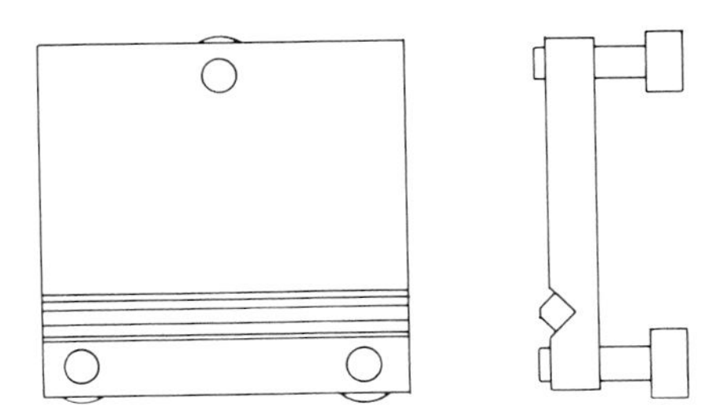

5.21 Jig for hand-holding eight-sided work.

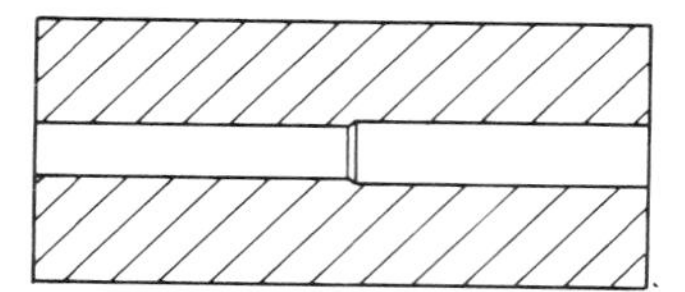

5.22 Two-diameter jig for circularising rods.

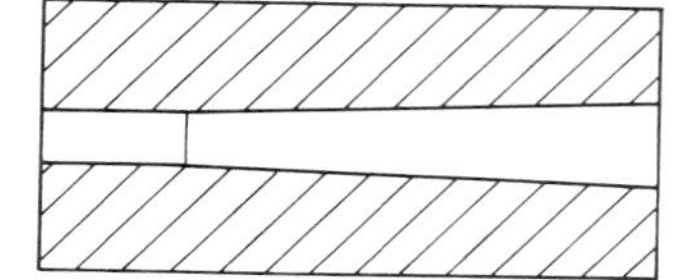

5.23 Taper jig for circularising rods.

Possibly the simplest of all laps for producing cylinders is one with a very slightly tapered hole finishing in a short parallel section of the required diameter (figure 5.23). Typically, for a 3 mm diameter rod, a piece of suitable metal of about 12 mm outside diameter and 20 mm in length is drilled axially with a 3·1 mm drill. Tapering can be done with a taper reamer which opens and enlarges the mouth of the hole sufficiently for the octagonal bar to enter.

The lapping operation is carried out by rotating either or both rod and lap and constantly replenishing the hole with an abrasive slurry. Relative rotational motions must be slow otherwise the contact area quickly becomes devoid of abrasive. With suitable lengths of lap and rod, the protruding finished cylinder can be grasped when it is no longer possible to hold the in-fed octagonal end.

The success of producing cylinders via tapered hole-laps lies in restricting their use to finishing processes with 600 carborundum—not attempting to remove large amounts of material. An unexpected ùse of the technique is that of removing slight lobes from centreless ground work: lobes which may be either inherent in the performance of the machine or stem from different hardnesses around the circumference of the crystal.

Cylindrical work produced by these methods does not usually suffer from lobing or other departure from circular form. Many crystals differ in hardness in their length and the final sizing operation minimises the effect due to variable erosion rates. Finally, a rare and delicate crystal is more safely lapped by hand than on the grinding machines.

Mechanical aid can be easily arranged for half-lapping, though the risk of breakage increases slightly. This takes the form of a rotating spindle (a small lathe or the output shaft of a motor–gearbox unit for example) coupled to the rod being worked with a short length of rubber tubing (figure 5.24). Both laps can be held on the smaller diameter work for the final stages (figure 5.25); but for larger cylinders, a single half-lap (a polisher, in fact) is used as shown in figure 5.26. The rubber tube is an effective torque-limiting coupling and, at the same time, a flexible drive. It is important to remember that when this mechanical rotation of the work is employed, the rotational speed must not be too great: corrugations may form since the fingers may be unable to provide a fast enough axial motion to the laps.

When lapping rods of about Mohs's hardness value 9 coarse diamond compounds are used in conjunction

5.24 Small lathe for mechanising a half-lapping process.

5.25 Two half-laps used simultaneously on a small diameter rod.

5.26 Single half-polisher in use on a stainless steel mandrel.

with cast-iron plates and half-laps. The final stages are performed more quickly by working a sixteen-sided bar initially and the necessarily higher pressures are best obtained by using the jig (figure 5.21), even though the cementing/uncementing steps have increased.

The same general practices apply where a cylindrical surface needs polishing. Polishers are made up from half-laps which have been lined with a suitable matrix like soft metal, pitch, wax and cloths. Again, if the rod is being rotated mechanically, the speed must not be too high since channelling may result. Figure 5.26 shows a stainless steel mandrel being polished.

5.4 Chamfering

Chamfering greatly decreases the edge weakness of any component and, in general, broad chamfers are used on work which may be subject to rough handling. It is usually a hand-operation (on small numbers of components at least) and the quality of hand-worked chamfers is as much the mark of craftmanship as the perfection of the optical surfaces. The aim is to produce a bevel with a fine grey finish, uniformly wide and free from chipping.

5.4.1 Straight edges

The control of edge-width is easily maintained on the longer edges of rectangular components but the shorter ones require more skilful treatment especially when they are only a few millimetres in length. The conventional practice is to draw the work over an ungrooved cast-iron plate using water or oil and 600 carborundum as an abrasive slurry. Most of the lapping motion should be across the chamfer and not along it, since fresh abrasive is more readily supplied by this action. As the contact width is very small at the beginning and tends to sweep abrasive grains along rather than riding over them, the first few strokes need making carefully. Suitably machined metal blocks can be used as guides to ensure, for intance, a 45° chamfer for two faces intersecting at 90°. No fixing between the work and the guide is needed as long as both hands are used to keep them in light contact. Polished surfaces may be protected during chamfering operations either by lacquering them or with PVC pressure-sensitive adhesive tape. Some of the more friable materials such as proustite require very careful handling to withstand chamfering on cast-iron laps. Extremely light holding in the fingertips helps to reduce the inertial mass of the work and hence the disruptive forces. Alternatively, the mass of the lap can be reduced by using thin metal plates or glass microscope slides either with loose abrasive or ones faced with abrasive papers. The latter can be used dry or with oil or water. For any advantage to be gained from lightweight laps they must be held sensitively: gripping the plate along its length between the thumb and forefinger should be avoided. A light hold by one end is recommended so that the remote end is stroked like a file across the edge of the work (figure 5.27).

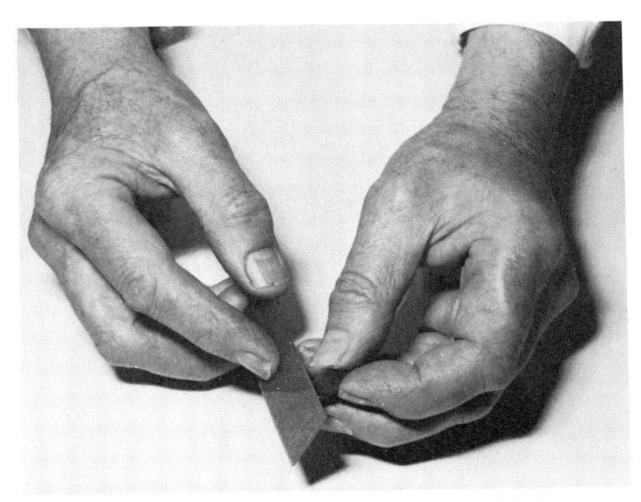

5.27 Chamfering a straight-edge with a lightweight lap.

5.4.2 Curved edges

Where a chamfer has to be produced on a curved edge that is not wholly circular, the coated plates (analogous to files) mentioned above are used. The majority of curves, however, are on the ends of cylinders and a simple method of chamfering one of these is with a lap made by machining a conical hole in the end of a metal rod, using a 90° centering tool. A thin layer of abrasive is brushed into the conical well and the work/lap revolved under light pressure (figure 5.28), and with frequent relief. If the ends of the component are already optically polished any loose abrasive must be washed rather than wiped away. The need for lacquering seldom arises, but chamfering is preferably done after lapping and before polishing.

Makeshift yet satisfactory chamfering can be carried out by coating the inside of a glass funnel with 600 carborundum and contra-rotating the work and funnel in light contact. Although the angle of the standard filter funnel is 60° instead of 90° it serves well enough for most purposes. Small funnels have the added facility of being usable on their outside surfaces so that internal edges may be chamfered with them.

In all the work described when loose abrasive slurries have been used it is important for the worker to develop an audio–tactile sensitivity in recognising the fine crushing sound and the feel of abrasive under a rotating lap. A hand operation develops this sense, but mechanisation once more requires that the operator should regard the machine as an extension of himself and be acutely aware of uniform abrasion. The only easily described symptom is that of squeaking sound which indicates approaching catastrophe, since either the cutting region may have dried-out; or too great a pressure has excluded abrasive from the interface; or too little abrasive is actively present—which can be due to too much or too little abrasive or water.

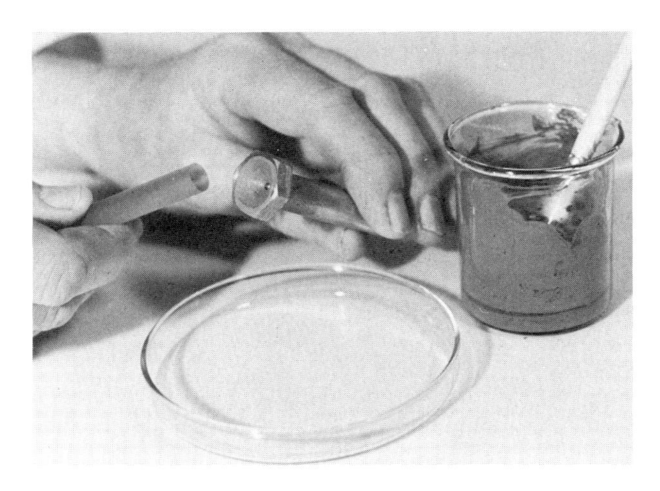

5.28 Hand-chamfering the end of a cylinder.

Consider the classic example of hand-working a chamfer on the end of a cylindrical component. At the beginning of the operation the edge of a precisely square cylinder contacting a glass or metal conical lap is virtually a line, effectively excluding loose abrasive. Excessive pressure excludes the abrasive, too, so that a very light pressure with frequent relief is needed until the mating areas are increased by material removal. Moreover, in this cylindrical chamfering process the rotary motion is along the contact line thus aggravating the abrasive-exclusion. When the component is being held by hand against a rotating conical lap the early stages of material erosion are aided by including a small amount of angular random movement to the work, which helps to introduce fresh abrasive slurry. Any positive rotation of lap or component must, of course, be at very low speeds to minimise sweeping away of carborundum from the interface.

The more robust materials can have chamfers diamond-ground on them with the centre-grinding technique described in § 5.3.1. Discs are cemented to metal rods for mounting in the three-jaw chuck and the wheel tilted at an angle for light cuts to be taken. Speeds and feeds are the same as those recommended for grinding cylinders.

In some special circumstances, chamfers may require a polished surface. The polisher usually takes the form of small soft metal, pitch or wood tools, conical—or more often radiused—in shape. The polishing medium is the same as that used for producing an optical surface on the component itself. Because the areas are still relatively small (though no longer sharp edges) the pressures likewise are almost uncontrollably light and the advice given for lapping is equally important.

5.5 Small Hole Graticules

The main problems to be overcome when making a pin-hole 50 μm diameter are the achievement of a regular circle, free from burrs, in thin material. This method results in several uniform graticules produced simultaneously in laminated, peel-off brass shims, each layer 50 μm thick. An attendant difficulty is that of preserving intact the small drill used during a delicate drilling operation and this is aided by hydraulically controlled cutting pressure.

The British Laminated Brass peel-off shims used comprise sheets of metal lightly soft-soldered together and, for the initial sawing-up into pieces from which about 1 cm diameter discs can be turned, they are clamped between steel washers. A Tufnol rod is cemented to this irregularly-shaped disc and forms a handle which is then held in a small lathe for turning the outside of the shims to size.

Drilling a very small hole demands not only light cutting pressures but that the drill must be held accurately at the centre of rotation, too. These requirements are met by the hydraulic feed mechanism (shown in figure 5.29) where a brass tube containing silicone fluid free from air bubbles is held in the tail-stock of a watchmakers lathe and feed pressure can be applied by lightly squeezing the rubber bulb. The shank of the drill and that of a similar small centering drill are precise sliding fits in a section at the orifice of the tube—its concentricity ensured by D-drilling the hole from the mandrel.

After the disc has been hydraulically centre-drilled, the 50 μm diameter drill is used to cut through about six shims. Individual graticules are carefully separated with a razor blade and the first two discarded because of oversize centre-drilling. A mild abrasive cleaning is advisable to removing the tinning and a superficial black-nickel electrodeposit (Field and Weill, 1946) can then be added. One drill has made many graticules; dimensionally uniform, precise in form and free from burrs.

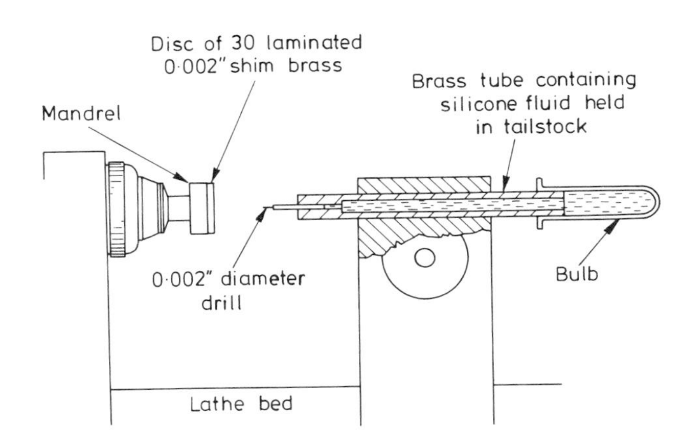

5.29 Hydraulic feed mechanism for drilling 50 μm holes.

References

Beale L S 1880 *How to Work with the Microscope* (London: Harrison) p 73.

Field S and Weil A D 1946 *Electroplating* (London: Pitman).

6 Measurement

'Do you know what it is to yearn
for the Indefinable, and yet
be brought face to face, daily,
with the Multiplication Table?' (W S Gilbert)

In terms of electro-optic components metrology there are five principal quantitative requirements that have to be dealt with: measurement of linear dimensions, of parallelism, of angle (usually orthogonality), of flatness and of curvature. When considering sawing and lapping processes, these five suffice; any polishing process requires some assessment of specularity, however, and it is an exception if such an assessment be other than qualitative. Even so, the topic is loosely dealt with under this chapter heading. Moreover, the slicing of many crystals has to be relative to an established crystallographic axis, but the techniques described in chapter 1 absolve the cutter from carrying out x-ray analysis.

Measuring instruments have proliferated and continue to do so. The lesson of the wheeltapper's cracked hammer has been taken to heart and many devices are inspected, gauged, labelled and documented by the user with a meticulousness which is unarguable. The technician has to be equally sure of his facts and familiarity with a variety of techniques is desirable. We intend to deal relatively briefly with the commercial equipment that we use and which are covered by their manufacturers' handbooks; but more fully with techniques, together with the jigs and fixtures which have been designed or adapted to carry out specific jobs and facilitate measurement generally.

It is hoped that this book will find its way into the hands of the technicians carrying out the work as well as being consulted by those whose function is mainly advisory. Many papers are written by highly qualified people one or more steps removed from the work they report. Their readers are equally gifted but themselves sometimes remote from direct implementation. The result is often the attempted duplicating of a technique (warts and all) and not its perfecting.

All branches of art have their jargon—polishing is no exception—but the shorthand of the mathematician and physicist (especially in the field of measurement which is stocked with well trained theoreticians and experimentalists) sometimes baffles the technician. In fairness, the imperial thou's and microinches sometimes bemuse the scientist. To many, the following sentences state the obvious: however, all too often the obvious is only so when it has been stated. We shall expand a few simple relationships by example and consider the ways of expressing distance travelled in time. 'Miles (in) an hour' is common usage (but soon to be replaced with 'kilometres an hour'—the conversion of imperial to metric units and their confusion becomes a general problem). Now we can write:

$$\text{miles an hour} = \text{miles per hour} = \text{mph} = \text{m/h} = \frac{\text{m}}{\text{h}} = \text{m h}^{-1}.$$

Similarly,

$$\text{micrometres per cm} = \mu\text{m per cm} = \mu\text{m/cm} = \frac{\mu\text{m}}{\text{cm}} = \mu\text{m cm}^{-1}.$$

and,

$$\text{grams per square cm} = \text{g per cm}^2 = \text{g/cm}^2 = \frac{\text{g}}{\text{cm}^2} = \text{g cm}^{-2}.$$

Table 6.1

Centimetres (cm)	Millimetres (mm)	Microns (μm)	Ångstroms (Å)	Inches, (in) Accurately	Trivial approximation
2·54	25·4	$2{\cdot}54 \times 10^4$	$2{\cdot}54 \times 10^8$	1	1
1	10^1	10^4	10^8	0·3937	0·4
10^{-1}	1	10^3	10^7	$0{\cdot}3937 \times 10^{-1}$	40 thou
10^{-2}	10^{-1}	10^2	10^6	$0{\cdot}3937 \times 10^{-2}$	4 thou
10^{-3}	10^{-2}	10^1	10^5	$0{\cdot}3937 \times 10^{-3}$	4 tenths
10^{-4}	10^{-3}	1 (10^0)	10^4	$0{\cdot}3937 \times 10^{-4}$	0·4 tenths
10^{-5}	10^{-4}	10^{-1}	10^3	$0{\cdot}3937 \times 10^{-5}$	4 microinches (μ″)
10^{-6}	10^{-5}	10^{-2}	10^2	$0{\cdot}3937 \times 10^{-6}$	0·4μ″
10^{-7}	10^{-6}	10^{-3}	10^1	$0{\cdot}3937 \times 10^{-7}$	0·04μ″
10^{-8}	10^{-7}	10^{-4}	1 (10^0)	$0{\cdot}3937 \times 10^{-8}$	0·004μ″

Table 6.2

	Angle subtended		Linear dimension	
	Seconds of arc		Metric	Imperial
Microradians, 10^{-6} rad (μrad)	Accurate to 4 decimal places	Approx.	Microns/cm (μm cm^{-1})	Microinches/in (μ″ in^{-1})
1	0·2063	0·2	0·01	1
5	1·0313	1·0	0·05	5
10	2·0626	2·0	0·10	10
15	3·0939	3·0	0·15	15
20	4·1253	4·0	0·20	20
25	5·1566	5·0	0·25	25
50	10·3132	10·0	0·50	50
100	20·6265	20·0	1·0	100
150	30·9397	30·0	1·5	150
300	61·8794	60·0	3·0	300
Milliradians, 10^{-3} rad (mrad)	Minutes of arc		Microns/cm (μm cm^{-1})	Mils/inch (0·001″ in^{-1})
0·291	1·0	—	2·9	0·291
0·582	2·0	—	5·8	0·582
0·873	3·0	—	8·7	0·873
1·0	3′ 26″	—	10·0	1·0
5·0	17′ 11″	—	50·0	5·0
10·0	34′ 23″	—	100·0	10·0
17·453	60	—	174·5	17·453

The last entry in each example is in the inverse notation.

The idea of moving the denominator to the numerator line and changing its sign thus:

$$\frac{\text{numerator}}{\text{denominator}} = \text{numerator denominator}^{-1}$$

is very useful, not only for printers, but where large and small numbers are involved and powers often may be dealt with:

$$1000 = 10 \times 10 \times 10 = 10^3 \text{ and}$$

$$\frac{1}{1000} = \frac{1}{10^3} = 10^{-3}.$$

Again $300\,000\,000 = 3 \times 10^8$ and $0{\cdot}00000006 = 6 \times 10^{-8}$ etc. Sawing and polishing instructions from various sources can dimension specimens in the same, different ways, for example, $2 \times 2\,\text{mm} = 2 \times 10^{-1} \times 2 \times 10^{-1}\,\text{cm} = 2 \times 10^3\,\mu\text{m} \times 2 \times 10^3\,\mu\text{m}$ ($1\,\mu\text{m} = 1$ micrometre $= 1$ micron $= 1$ metre $\times 10^{-6} = \text{m} \times 10^{-6}$).

The tables which we find useful in linear, angle and flatness measurement have been placed at the beginning of the chapter rather than with their relevant sections so that the inter-relationships are more obvious and the opening explanatory sentences are near to them.

Table 6.1 covers the relationship between the more common linear dimensions with the trivial names prevalent in engineering practice. The fundamental conversion figures used are 1 cm = 0·3937 in and 1 in = 2·54 cm. A 'thou' is 0·001 in and a 'tenth' 0·0001 in.

In table 6.2 the linear displacement which subtends a small angle is given both approximately for quick calculation and more precisely for those who are interested. The figures are based on the radian—that is $180°/\pi$. The value of pi has been calculated (and published) to the 500000th decimal place but our less critical value gives one radian equal to 57° 17′44·8″; in seconds alone, R = 206264·8″ and 1 second = $4{\cdot}84813695 \times 10^{-6}$ R.

Table 6.3 relates the departure from a flat surface for a given interference fringe, or fraction of a fringe when viewed under a monochromatic light source. The gas discharge lamp is helium filled with its strongest emission line at 5876 Å (Henderson, 1970). Strictly, there are a family of emission lines but it is convenient to specify only the strongest and calculate from that.

Table 6.3

Interference fringes or bands at He 5876 Å		Wavelengths, $\lambda_{(He)}$ 5876 Å		Departure from plane			
Fractional	Decimal	Fractional	Decimal	Approx. (μm cm^{-1})	Accurately (Å cm^{-1})	Approx. (μ'' in^{-1})	Accurately (μ'' in^{-1})
2	2	1	1	0·6	5876	24	23·13
1	1	$\frac{1}{2}$	0·5	0·3	2938	12	11·56
$\frac{1}{2}$	0·5	$\frac{1}{4}$	0·25	0·15	1469	6	5·78
$\frac{1}{4}$	0·25	$\frac{1}{8}$	0·125	0·075	734	3	2·89
$\frac{1}{5}$	0·2	$\frac{1}{10}$	0·1	0·06	588	2·4	2·313
$\frac{1}{10}$	0·1	$\frac{1}{20}$	0·05	0·030	294	1·2	1·156
$\frac{1}{20}$	0·05	$\frac{1}{40}$	0·025	0·015	147	0·6	0·578

6.1 Basic Equipment and Techniques

There are many items of equipment which are fundamental in engineering metrological practice and although not all these are necessarily applicable in optical polishing, some require a special mention. They are generally useful in two or more of the types of measurement enumerated.

6.1.1. Surface tables, angle plates and stands

Many of the surface tables which we use are cast iron; 90 × 60 cm, skimmed, weathered and finally planed on a moderately accurate machine. No additional surface treatment has been given. Their use can give one psychological advantage: should it be necessary to drill and tap holes in the surface (a practice amounting to sacrilege) for affixing such components as optical benching or rotary tables, one's engineering conscience is not affected in the way it would be if they were hand-scraped plates. The method used for drilling the roughly planed plates is to stand a fairly heavy bench drilling machine on the surface table and to reinforce its mass by adding sufficient iron or lead weights on the drill table. This maintains stability during machining operations which necessarily involve the shaft of the drill's being swung over the edge of its own table.

It will be evident at this stage that we have collected a range of equipment, much of which is of our own design and of in-house manufacture. It would appear from our conversations with other workers on a wide scale that equipment generally is not always readily available so a few substitutes for the conventional engineering solution to solid bases will be mentioned. Again—as in the case of polishing machines—thick, laminated wood has very good stability. It might be argued that the softness of the material would result in poor wearing properties. Sheathing the important surfaces (both upper and lower) with resin and glasscloth provides a durable surface. Where this has been built up to a sufficient thickness, it is possible to machine the glass-reinforced resin with tungsten carbide tools. Box sections, suitably faced and machined, make presentable substitutes for cast-iron plates. Concrete has been explored for the purposes of machine tool beds and it should by the same logic make good surface tables for metrological purposes when it has been sealed to prevent dust forming on the surface. All moulded materials offer the opportunity for insetting machined components which can be in the form of kinematic abutments (see §6.1.3.1) or simple lapped flat plates where required.

Conventional bricklaying is not far removed from concrete manufacture and we have seen an excellent structure laid by a research graduate who needed an optical stand with the minimum of delay. A small quantity of bricks, sand and cement was used to complete the stand over a weekend. It is worthwhile working in some metal fastenings with the top layer of bricks—some short lengths of rolled steel joists (RSJ or I-section girder) are useful insets on which to mount equipment. Granite slabs, obtainable from a monumental mason's yard, have excellent stability.

Where the layout of components is more or less static and linear, RSJ is satisfactory. Although the material is very straight, it is often a little too rough on the faces for metrological purposes. Either surface grinding or planing remedies this and levelling feet can be welded or bolted to the lower limbs. It is especially suitable for long test benches which are used, for example, to space two optical stations very rigidly as are required in the mirror tests of Foucault (1859) and Ronchi (1925) (see §6.6). The RSJ arrangement of optical benching will perform equally well for prism testing, using a single auto-collimator.

Of the many precision angle-blocks and clamps readily available in the fields of engineering tool making and metrology, we shall list but a few and their uses. One of the more useful is the box angle, so called, which is a five-sided cast-iron box machined on external faces and provided with both through and tee slots on opposite faces (figure 6.1). It is valuable in packing-up components to a required height. When it is bolted to a surface table, work can be clamped to the vertical sides via the slots and thus a measure of vertical height adjustment is available. The simple right-angled bracket, too, may be used for the latter function. Magnetic bases, positioned in the demagnetised state and locked in place on aligning the poles, serve variously as direct and indirect clamps, abutments and registers.

6.1 Box angle.

Whatever the method of packing-up or holding a piece of work relative to a measuring instrument (or vice versa) the golden rule when precision is sought is that the smallest number of mating components must be used. These, in their turn, must be solidly co-related. All the common ways of achieving alignment which rely on sliding rods and clamps in the form of, say, a grub screw on a soft shaft fall short in respect of rigidity when working to seconds of arc. They are admittedly easy to use and provide rise, fall and rotation but they require the equivalent of the gardener's green fingers to minimise uncertainty. The skill that such an operator has patiently to acquire is in the 'letting go' allowances; knowing intuitively just how much to allow for the spring of an instrument on the removal of the adjusting fingers. It is for this reason that we generally use very solid arrangements.

An interesting problem occurred recently the solution to which showed what interactions between apparatus and users can happen, and the strange mixture of willowy instrumentation, touch and environment which prejudices measurement. The system involved neodymium-glass lasers with external resonators. Because of its relatively low optical gain (when compared with neodymium–yttrium aluminium garnet lasers) and the very high optical perfection of the glass, it was extremely sensitive to alignment at threshold levels. The apparatus was completely mounted on a conventional thin stemmed, telescopic rise and fall arrangement. The micrometer-driven, non-kinematic mirror mounts (cf §6.1.3.3) were normally capable of being set to a few seconds of arc when held on a solid foundation. Their relatively stiff and very positive micrometers had not seriously limited their performance. However, when they were bolted to the top of a 7·5 cm brass stem of 0·7 cm diameter it became impossible to set the mirrors with anything approaching the essential accuracy. PTFE sleeved/stainless actuators described in §6.1.3.3 were used to replace the stiff micrometer drives and, since they could be adjusted with very light finger pressure it became possible to set the system without difficulty. The long-stemmed rise and fall device still prejudiced the new mirror mounts which were designed to be stable under some few g's, but were literally waving in any draught in that particular situation. However, it demonstrated forcibly the advantage of having freely moving but firm actuators with reasonably large control knob diameters; for the forces which affect a whippy arrangement to a large extent must also affect a more rigid one to some lesser degree. Any cylindrical column which supports a metrological device should have a length-to-diameter ratio of not greater than 8:1; where heavy gauge heads or mounting assemblies are in use, a ratio of 5:1 is often advisable (i.e. a column 25 cm long 5 cm diameter).

6.1.2. Dial test indicators (DTI s) or dial gauges

These are possibly the most useful metrological instruments in the entire field of materials processing (figure 6.2). They can be applied in a variety of ways, the most usual of which is as direct-reading instruments (figure 6.3), where work is slid under the stylus and the readings taken from the scale. The inbuilt accuracy and linearity of the instrument is its limiting factor. Baker (1958) assessed DTIS and found them

6.2 Stylus and plunger-type dial test indicators (DTIS).

6.3 DTI in use for measuring laser rods.

suspect over a range of travel, but latter-day models on which we have carried out calibration checks have been proved to be accurate to better than 2 μm. Even so, an old 'clock gauge' should be carefully checked and the accuracy of new ones confirmed before any definitive figures are quoted.

An indirect technique is to use one as a zero-indicator, mounted on a kinematic carriage system driven by a large-diameter drum micrometer (figure 6.4) (see §6.1.3.2). In this function it gives the highest accuracy (repeatability in essence) and any required readings are taken from the scale on the micrometer.

Most of the dial gauges of the lever type have fittings for attachment to scribing blocks and to the vertical pillar of a magnetic base (figure 6.5). Both can be either slid over a surface table or, in the case of the magnetic base, locked to it. A feature of the scribing block is the pair of pegs in its base which can be pressed to protrude through the mating surface. Registration of these pins against the edge of a table or angle block is used to produce a linear movement of the dial-gauge stylus during crystal alignment arrangements (figure 6.6).

6.5 DTI on a magnetic base.

6.4 DTI on a kinematic carriage, driven by a large drum micrometer head.

6.6 DTI on a scribing block, showing pins registering against the edge of the table.

The more common application of the small lever DTI is for measuring specimen height vertically. In our experience it is possible to estimate the thickness of a specimen to better than 3 μm using a 25 μm/division scale. A × 4 or × 5 magnifier helps considerably. Perhaps unexpectedly, this arrangement seems to be more critical than using the more sensitive DTIs without optical magnification. The small plunger-type gauges used for monitoring specimen height and thus thickness are light enough to be mounted on an adjustable-angle polishing jig (figure 6.7). Although the stylus pressure is on the upper end of the central, spring-loaded shaft, it does not interfere with the auto-collimated setting. A continuous indication of specimen erosion rate is obtainable. Some allowance has to be made for the wear of the conditioning ring of the jig, but this is a constant for any jig once the polished face becomes grey. It is a very different and more complex wear-process when the jig is on a polishing machine, but, relative to specimen, it is still predictable. Even so, the wear can be reduced to a negligible amount by facing the existing ring with tungsten carbide or high alumina ceramic. Whether or not special ring facings are worthwhile depends on the relative rates of polishing of ring and specimen: proustite, for example, polishes so speedily that there is very little loss of the surrounding jig face. Whatever the technique, the use of the gauge in this way has proved itself to be a valuable economy in cleaning and measuring time.

6.7 DTI stylus registering on a jig-centreshaft for monitoring lapping and polishing progress.

When using a DTI to indicate work height on a porous substrate), then a reading of specimen height jig, care has to be taken either to pack up the the jig stand so that the conditioning ring is parallel to the surface table; or to use a dial-gauge mounting which relates to the ring itself, for the purposes of sliding and measuring (figure 6.8).

Many electro-optic materials are too delicate to withstand the pressure of a stylus sliding on their surfaces and a successful compromise practice with all but the most friable is to lift it clear of the specimen before moving the gauge. The stylus is lowered only when the components are stationary. Careful lifting can be done with an elastic band (figure 6.9) but the DTI body must not be moved, otherwise the reading would be meaningless. For the protection of the most delicate surfaces a thin sheet of polyester film (which is sufficiently uniform in thickness not to confuse the readings) can be placed beneath the stylus tip. The film should be 25 μm or greater in thickness since anything thinner is difficult to manoeuvre into position.

In metrological practice, it is usual to take three readings at each operation: one on the block (or porous substrate), then a reading of specimen height and finally a return to the first position as a check that nothing has moved during the previous operations. Again, it is sound practice to relieve static-to-sliding friction in mechanical instruments by tapping the surface table on which they are standing with a soft mallet, usually Tufnol or nylon faced. This should be done at each new setting of the components. The technique is recommended with mechanical, electro-mechanical and pneumatic instrumentation since the complete arrangement of jig (or block) surface

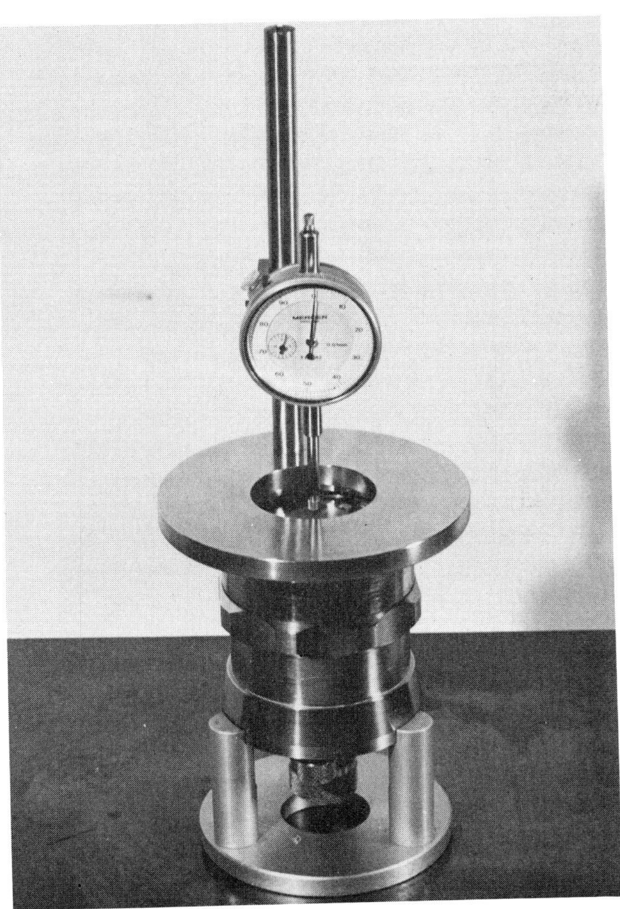

6.8 DTI registering on a jig-held specimen, referenced from the conditioning ring.

table and comparator stand is a series of complex mating surfaces. In spite of the fact that these may be clamped—some by bolting and others magnetically—small unaccountable movements sometimes take place between the gauging instrument and the workpiece.

6.1.3 Kinematic planes and mechanisms

Uncertainty in performance is a major snag for the user of a measuring instrument since it may necessitate constant checks, realignment and recalibration. This uncertainty, relative to a specific application, must be reduced to negligible proportions and the maximum constancy obtained—that is, the instrument must be precise.

Kinematics may be taken as the consideration of pure movement independent of applied forces. Thus, in a mechanism it depends on the achievement of translation (linear travel) and rotation where each motion does not influence any other, and constraints that are necessary without being superfluous. Designing with scrupulous regard to the kinematic principles which have been stated by Thompson and Tait (1863) and elaborated by many authorities since (e.g. Maxwell, 1890; Whitehead, 1933; Pollard, 1929; Strong, 1938; Braddick, 1960) fulfills the above requirements.

A rigid body, able to move in space, has six degrees or modes of freedom: three of translation along orthogonal x, y and z axes and three of rotation around each of these axes (figure 6.10). A ball at rest on a flat surface

6.9 Elastic band in use for lifting the stylus out of contact.

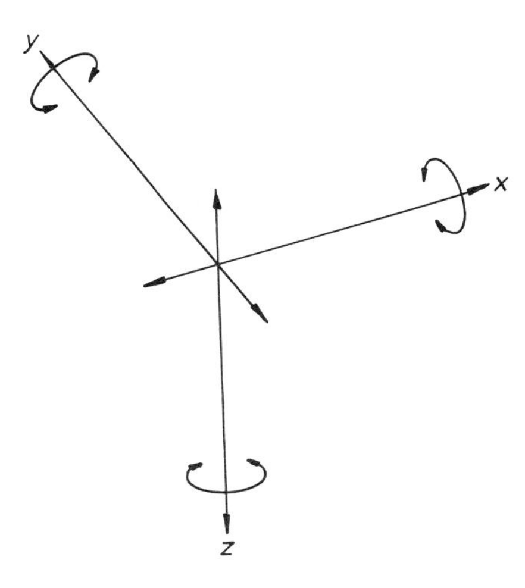

6.10 Six degrees of kinematic freedom.

has lost, because of the force of gravity and its position on a solid plane, one degree of freedom, but it still has all three rotational degrees and two translational. Place it in a dimple under the same conditions and it has only the rotational modes left to it. Placed in the dimple with a pin through it, protruding both sides, parallel to and contacting the flat surface, it has lost one more degree. It can still rotate round the shaft and spin like a top.

A bead threaded on a taut wire, as in an abacus, has only two degrees of freedom (if loose fits are ignored). It can translate along the wire and revolve around it. For the bead to be and remain a good fit on an accurately cylindrical wire it should make bearing contact at each end of its central hole. Its middle is relieved. A third bearing is a third constraint and is superfluous, while one long bearing is more likely to be a series of good and poor fits even if it has been reamed. Now if the tendency for the bead to rotate be removed by attaching an arm at one side, say, which abuts the surface in the horizontal plane, the bearing can become point contacts. Figure 6.11 illustrates the progression. The result approaches rigorous kinematic design, especially when the anti-rotational arm itself exerts no undue influence on the translation of the bead (bearing arrangement) on the wire (shaft). It is generally true that kinematic design has its main application in the small instrument field and that when loads and movements become large it is increasingly difficult to apply rigorously. Even so, in the evolution of any mechanism there must be scope for considering these precepts and applying them whenever practicable, and yet their employment is ignored by many design drawing offices. There are, of course, famous companies who have used kinematic designs: for example, the Cambridge Instrument Company from the early 1900s. The following mechanisms are, we feel, simple and precise. Points of design will be noted in them which were lectured upon by Professor Willis as long ago as 1840. We can claim very little originality of thought, but, perhaps, the merit of application to readily obtainable, cheap components. Thus extensive use has been made of ball, needle and roller bearings because of their correct hardness, high surface finish and close dimensional tolerances. These cost less than a few pence each and precision has been obtained by their location rather than by primary machining accuracies.

Other factors, too, are important to the user: spatial freedom to adjust the controls is an obvious advantage, but equally important is the 'feel' of the control itself. In a mirror mount when the movement of the actuator is not monitored but its effect on the optical components is observed, the feel of the control must be unobtrusive in order that the primary dependence is upon the eye/fingertip coordination and not vice versa. The actuating thread must not be so loose that it is imprecise since in this condition a smooth sensation achieved by packing it with a viscous lubricant results in drift after setting-up. It must not be so tight that restoring moments after adjustment distort the setting either of the instrument or of its attachment to a stand or to an optical bench. Thus the motion achieved via the screw thread must be fine, smooth and free from bounce and drift. Large knobs and keen knurls are important factors in the translation from flexible fingertip to steady torque thread adjustment. This is a facet of the kinaesthetic sensation discussed by Whitehead (1933).

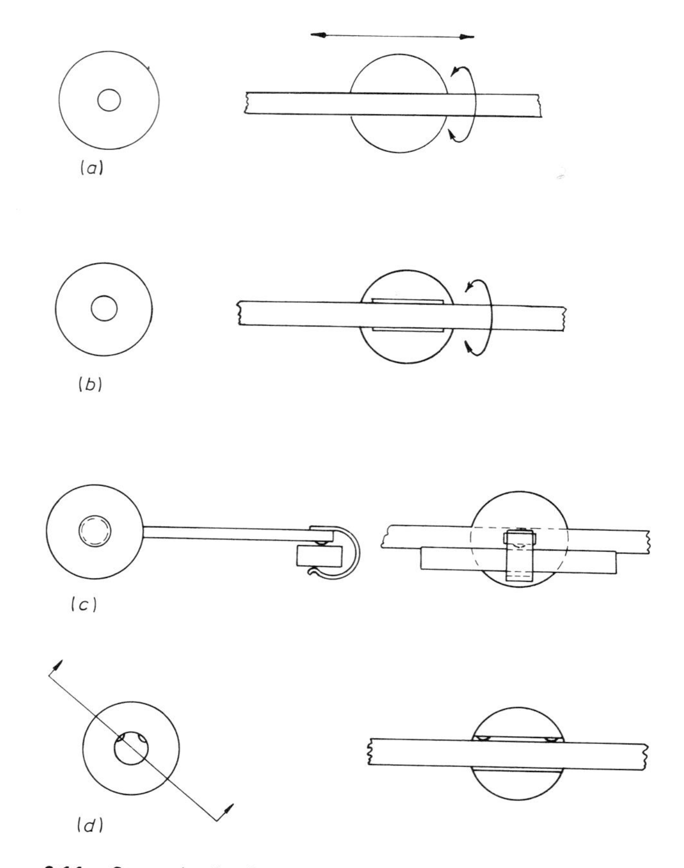

6.11 Stages in simple constraint: (*a*) two degrees of freedom, simple case; (*b*) two degrees of freedom, improved by bearing relief at centre; (*c*) removal of rotational degree of freedom; (*d*) bearing contact made into point contact.

6.1.3.1 Kinematic planes, or three-ball planes

That a table or chair standing on a firm surface usually makes contact with only three legs is a common example of kinematic location. Equally, sitting on a three-legged stool with one's weight outside the triangle formed by the feet points to a user limitation. Given a flat, solid surface table, the difficulty of registering another smaller flat on it which can be repeatedly removed and replaced without changing the angle of contact would seem to be simple; but when repeatability to better than a few seconds of arc is needed, the flat should contact unambiguously at three regularly spaced points. The plane defined by the points has been termed a kinematic plane and, since it is most easily obtained by holding three ball bearings on a surface (flat or otherwise) it will be referred to as a three-ball plane. One cautionary point should be emphasised from the beginning: when work is placed on a plane and monitored either with a DTI, comparator or pneumatic gauge head, the applied measuring pressure must be either immediately above a ball or within the equilateral triangle formed by the balls. If the pressure should be applied outside this area the tipping force, though not always apparent, can give rise to errors.

Since they are such important components in both our optical and mechanical measuring arrangements, the various types and their applications will be described in some detail. Those for use with solid blocks—when, for example, single laser rods are polished—take the form shown in figure 6.12. Each ball is fitted against a shoulder formed by drilling a two-diameter hole, the larger one fits the ball and the smaller forms a pilot hole right through the steel base. The purpose of the latter hole is to allow worn balls to be driven out and replaced with new ones. Ball replacement occurs more frequently than may be imagined. Most of the wear seems to occur in the constant testing of surfaces which have just been lapped. The work has a surface roughness of about 0·2 μm CLA and it probably has some abrasive embedded in the fractured surface. Using a three-ball plane which has large flats worn on the balls to some extent degrades its performance as a plane-defining device. The steel balls are fixed in the base with a semipermanent cement such as a cyanoacrylate resin or a hard wax. The underside is relieved to leave a rim which is lapped flat. The block can then be easily wrung to a collimator base.

Some means of centering the work block over the triangle defined by the balls is therefore needed and it usually takes the form of two steel pins situated

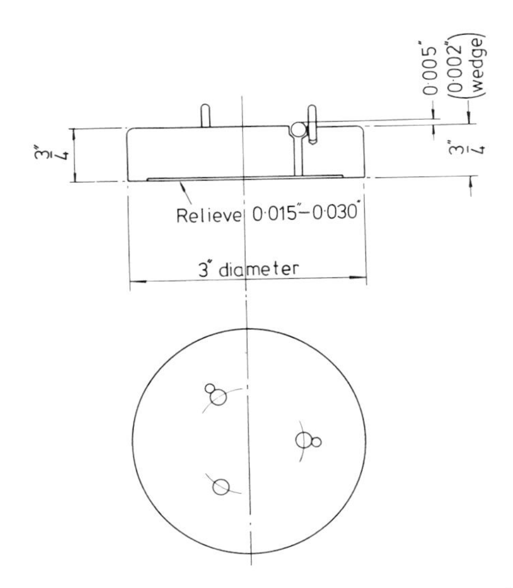

6.12 Kinematic (3-ball) plane. Three 1/4" balls on 1 1/2" PCD (120°) press fit or loctite. Two 1/8" diameter pins 3/8" high (120°) 0·015" spaced from ball holes, closer if possible.

at 120° and positioned to register the block as shown in figure 6.13. Plastic collars, usually of PTFE, are fitted to the pins to enable work to be turned smoothly by hand. In operation, the block is subjected to a slight downward pressure combined with a slight lateral force which maintains it in contact with the pegs. This ensures that both rotational and thrust factors are defined and in this way any slight curvature of the block facing has a negligible effect on the parallelism readings.

A further important constructional feature is in the arrangement of the ball heights above the base: one ball is set to be 0·05 mm above the other two, relative to the underside of the base. This difference is made necessary because most Angle Dekkors and auto-collimators are restricted to fine adjustment in angle in one direction only. The second axial adjustment is normally obtained by setting the auto-collimator tube correctly in its stand—a method that we find somewhat crude. By making the three-ball plane effectively a slight wedge, its rotation provides a nice control in any desired direction of tilt within the approximately 2′ capability of a 10 cm plane. It might be commented that such resetting is unnecessary since a specific reading can be taken from the auto-collimator scale. In practice, we are often using Angle Dekkors well outside the normally accepted limits of sensitivity, purely as coincidence

6.13 PTFE sleeves and 3-ball plane in use.

indicators sensitive to a few seconds of arc (see §6.1.4). Once a coincidence setting has been established for a specific check or measurement, the plane must be prevented from further unwanted rotation. If this is not done, any recheck would be potentially meaningless, since an unnoticed excursion of the plane would alter the angle. Magnetic bases can be used to hold a vee-block in place on the surface table or collimator base, and in its turn, to hold and register the kinematic plane (figure 6.14). This serves to prevent movement during any rotation of a block or jig on the plane and gives some guidance to the outside diameter of the work block if the plastic-sleeved pins have not been fitted.

Where a three-ball plane is being used in conjunction with an electromechanical comparator, rotation of the item undergoing test is a necessity with the two pins (or vee block) serving as guides to control the rotational axis. Various jigs, too, make use of rotational freedom in order to provide effective doubling of the image excursion in an auto-collimator. Because of the milled, knurled or fluted profiles of the conditioning ring's holder and locking nut, a brass collar is often slipped over the three-ball plane, with a smaller diameter a loose fit on the jig outside diameter (figure 6.15). This guides the jig, prevents it from slipping off the balls and makes the fringe pattern of the ring less important in its effect on determining parallelism.

With the increasing use of stainless steel (EN58J) as a conditioning-ring facing material on heavy devices such as the Mk II adjustable-angle jig, their rotation on a kinematic plane causes annular marks on the polished ring-face. If rotation is necessary, in order to secure the error magnification effect then greasing the stainless steel face with a very thin layer of heavy duty grease (such as one based on an aluminium stearate) reduces the marking from the ball contact points. Of course, if the face is brought sharply into contact with the balls, both ring and plane suffer. The effect is more damaging with glass.

6.14 Kinematic plane, registered against a vee-block on an auto-collimator's baseplate and held in place by a magnetic base.

6.15 Brass collar to retain jig on balls during rotation.

There are considerable variations in construction of our three-ball planes. Some have been very hurriedly made for a specific diameter of work by grinding three ball bearings hemispherical and cementing them to a flat base with Tan wax. This arrangement is readily modified to suit different ball-circle-diameters.

We have used ordinary hardened steel ball bearings because they are precise, cheap and plentiful. There is, of course, no reason why other materials should not be used. Balls are readily available in bronze, from the major ball-race firms; sapphire, from F Lee and Co. and tungsten carbide from Spheric Engineering. The latter should make excellent planes from the standpoint of wear. Sapphire, however, brings brittleness with its hardness and would have to be used carefully to avoid chipping on contact with the work. Some of the high alumina ceramics may well have advantages, but they have a predictably abrasive surface due to their microcrystalline structure.

6.1.3.2 *Kinematic slide*

A brief look at the instrument field reveals a wide range of slides and movements. Few sliding mechanisms perform their primary function of giving an accurate exchange of the rotary movement of a micrometer head for the linear motion of a slide. A commonly used and at the same time a particularly unsound arrangement is one in which the carriage moves along double or single parallel rods. The fits between these and the sliding bearings must be extremely accurate which entails very close machining and individual assembly. Such mechanisms are often mediocre in performance and sensitive to deposits of atmospheric dust which accumulate in the sliding ways.

An improved design is one where vee-ways are ground in hardened steel plates in which a triangular system of balls is arranged to run. This requires the precise grinding of ways in large, hardened components and has, in our experience, given trouble in manufacture. Excellent performance is obtained from systems using flexural constraints (Jones, 1962) but when the wanted linear movement exceeds a few millimetres, the mechanisms become disproportionately large.

The mechanism described here was designed in 1956 for a Grade I J-band waveguide attenuator; however, since this application the basic mechanism has found use in metrological, servo and allied fields (Fynn and Hartwright, 1969). The slide arrangement is based on one described by Wickman (1923). It is not fully kinematic in the strictest sense since it makes use of rolling balls. Commercially available clustered needle rollers provide a cheap and relatively easy way of achieving precise vee-ways. Their use in machined slots require less skilful machining and finishing too. Short-term errors are reduced and any long-term ones caused by slight misalignment would produce a smoother curve which is more easily allowed for in calibration.

The standard translation is 1·5 cm (0·6 in) but much longer travel is obtainable by using the Dexter linear-bearing system where a continuous series of square ended rollers form the way. Long bearing systems may be built up by this means and the roller end junctions are said to have negligible effects.

The instrument (figures 6.16 and 6.17) consists of two main components, a fixed body A and a moving carriage B, which runs on three steel balls arranged at the corner of a triangle. Two of the balls run in ways set in the long side of the carriage and the third in ways set in the short side: corresponding ways are arranged in the body for the two balls while the single ball runs on a hardened steel flat which provides an anti-rotational constraint. The ways are made by insetting 3 mm diameter needle rollers in parallel slots milled in the carriage and body. The rollers are of high dimensional tolerance and surface finish: $\pm 1{\cdot}25$ μm (50 microinches) on diameter and 0·05 μm (2 microinches) CLA. An improved performance is obtained by lightly polishing the ways with a short rod of the same diameter as the balls. It is worth

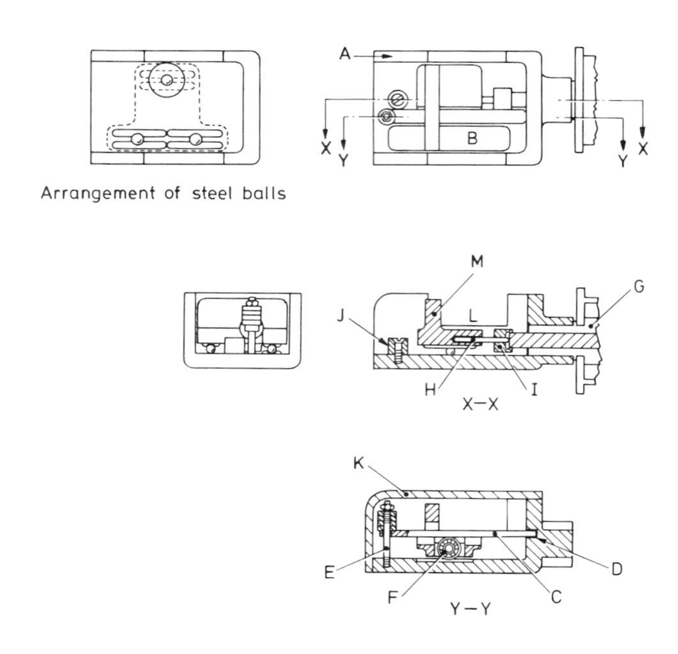

6.16 Diagram of kinematic carriage.

mentioning that polishing of the ways with a soft metal (e.g. brass) rod can be equally well applied to ground vee grooves. The improved surface finish and line contact for the balls result in a much less gritty feel and better rolling performance.

The steel flat is surface ground and then polished with 8 μm diamond paste on an optically flat Tufnol polisher, to within $\lambda/4_{(\mathrm{He})}$. The pressure bar C is of carbon steel and is polished on its working face. To provide a fulcrum, one end is barrel shaped and locates in a clearance hole D in the body. When the rubber spring assembly E is loaded by the adjuster nut, the load is transferred by means of the pressure bar to the ballrace F. Due to the 2° incline on the bar a small proportion of the load appears as a force along the slide axis and serves as a return force against the micrometer, thus obviating lost motion (backlash). The remainder of the load provides a retaining force to the carriage.

The movement of the carriage is effected by pushing it against its return action with a suitable

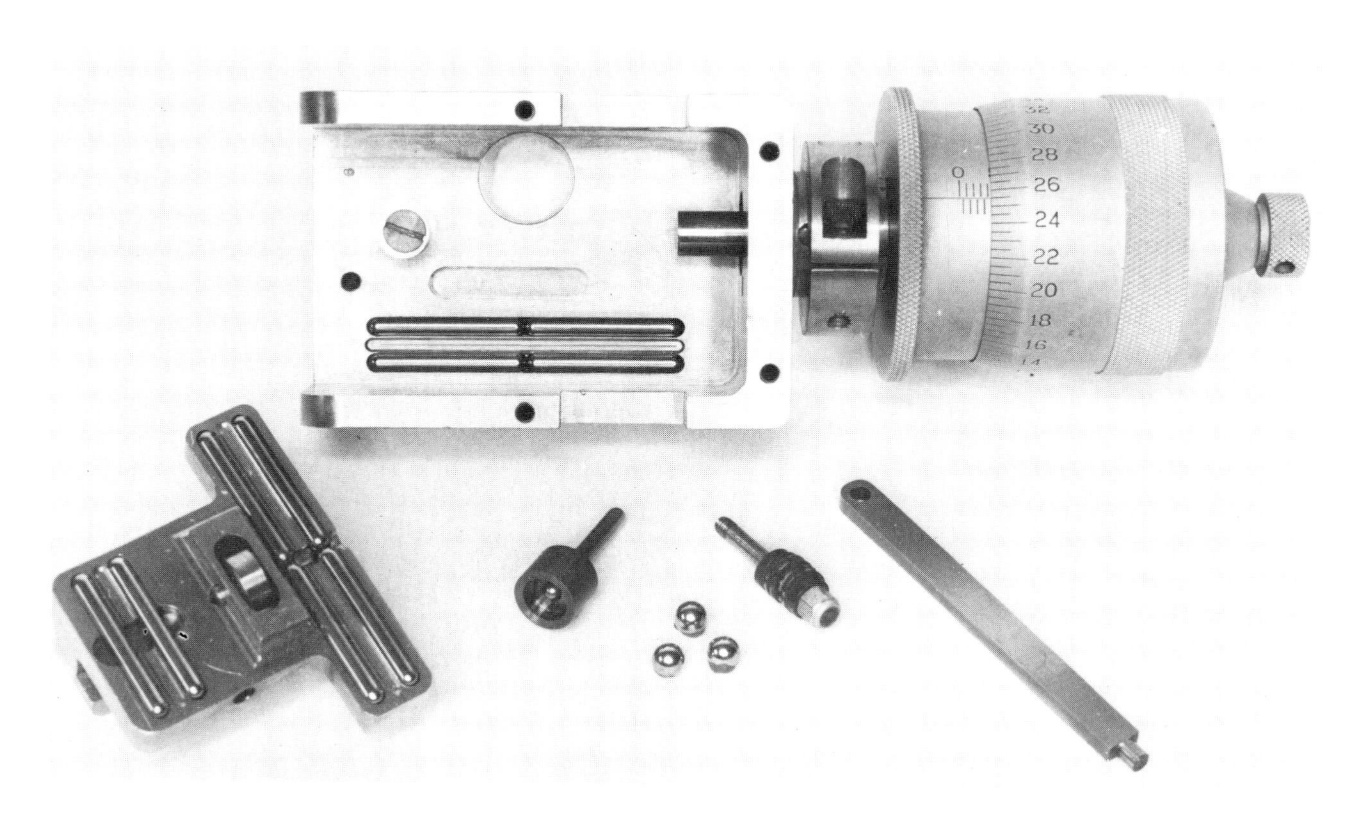

6.17 Components of kinematic mechanism.

micrometer head G held in the body. To remove the rotational component from the micrometer travel, a 3 mm diameter roller H—a wobble pin—is inserted in an approximately 1 mm deep hole in the carriage. The end stop J prevents over-run and the mechanism is protected by the cover K. Access to the carriage is obtained through cut-outs L in the body. The vertical face M is a suitable attachment point for extending the movement outside the body. There is an absence of detectable lost motion, certainly to better than 0·125 μm (5 microinches).

6.1.3.3 Kinematic mounts (Fynn *et al* 1974, 1976)
Where the limited movement of one plate relative to another is being considered, three connections between two plates are required in essence: a universal joint and actuators for travels along the *X* and *Y* axes. It is from the method of achieving these connections that precision must result. Thus of the six degrees of freedom possessed by a rigid body, the three independent motions along the rectangular axes are fixed at the universal joint. Of the three independent rotations, one is fixed by an anti-rotation constraint. The two remaining provide X and Y movements.

Few commercial mounts conform to kinematic design principles—for example, many of them employ ball-ended micrometer shafts registering in vee-ways for combined drive, hinge and anti-rotational constraint. Inevitably, a ball perfectly aligned on a micrometer-drive axis is not obtainable and side forces impart slight orbiting to the driven plate. This can be remedied by decoupling the rotary-to-linear motion using a wobble pin (as noted in §6.1.3.2) which is, for convenience and cheapness, a hemispherically ended needle roller. Wherever an abutment is required roller bearings are used endways and ball bearings provide rolling constraints, usually sandwiched between the flat ends of two rollers.

Figure 6.18 illustrates a precision mirror mount (PMM) mechanism. Two plates, a fixed one A and a free one B are connected by the units, X, Y and U and springs S1 and S2 (not shown). Movement along the X axis is obtained by rotating the large knurled knob and its fine thread Xm in the threaded insert Xt. This rotary motion is changed into translation (linear travel) by the wobble pin Xn, one end of which, Xna, is registered in the conically ended bore of Xm. The opposite end Xnb abuts the plane end of a roller bearing Xr which is cemented in the moving plate. It is registered thereon by the collar Xc which allows slight but necessary angular freedom to the wobble

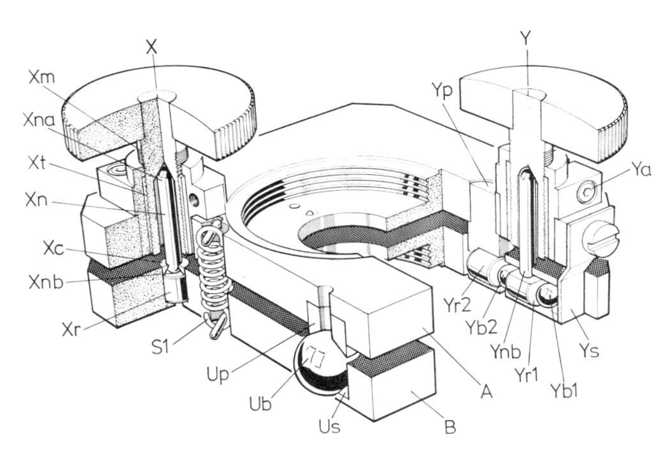

6.18 Precision mirror mount (PMM).

pin yet restricts unwanted excursion across the abutment face. An adjuster like Ya is provided so that the thread tightness may be altered to suit the kinaesthetic requirements of the operator.

Y movement is similarly obtained by rotating Ym in Yt and the consequent travel of the needle roller Yn (Y subscripts are as for their X counterparts). Since this part of the mechanism contains the anti-rotation loop, the roller Yr1 is cemented in the free plate B with its plane ends contacted by two steel ball bearings Yb1 and Yb2. These provide rolling contacts rather than frictional slides. The needle roller end YnB abuts a flat ground on the side of the roller bearing and it is registered thereon by a small clearance hole in B. The anti-rotation loop is formed by the chain of components: (A) Ys–Yb1–Yr1(B)–Yb2–Yr2–Yp(A). Ys is a leaf spring attached to and projecting from A. It applies compression via the ball Yb1 to the roller Yr2 fixed in plate B and from this via the ball Yb2 to the flat end of a roller Yr2 solidly attached to the post Yp which is projecting from A and held by a cap-head screw. This combination of fixed and rolling constraints results in easy X and Y adjustments of the plates even though they are held together under considerable spring tension: typically 1 kg.

The corner hinge unit U allows a limited joystick action to the plate before the individual constraints are applied. A ball bearing Ub is cemented to a post Up projecting from the fixed plate A and engaging in a hardened conical seating Us cemented in B. The centre of this ball and the needle roller abutments at Xnb and Ynb define a plane parallel to the flat surface of B. A refinement which may be made to the universal joint is to grind three flats on the mating bearing surface of the ball so that the contact is localised at three small areas. This concept is mentioned by Thomson and Tait (1863) and credited by them to Professor Willis.

Two tension springs S1 and S2 are used to pull the plates together. The tension is great enough to keep the wobble pins in contact with their abutments when optical components are screwed in place against the self-contained three-ball plane on the moving plate.

Limited angular adjustment of a heavy component is frequently required and it is impracticable to overhange these from the PMM for two reasons: firstly, the cantilever load overrides the standard tension springs, secondly, any cantilever arrangement decentres the component at the end remote from the hinge when it is adjusted in angle. For these reasons, a mount with a heavier load capability is used with a limited rotation of $\pm 2{\cdot}5°$. Many of the features of the standard mount have been incorporated in the turntable or tilt and rotate system displayed in figure 6.19. The main load of the device to be adjusted is placed above the universal coupling 3, constructionally the same as Uc (figure 6.18). The tilt control simulates Xm adjustment while the rotational drive approximates to the Yr arrangement since it is a system driven by a fine screw thread.

In order to leave a large, uncluttered turntable space the tilt and rotate mount (TRM) is in the form of two L-shaped plates, 1, 2 spaced by three couplings: a universal coupling 3, tilt control 4 and a fixed length spacer 5 in the form of a wobble pin. This in line with

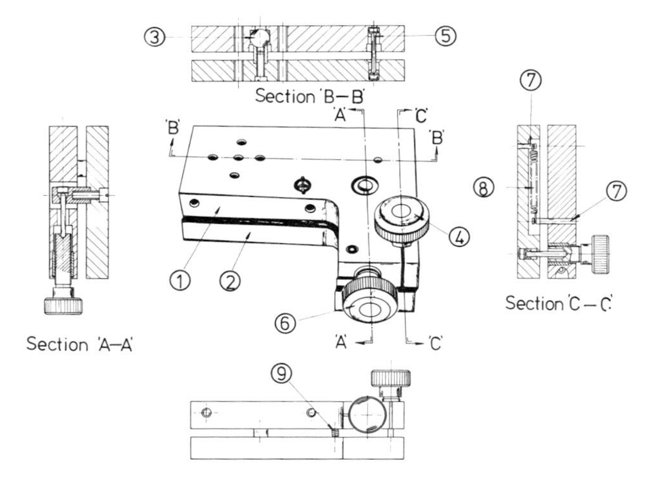

6.19 Tilt-and-rotate mount (TRM)

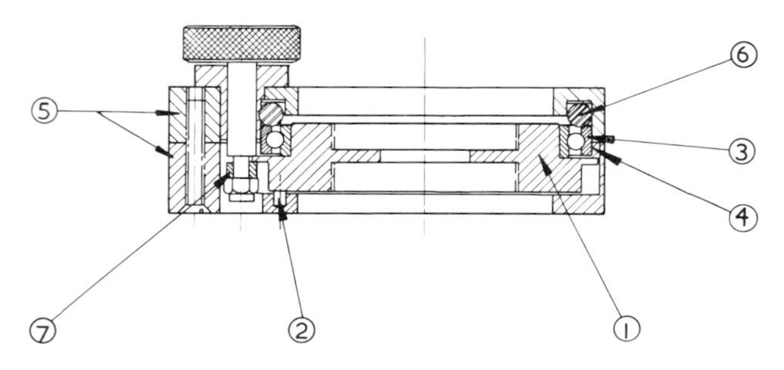

6.20 Precision rotation mount (PRM)

the universal coupling and together they define the tilt hinge-line. Finally the micrometric rotational control 6, like the previously described X and Y actuators, has a wobble pin output. The pin abuts a hardened steel face on a lug 7 which protrudes from the fixed plate into a large clearance hole in the moving plate. The restoring force is provided by a tension spring 8 anchored to the upper and lower plates and parallel to the rotational micrometer axis. Another short spring 9 provides the tensioning between the hinge line and the tilt micrometer. It is anchored to both plates and disposed centrally amongst the three plate-spacing elements.

The precise rotation mount (PRM) is intended to give accurate 360° rotation around the optical axis and can be used to establish angle. The plane of rotation is constant to one second of arc (with respect to the body of the device) throughout the complete revolution. In those applications where angular adjustment of the PRM is required, provision is made to bolt it to the side of the TRM. An optical component can be arranged to overhang by circa 2·5 cm so that its optical centre is above the universal coupling of a TRM. This reduces the decentering effect when the TRM controls are adjusted.

The design of the PRM is very simple in concept and depends on the rotor 1 (figure 6.20) having one face optically flat to $\lambda/4_{(He)}$. This face is registered by three low-friction pads 2 spaced at 120° and cemented into the stationary housing of the unit. An aircraft torque tubular race 3 is used to provide both the radial and thrust bearing requirements of the rotor. The outer of the ballrace is a sliding fit in a PTFE-lined bore 4 in the body of the mount 5 and an O-ring 6 is used to provide a small thrust-load to the rotor, via the ball race. Thus the rotor is held in constant contact with the three pads and in so doing defines the plane of rotation. Hand control is by means of a conventional friction-drive 7 actuating the rim of the rotor.

The plates of each of the mechanisms described above are made of aluminium alloy. The choice of this material is on grounds of cheapness and ease of providing a durable finish by dense black anodising. Although most aluminium alloys are known to undergo dimensional changes with time due to hardening and some stress release after machining them, quantification is not easily obtained. A feature of kinematic design is that the precision of the instrument is minimally affected by wear. Moreover, unless any warping which takes place is sufficiently great to cause components to rub together—such as a needle roller in a clearance hole—the actuators, abutments and rolling constraints themselves are little affected.

The relatively high coefficients of thermal expansion of aluminium alloys when compared with stabilised steels raises the possibility of warping due to unequal temperature distribution caused by draughts, radiated heat etc. This in its turn is minimised in a plate by the high thermal conductivity. An extreme experiment was carried out by applying heat to the fixed plate and raising its temperature by 7 °C. The moving plate remained at ambient and the graticule drift was observed on an auto-collimator arrangement to measure a mirror on the mount. When the temperature equalised the optics regained alignment. No permanent warping took place, but it is predictably advisable to minimise any non-uniform heating or cooling.

Notwithstanding the excellence of commercial micrometer threads they have been considered to be too coarse for the X and Y drives on the mounts for high precision work. The models have threaded stainless steel shafts, 10 mm diameter with a pitch of 0·212 mm (120 TPI). Since the distance between the centre of the universal joint and the X and Y movements respectively on the PMM has a mean value of about 51 mm, then one complete revolution of a knob tilts the plate through about 14′. It is a demonstrably simple matter to adjust a knob in increments of 15″, and 5″ can be achieved with moderate care. The tilt resulting from these inputs is calculated to be about 0·0083″ and 0·0027″ respectively. However, the measurement of tilts of this magnitude is beyond the scope of our available instruments and our checks can confirm the calculations only to 0·1″.

The threaded female components are continuous sleeves of polytetrafluorethylene (PTFE) in split housings. Merton-type nuts (Merton, 1950) consisting of three PTFE pads arranged at 120° in a split housing have been tried but their performance is indistinguishable from that of continuous sleeves. It is possible that test equipment more sensitive than that giving 0·1″ resolution might differentiate between the two, but unless a worthwhile difference is established the ease of manufacture is the arbiter.

The needle rollers used as wobble pins for transmitting drive to the plates are of 3 mm diameter and their clearance holes in the threaded shaft are about 6 mm diameter. For the PMM, the location collar Xc allows about 0·05 mm clearance to the wobble pin Xn. The clearance hole in B performs the function of Xc for the roller Yn. These precise mounts are commercially available from Malvern Instruments.

6.1.4 Auto-collimators

The auto-collimator is a very useful optical metrological instrument since it can be applied in situations where comparators can only be used with difficulty in addition to the majority of situations when they serve well. Figure 6.21 shows a typical arrangement of

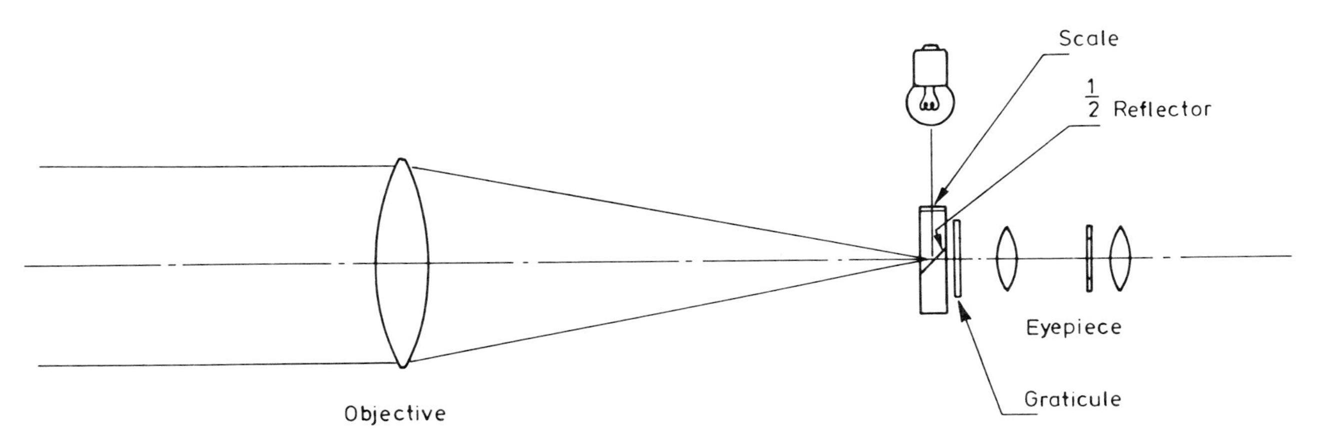

6.21 Auto-collimator: simple optical system.

the optics schematically. In some of the optical systems, an illuminated cross is viewed against two right-angle scales; in others, an older and often better arrangement provides a movable illuminated minute scale viewed against a pre-set illuminated scale. Nearly all the arrangements have the pre-set scale at the eyepiece focus. An instrument with greater angular sensitivity such as the 'Microptic' has the scales replaced by a fiduciary mark, traversed by a micrometer drive. Readings are taken when the fiduciary marks and the return image of the crosswire coincide.

Various modifications have been devised to reduce the eye fatigue which occurs with frequent readings. The earliest arrangements were projection instruments with graticule images presented on ground glass screens, while later versions utilised a photoelectric system to indicate coincidence.

In the use of auto-collimators, the main difference between the field of engineering metrology and that of crystal working is in the very much smaller size and lower reflectivity of the observed faces. The largest possible reflectors ($\sim 90\%$ reflectivity) are used in engineering, whereas crystal surfaces are often a few millimeters in diameter with reflectivities 10% or less—typically, transparent specimens.

All auto-collimators depend on some beam splitting system which can be a single prism, double prism or a bevelled-edge plate. It can be achieved, too, by means of a semi-reflecting thin plate, as in the model made by Optical Tools for Industry. Another useful feature of this particular instrument is the interchangeability of the eyepiece with the lamp and its associated graticule, allowing the great convenience of right-angle viewing when the instrument is in use in the horizontal position.

Those instruments with lower sensitivities such as the Angle Dekkor (1′ divisions, 1° scale) are prepared by using their pivoting mountings either horizontally or at any angle to the vertical. One with higher sensitivity such as the Microptic (0·2″) usually has horizontal mountings and thus takes advantage of a more stable support. Figure 6.22 shows both types: the higher sensitivity instrument has a micrometer driven graticule and is mounted horizontally while the Angle Dekkor is used in a lower sensitivity situation with its standard mountings vertical.

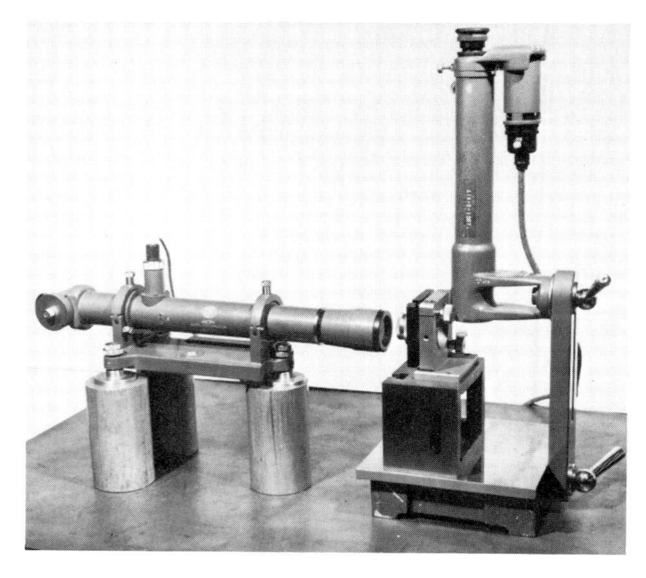

6.22 Mirror mount tilt mechanism under test by two auto-collimators.

Adjustment to the level of illumination at the graticule is done electrically in the transformer unit, except in some earlier models when the lamp is powered directly from the 240 V mains. In the latter case, no easy adjustment is provided; however, there is a slit between the lamp housing and the body of the instrument where a neutral density filter may be conveniently inserted in order to obtain the all-important optimum level of illumination.

Finding an image on a reflecting surface can be time-consuming if scanning has to be carried out along both X and Y axes. Some of the more sensitive instruments have spirit levels for this purpose so reducing the scanning to one direction. On those which have none, an ordinary spirit level can be applied to the tube of the instrument. The general technique is enlarged upon in §6.3.2.

6.2 Linear Measurement

The tolerances placed upon the linear dimensions to be measured to some extent determine the technique to be used in both contact and non-contact methods. Accuracies less precise than say 0·5 mm require no great discrimination on the part of the measurer and correspondingly little niceness in the scale or instrument, whereas the determination of length or diameter to better than a micrometre can be demanding in all ways.

6.2.1 Contact methods

Standard steel rules, maintained at the specified temperature (usually 20°C) can be read to an accuracy of about 0·05 mm (0·002 in) with the aid of a ×4 magnifier. However, for such a precision to be achieved the work must have sharply defined limits. Chamfering of an edge and turndown reduce discrimination. A rule can only be used effectively in one dimension, that is, against a side or across a diameter or diagonal.

A refinement to measurements made with a rule can be obtained by adding a vernier. This device, invented by Pierre Vernier, enables the user to read linear or angular scales to a degree not easily estimated on a single scale alone. The most-used type is termed a direct vernier and, when added to a caliper (as in figure 6.23) measurement ot 0·025 mm (0·001 in) across the body of a specimen is readily obtained. The vernier itself consists of a short scale which has one extra division relative to an equal length of the main scale for example 10 graduations on the former occupying 9 graduations on the latter. It is common practice with vernier calipers ruled in imperial units to have each main scale division equal to 0·025 in: the vernier has 25 divisions occupying 24 on the main scale and thus each is 0·001 in smaller. In metric units, a main scale is graduated in millimetres and its auxiliary vernier scale has typically 50 divisions occupying 49 mm. Each vernier division is thus 0·02 mm shorter than a main scale division. By using an eyeglass it is not difficult to discriminate to half a vernier division: 0·0005 in and 0·01 mm respectively on the scales described above.

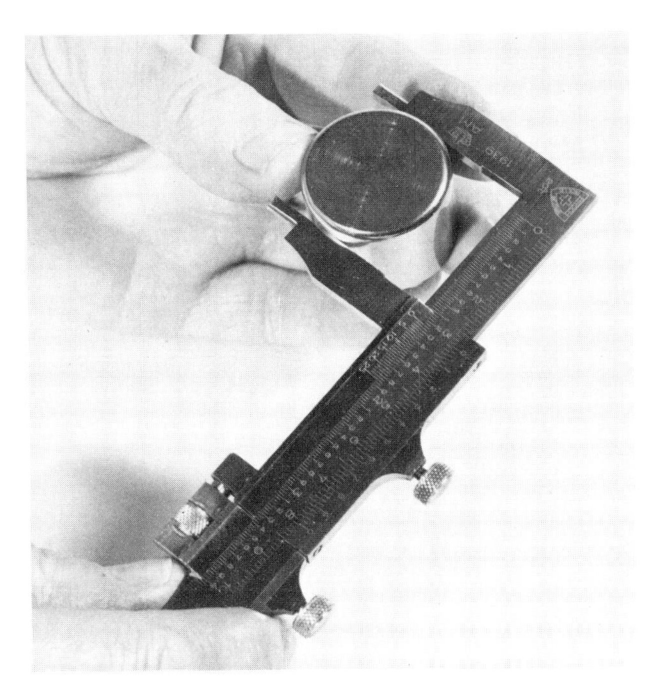

6.23 Vernier caliper.

6.24 Vernier caliper scale.

Figure 6.24 shows a vernier caliper setting with both imperial and metric graduations. A complete measurement is made by summing the readings of the main scale immediately before the zero of the vernier, and the vernier scale reading which coincides with a main scale graduation. In the figure, the scale reading is 0·021 in or 0·054 cm.

Vernier calipers are essentially handy instruments which may be used quickly. Many of them read from 0–9 in (0–23 cm) on a single scale and may be regarded as the equivalent of nine separate micrometers. They can be used to make both internal and external measurements. In laboratories, they are used mainly for rapid measurement of glass tube diameters; however, when sensitively applied they are capable of much more precise work. Since they can mark even quite robust surfaces, it is advisable to protect the specimen with polyester film and subtract its thickness from the final measurement. However, a degree of pressure has to be exerted on the work via the light clamping actions of the jaws and this must be borne

6.25 Vernier height gauge.

in mind when using calipers. The errors introduced by jaw pressure are minimised by consideration for the specimen.

A valuable variant of the caliper is the vernier height gauge (figure 6.25), on which the linear vernier scale indicates distance above a reference surface table. A scriber can be fitted to the vernier slide in order to engrave lines of a given height above the reference plane. A dial gauge, too, can be clamped in place of the scriber and performs as a zero indicator. The gauge may be used for engraving bright-line graticules, for optical measuring purposes, by scribing through aluminium films deposited on glass substrates. For making 'instant' graticules, the aluminised disc is mounted in a dividing-head and a scriber with a fine point fitted to the height-gauge. Lines are drawn at 90° or as required. A limiting guide is needed for fine work and the scriber should be mounted on a spring-hinge in order to apply a predetermined pressure. The process in no way replaces conventional graticule making techniques but it is a simple way of producing bright lines—for instance, for collimator use.

The vernier caliper provides a simple test for assessing an individual's ability to discriminate within a division and thus the ability to estimate a plain, unverniered rule. It is not difficult to make fairly precise readings when the edges of the work coincide with the graduations, but when the length falls either side of centre in an open space—say a 0·025 in division—mental subdivision is needed. If this is practised by reading the main scale of a vernier caliper as accurately as possible with the vernier scale covered and the 'guestimation' checked on the vernier, the feedback of measuring biases rapidly improves confidence and also indicates personal tolerances.

The ordinary screw micrometer typifies the firm contact method of measurement. They are generally unsuitable for use on either polished or otherwise delicate material but are suited to robust, unpolished (grey) components. The forces applied in making a measurement, and consequently the damage caused, depend greatly on the skill of the operator. There is a ratchet or friction drive provided on most micrometers, arranged to limit the torque applied to the screw, and in consequence replicate the pressure at the measuring faces; even so, it is dangerously high for many optical purposes. Generally, the user develops sufficient skill with the fingers driving the screw directly to produce a safe pressure. Because the pressure applied is in line with the scale on the thimble, the micrometer is less open to abuse than the vernier caliper, that is the 'feel' is more sensitive. At the smaller end of the range, 5 cm or 2 in micrometers can be read to 0·003 mm or 0·0001 in with well developed operator dexterity. Micrometers with capacities greater than 5 cm become progressively more difficult to handle with a correspondingly reduced reading accuracy.

A worthwhile variation in micrometer use is achieved by mounting a micrometer-head and a dial gauge coaxially (figure 6.26). The latter is set to read a convenient zero when in contact with the micrometer-screw, itself set at zero. Work is inserted between

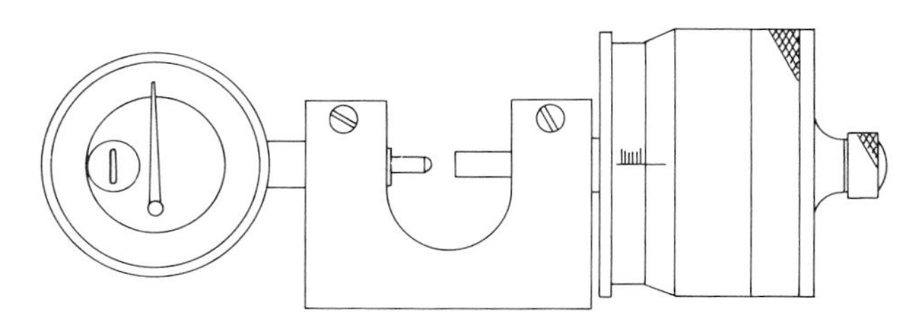

6.26 Coaxial mounting for micrometer head and dial test indicator.

the dial-gauge stylus and the micrometer face and the screw advanced until the gauge reads its arbitrary zero: pressure is thereby limited by the DTI. Because the spindle has a rounded end, non-parallel work is the better accommodated during measurement. Many variations are possible, from the extremely basic zero indicator of a spirit level mounted on a spring hinge arranged to be tilted by the movement of the micrometer to the use of an electronic or combined electronic and pneumatic comparator. Comparators are more fully dealt with in §6.3.

The success of screw micrometer arrangement depends on the extreme accuracy of the threaded parts: since this is well beyond what can be read on the scale of a conventional thimble, errors are mainly those made in taking readings. The thread-accuracy of the standard 2·5 cm head may be used to greater advantage by fitting large-diameter drums engraved with longer scales. One 7·5 cm diameter on a standard 0·5 mm pitch threaded spindle gives a great gain in resolution. It is with these large drum micrometer heads that reading accuracies of about 1 μm are directly obtainable.

When greater accuracy in a linear measurement is needed, the standard metrological practice is to compare the length of the work with Grade 00 slip gauges wrung together to give approximately the same dimension. Since the departures from specified values are certified for these slips, usually to a few microinches (when 1 microinch = 0·025 μm) a calculation of their combined length should be precise to a few parts in a million. It depends then upon the flatness of the reference surface—say a kinematic plane—and the limits of the comparator method to what accuracy a measurement may be made. A DTI may discriminate to better than a micrometre, and many capacitive, electronic and pneumatic comparators are both reliable to a few millionths, and linear over a range of several thousandths, of an inch. Figure 6.27 shows a number of slip gauges wrung together and compared with the length of a component.

In the absence of any more convenient means a very sensitive form of length comparator can be made which incorporates a sensitive spirit level. The work is positioned on a precise abutment such as a three-ball plane and a suitable gauge block or blocks wrung to the surface table near to the work. The spirit level is arranged to straddle the two elements and is spaced from them by small ball-bearings, two on the gauge block and one on the surface undergoing test. Any small variation in the height of the work will be seen as changes at the spirit level.

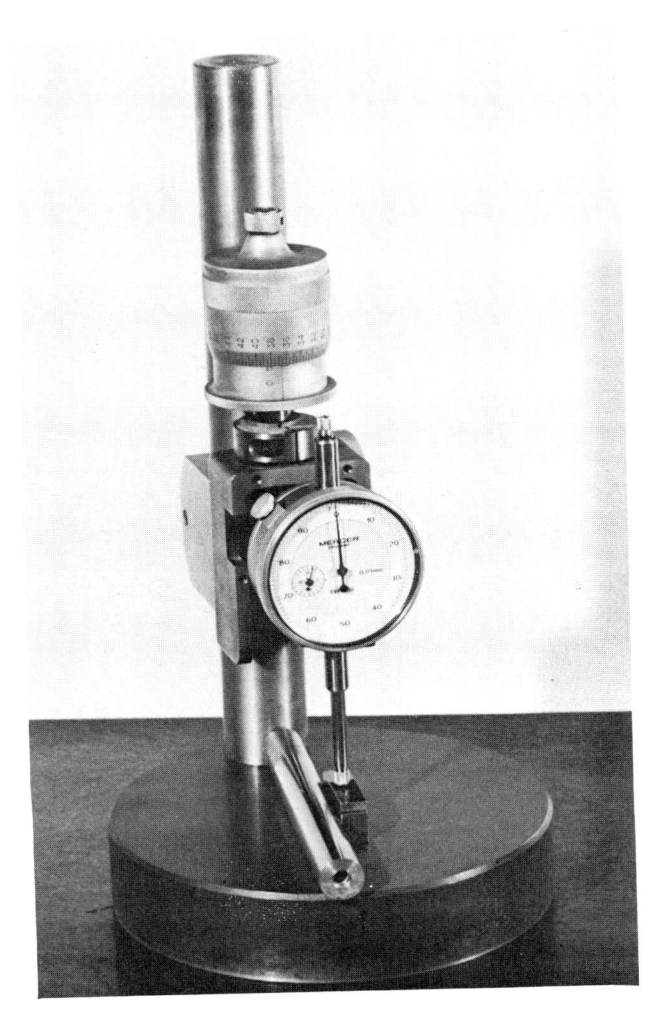

6.27 Wrung slip gauges for accurate diameter measurement.

6.2.2 Non-contact methods

In cases where the material is too soft to allow measurement by contact methods, a simple technique is to arrange a steel scale on black velvet under a very low power binocular magnifier. The specimens can be slid on the velvet into near contact with the scale, using soft brushes to manoeuvre them into place.

Magnifiers complete with graticules are valuable aids for determining the linear dimensions of small components. A very cheap one which we use has a focusing control to compensate for individual eyesight. They can be used equally well in contact with specimens since a section of the body has a window which allows direct illumination (figure 6.28). Versions containing imperial and metric graticules are available

6.28 Adjustable focus magnifier with integral scale.

6.29 Travelling microscope, in use for real/apparent depth measurements and calculation of refractive index.

from Flubacker and Co., sold in the UK by Matchless Machines Ltd. The divisions are 0·005 in and 0·01 cm respectively and it is possible to estimate to one-fifth of a division. With these simple instruments, as with many others described, sensitivities beyond what might normally be expected are obtained through developed observational skill and attention to details of colour and levels of lighting. The main difference between the graticule magnifier and the travelling microscope is the speed in operation of the former.

Travelling microscopes are particularly suitable for the more precise non-contact measurements. The instrument with which we have the most experience is the Cambridge Instrument Company's kinematic model, a single-axis design manufactured in about 1940, with 0·01 mm graduations. Its discrimination to less than 2 μm enables the user to estimate the thickness of delicate slices both on and off the substrate. The technique is found to be very valuable in observing the progress during the polishing of composite specimens—composite, that is, in the laminated sense.

Another application of the Cambridge instrument is for the estimation of refractive index by the real over apparent depth method. Since it involves good resolution in the vertical or microscope axis, an accurate micrometer head (or dial gauge) is required capable of discrimination to one or two parts in 10^4. A simple modification which allows measurement along the microscope axis is made (figure 6.29) by bolting the micrometer head to the body of the instrument and clamping an adjustable abutment to the lower half of the lens tube. A steel ball bearing is cemented to the abutment and forms a hardened contact point for the micrometer spindle-end. After focusing the microscope on the upper surface of the transparent material undergoing test, the micrometer is used to indicate the microscope position by 'feel'. The microscope is refocused on the lower surface and the micrometer spindle contacted again. The difference between the reading gives the apparent depth, while the real depth can be measured separately.

Estimation of refractive index by this technique is of course very dependent upon the accuracies of the apparent and real depth measurements. A main source of potential error lies in the vertical microscope tube where the great enemy of accurate measurement, spring, has been inevitably introduced by the attachment of the micrometer and abutment bracket to an already lightly constructed focusing arrangement. An accurate reading requires considerable feel, or tactile sensitivity. Two aids can be employed: a DTI as a zero indicator, or electrical contacts at the abutment in place of a steel ball.

There are several methods of determining length precisely by non-contact means which are outside the scope of this book—by the use of lasers, for example. However at the fine end of metrology, and readily available, is the use of slip gauges as described in §6.1 in conjunction with a pneumatic/electronic gauge effectively floating over the specimen. Mercer's Hover probe can be calibrated to be linear over a range of several hundreds of micrometres and still be precise to 1 μm. Thus it performs not only as a comparator but also as a direct reading instrument for specimens lying within its range: normally 2 mm ± 1 μm and 0·4 mm ± 0·1 μm.

6.3 Plane-Parallel Measurement

Subdividing this section into contact and non-contact techniques and instrumentation is not strictly pertinent, since many measurements are carried out by contacting a disc or waster co-planar with the specimen without actually contacting the specimen itself. Only optical, capacitive and electro-pneumatic measuring techniques can be non-contact, though the last-named is again more frequently related to an extension of the polishing plane at the specimen; whereas our optical assessment relies on an inverted contact system. Where the work itself is monitored, the method has been classified as direct; where an adjoining co-planar surface is measured and the readings extrapolated to indicate the parallelism of the specimen, the method is classified as indirect.

Consider the boundaries of the problem: from table 6.1 it is apparent thay any dimensional measuring system (as distinct from an angular one) can resolve a departure from plane parallel of 2″ of arc when two conditions are satisfied. Firstly, it must be capable of measuring to 0·1 μm; secondly, the minimum of two measurements made, diametrally opposed on a surface, must be therefore at least 1 cm apart. Thus a specimen 3 cm in diameter can be directly checked with a precise comparator to a parallelism of less than 1″, but a small laser rod of 0·3 cm diameter is too small to be so precisely measured over a baseline of about 0·15 cm by probes or styli which are similar to it in size. The options are either to extend the polishing plane of the work as noted above and carry out mensuration on a much longer baseline, or resort to optical measurement. The former technique will be dealt with first.

6.3.1 Indirect measurement

It is a fortunate coincidence that the polishing of a large proportion of laser optics and to an extent semi-conducting compounds, poses plane parallel measuring problems which can be solved by techniques readily available in the field of engineering metrology. Highly developed gauging systems resolving less than 0·1 μm have been in existence since the early part of this century, notably the 'Millionth' comparators of Eden and Sears. Such was the quality of those instruments that it would have been possible to have produced the highest quality laser optics then, using the comparator as a parallelism indicator. One of the features of the Millionth was a cluster of three hardened steel ball bearings, forming an artifical plane, and an abutment with properties of excluding dirt and obviating the uncertainties of the thickness of a wringing film. The three-ball plane as it is now called has found many applications in both the mechanical and optical fields.

Mechanical indicators have developed into more compact devices where it is no longer necessary to utilise light-beams for the final stages of magnification. The earlier stages of magnification still rely upon bending or twisting steel tape, as did the old instruments. Even the standard dial gauges now have sensitivities which allow the measurement of quite small optical components to be made to a tolerance of better than 2 μm. However, they should be used with caution, since the spring pressure which they exert can be many hundred grams cm^{-2}.

Modern electromechanical systems have advantages because they apply very low stylus pressures and have relatively high magnifications, typically Rank Taylor Hobson's 'Mitronic' or Mercer Electronic's 'Magna Gauge'. Where conducting work-faces are being measured, very high magnifications are possible with high frequency capacitive gauging, though these have tended to give way to the commercial versions noted above.

In the original capacity micrometer of Whiddington (1922) small changes in a capacitance between an electrode surrounded by a guard ring and an ohmic specimen were monitored while a medium frequency was applied. Later developments of this system used much higher frequencies with an adequate guard-ring size which led to a very stable instrument (Fynn, 1949). The system was used in most of our early laser work for parallelism indication with a device which enabled it to read on non-conducting surfaces such as glass wasters. The probe was fitted with a stylus, the capacity plate mounted on a spring hinge and required only a few grams to move it through the operating range.

Measuring and correcting the parallelism of block-mounted laser rods typifies the indirect method on any plane parallel specimen and it will therefore be described in this context. When the packing rings (see chapter 7) have been worked during earlier stages to about 0·0001 in (2·5 μm) departure from parallel—measurement being made with an ordinary DTI—it is at the latter stages that the final parallelism measurement and compensation is applied. The capacitive comparator has a 100 MHz oscillator in the sensitive head, the tuned circuit of which is controlled by the varying capacity produced by movement of the indicating stylus. The output of the oscillator is fed using a single gain stage to a simple tuned circuit and rectifier, followed by an amplifier with a backing-off control. The final display is shown on a 0–1 milliammeter, rapidly changing movements can be shown on a cathode ray tube. With this arrangement, linear gains of 100 000 are readily obtained, but for the laser work a gain of 10 000 is adequate. This means that the pointer of the meter indicates a positional change on the scale, of say, 0·02 in (0·5 mm), that is two microinches (0·05 μm) on the specimen.

Horizontal measuring systems (figure 6.30) which make use of vee blocks to control the rotational axis of the lapping block during testing require the exterior of the block to be truly cylindrical. This avoids the introduction of length errors due to poor axial control. If the blocks have been well machined a small lapping operation—such as the one for hand circularising of rods (chapter 5)—is sufficient to bring them circular. A convenient test for this parameter is to revolve them in a vee block with a dial gauge registering on the vertical axis. A better arrangement would be to examine the block diameter on a Rank Taylor Hobson 'Talyrond'.

The horizontal measuring system consists of an adjustable length vee-block B on which the block has freedom to rotate. One end of the block A is lightly loaded against an abutment C, while the stylus D of the linear magnifier is at the opposite end. Thus when the laser rod and block are rotated, any changes in length indicate errors in parallelism. The usual practice during correction stages is to mark the block at the maximum length position and return it to the polishing machine with the push-rod set eccentrically by about 3 mm and coincident with the marked high region.

Where it is desirable to dispense with the precise exterior on the block, a different system of measurement is employed. This involves running it on a three-ball plane with the block axis vertical (figure 6.31). Length changes are indicated by a comparator with its transducer mounted on a kinematic slide carriage (§6.1.3). Since this is free from lost motion to better than five microinches, it provides a ready means of calibration. In this case, the hand manipulation of the block demands extra skill in providing the smooth rotation and maintaining continuous contact with the three balls. The action is one of a slight downward pressure combined with a slight lateral force to keep the block in contact with the pegs (figure 6.13).

6.30 Horizontal measurement of plane parallelism.

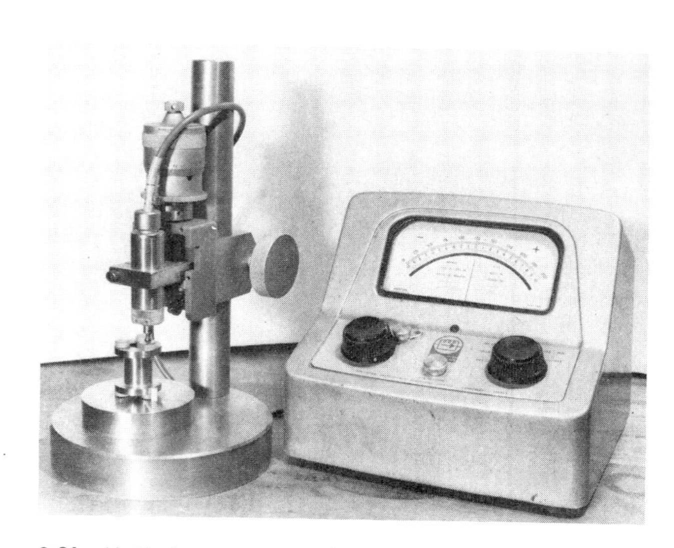

6.31 Vertical measurement of plane parallelism.

6.3.2 Direct measurement

The indirect methods described in §6.3.1 become direct when the stylus or probe of the comparator is applied to a suitably large and robust specimen. This is equally true when a non-contacting pneumatic or electromagnetic gauge is employed. A great advantage of these non-optical methods is that they can be carried out on both grey and polished surfaces. The technique which is generally applicable for direct measurement of specular surfaces makes use of the auto-collimators described in §6.1.4.

The basic concept is that a collimator can be set very accurately above a flat surface and conveniently at right angles to it. A kinematic plane is used in conjunction with a 10 cm diameter, semi-polished stainless steel disc which replicates the plane of the balls. Section 6.1.3 describes the method of setting-up the three-ball plane and disc. Since angular accuracy depends upon transferring the measuring plane, in effect, to that of the block or adjustable-angle jig using the medium of the disc, it must be made very precisely, preferably to better than one second of arc in parallelism—dimensionally, parallel to 0·5 μm. A disc is made on a high-grade surface grinder. It is rotated at intervals and the final cut repeated a number of times when the second face is being ground. This procedure helps to produce a profile which gives a good start in producing parallel plates when subjected to flat-lapping on the 60 cm machine for instance.

Once both surfaces have been lapped to a state of uniform greyness, the thickness of the plate is measured on a three-ball plane with a sensitive electromechanical comparator operating above one ball. It is usual to true the 10 cm discs to one second of arc by hand-lapping or by bias-weight loading on a large ring-lapping machine. A compromise between the two techniques can be made by applying hand bias to improve the state of parallelism of the disc while using the machine to restore flatness to the worked surface. For collimator use, an adequate polish may be obtained by burnishing the disc against a cast-iron plate, or rubbing it with a flat Tufnol polisher or using a diamond impregnated polishing cloth. As a general rule, if fringes can be seen between an optical flat and the metal surface, an adequate image can be viewed in the auto-collimator.

Any fault in the manufacture of a disc is apparent by a shift in the reflected image when it is rotated on the kinematic plane. The very small degree of surface polish is usually arranged to give a comfortable intensity of the returned image. Plane parallel glass and quartz plates could be substituted by stainless steel ones, but a reduced voltage on the auto-collimator lamp would be needed in order to reduce the image-spread.

Once the collimator has been set precisely square relative to the disc, then the kinematic plane, whatever flat surface is monitored, can be worked and corrected to the plane parallel state. Specimens, blocked or free, can be checked and rechecked after each corrective biassing until they come within the specified tolerances. However, the indirect mode is more often applied to such work. It is in the measurement of specimens mounted in adjustable-angle jigs that the auto-collimator has the greatest advantages. The jig's conditioning ring is an extension of the specimen's polishing plane and when one surface has been worked (often unmounted) the tilt control knobs are used to achieve parallelism. Thus a component such as a window held in a workplate has a temporary polish worked on one side and is then collimated and adjusted to be parallel to the conditioning-ring face (figure 6.32(*a*) and (*b*)). When a specimen is mounted on a porous substrate, the rear side of the latter is viewed in the collimator looking down the jig's hollow shaft and the parallelism achieved is dependent upon the optical instrument, the accuracy of substrate and cement film (chapter 4) and overall jig precision. It follows that work mounted on a porous substrate need have neither side polished for collimator setting and measurement (figure 6.32(*c*)).

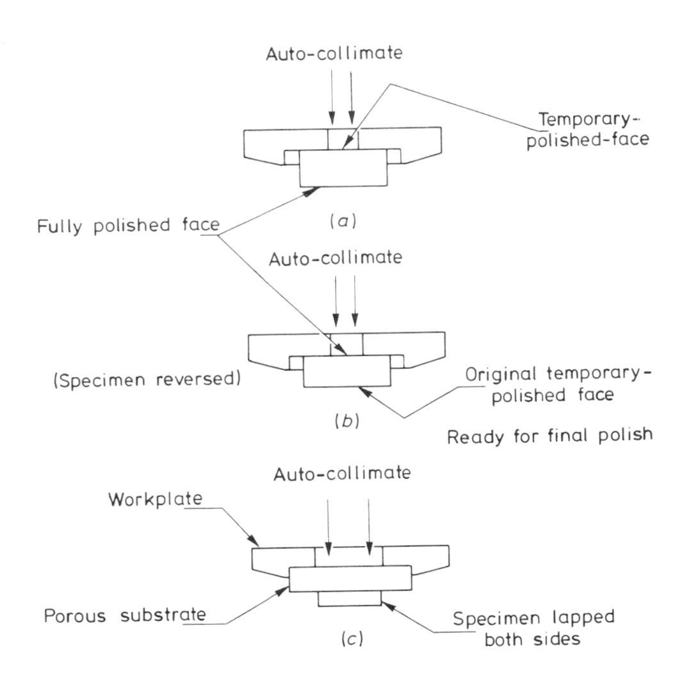

6.32 (*a*) Initial auto-collimation on face with temporary-polish; (*b*) second auto-collimation on a fully polished face, specimen reversed; (*c*) auto-collimation on porous substrate, obviating the need for a temporarily polished face.

The standard measurement steps in checking work in an adjustable-angle jig are therefore:

(1) check that the kinematic plane is held firmly in position by means of magnetic bases (or by a central bolt);
(2) place a cleaned stainless steel master plane-parallel disc on the three balls;
(3) establish that the collimator is set as required, for example 20–20 on its scale;
(4) remove disc and carefully place the jig on the balls;
(5) adjust the tilt control until the collimated image on the rear of the specimen (or porous substrate) repeats the setting seen on the disc (20–20);
(6) for increased sensitivity, rotate the jig smoothly on the plane through 180° and check settings;
(7) repeat until setting errors have been halved;
(8) finally, remove jig, replace disc or recheck the validity of the initial setting. This shows whether or not anything has been inadvertently moved.

Because of the difficulty which is often experienced in initially finding an image when a very sensitive instrument is used, the tendency is to employ a less sensitive one. It is then required to indicate departures from parallelism at the very limit of, and beyond, its normal discrimination. When functioning purely as a coincidence indicator, it is necessary to operate the bright line of the illuminated source directly over some specific division on the adjustable graticule. In so doing the effect we call reddening or line-blistering is used: it is observed when the bright graticule line is traversed across a stationary graduation and red fringes develop on both sides of the black line. Very critical alignment is obtained when these are balanced in magnitude and colour. Having mentioned the value of line-blistering, it must be added that some types of auto-collimator—particularly the very old Hilger and Watts T223 shown in figures 6.15 and 6.22—are excellent in this respect. Some of the more modern instruments do not have sufficient brightness of line to blister on weak reflections. However, in all cases a change in the graticule will give the necessary intensity. New ones have been made by taking a glass disc identical in physical size to the graticule plate, evaporating aluminium on one surface and engraving two lines at 90° with a fine scriber—see §6.2.1.

In the case of some work where the ends are nominally parallel on an imperfect optical medium, a combined sharp and diffuse image will appear in the auto-collimator. A minute trace of oil of cedar wood will destroy the image from the lower surface. The technique is applicable to many other measuring situations when more than one image obscures the sharpness of the reflected graticule and is most common in laser rod manufacture by jig methods. More or less any oil or grease serves to destroy the image from the remote face.

Rear alignment of a specimen in a jig is a valuable means of measuring and achieving parallelism. However, sometimes it is sufficient to polish a single surface or repolish a face of a specimen without any parallelism or angular requirement. It is lapped and given a temporary polish which must be flat but need not be highly specular, then mounted on a standard worktable, fitted to the jig and the conditioning-ring set at the correct height. The jig is mounted face upwards on a tubular stand, beneath the auto-collimator. Two reflected images will appear if both specimen and a sector of the ring are brought within the field of view of the instrument. Using the jig tilt controls, it is a simple matter to superimpose the specimen image on that of the ring. Since the technique is not a very precise one, the alignment can be often improved by

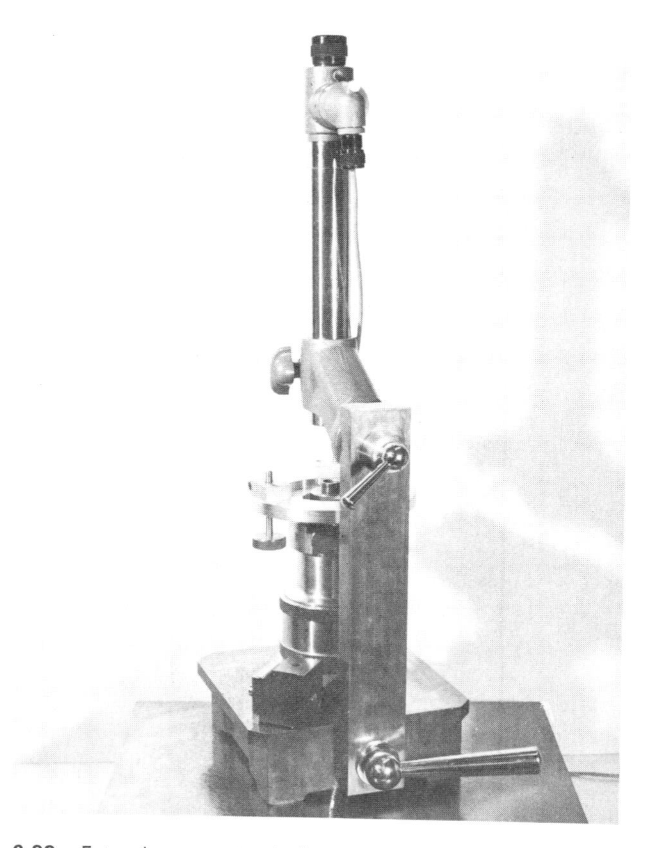

6.33 Extension-arm attached to an auto-collimator.

placing an optical flat on both the jig-ring and specimen faces and viewing them under parallel monochromatic light. The fringes can be broadened to a maximum, firstly on the ring and then on the work. If such an alignment method has to be used on a grey surface, a small parallel optical flat has to be wrung or oiled on to the specimen in order to obtain a reflection. Alternatively, a temporary polish may be worked with diamond abrasive and a soft metal polisher.

In contrast to the situations where a multiplicity of images have to be dealt with there are those where scarcely discernible reflections are obtainable. Black paper apertures are valuable in decreasing the amount of scattered light and general background illumination. They vary from holes cut through the paper which is then placed above the jig load-control nut, to a rolled cylinder placed between the nut and the collimator which effectively unites the two bores. One of the traps associated with specimens in jigs which can confuse the operator during measurement on an auto-collimator, since it too can result in an indistinct return image, occurs when a droplet of water sometimes enters the bore of the jig during washing and gives a misleading picture: when the drop is large it is very obviously the cause of trouble, but a trace is not so easily diagnosed. The remedy is to dismantle the workplate and dry the components, rather than insert paper tissues or attempt to blow-dry.

A further modification may need to be made to on auto-collimator's stand since most of the adjustable-angle jigs standing on a kinematic planes are too tall to go under the optical system. It takes the form of a mild steel extension arm (figure 6.33) 5 × 2·5 cm, drilled at each end to clear the fixing bolts of the table and of the Angle Dekkor bracket.

6.4 Angular Work

6.4.1 Reference gauges and vee-blocks

6.4.1.1.

Although vernier protractors can give an accuracy of five minutes of arc, angular measurement usually involves reference to various standards combined with a means of alignment. The most conventional arrangement is to wring together the appropriate angle slip gauges and in turn wring these to the base plate of the auto-collimator which is then set at some arbitrary zero on the topmost gauge surface. The work—for example a prism—can be subsequently substituted for the gauges whereupon its polished or semi-polished face will show any error relative to the combination of slips. To ensure reasonable orthogonality both the gauges and the prism undergoing test must be 'swung' to produce the lowest image in the auto-collimator. If the slip gauges are set by this method initially, the lowest readings (indicating the steepest effective gradient) occur when the horizontal and vertical scales cross. It follows that these should be the same when testing the work. The accuracy is limited by the various wringing films and by the contact made between the components being checked and the auto-collimator's baseplate. The intensity of the return image should be adjusted to produce the least spread and the sharpest image.

A suitable standard polygon can provide a very convenient reference for work which has simple angles. An eight-sided standard is commonly used for a pair of 90° surfaces. It is placed on a tilt and rotate mount (TRM), raised or lowered to occupy the lower half of the auto-collimator aperture. The most convenient technique makes use of two auto-collimators at 90° to each other (figure 6.34). The prism which is being measured is positioned on top of the polygon and rotated till approximately aligned with the sides:

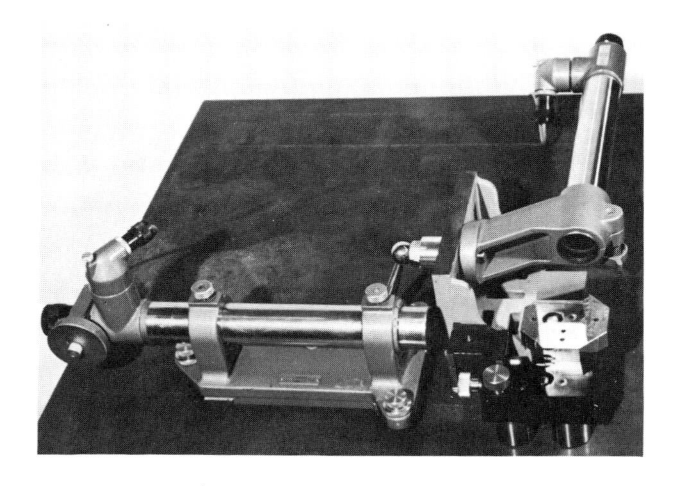

6.34 Optical polygon attached to TRM and viewed through two auto-collimators at right angles.

each collimator will show two images, usually different enough in character to be recognisable. Reading is made easier by setting the optical instruments with the two polygon-images on their respective auto-collimator zeros. The prism images then differ or coincide as the case may be, and the departures from square are noted from the readings on the scales.

When total-internal-reflection (TIR) prisms are measured with the hypotenuse face already polished, this surface has to be greased or in some way rendered non-reflective otherwise confusing return images will be seen. Moreover, measures have to be taken to prevent interference between the auto-collimator illuminators. Black paper masks suitably folded to keep them standing vertically are suitable for the purpose.

Where low refractive index materials are being worked, it is necessary to adjust the intensity of auto-collimator lighting for the maximum sharpness of the return image in order to obtain the highest accuracy. If the standard polygon is of polished steel very considerable reductions in illuminator brightness have to be made. It is normally easy to differentiate between the images returning from a glass prism and a metal polygon because the latter has a bluish tint by comparison with the warmer-image glasses. Optical elements being measured need have only the slightest specularity—in other words, a predominance of greyness is acceptable. This saves a great deal of polishing time while working surfaces and checking them when they may be outside the acceptable angular limits.

While the twin auto-collimator measuring system has obvious merits in convenience, a single instrument is adequate but requires a sub-turntable to rotate the polygon and prism through the required large angle. No great rotational precision is needed since final precision is obtained by trimming the TRM which is mounted on the sub-turntable. The complete assembly should be mounted on a surface table, preferably one with levelling adjusters. If the table is initially set level in two directions (at 90°) then as the various components are placed in position they too are levelled. When the images from the relative surfaces finally have to be found the operator has only to rotate the standard polygon slowly enough to find them. The procedure minimises tedious scanning. There are well known connections between eye-fatigue and accuracy in making instrumental measurements, so that anything which helps to ease observation is very well worthwhile. Just sufficient a level of room lighting is advantageous in order that the auto-collimator-micrometer may be read easily. There should be no direct sunlight: apart from the eye discomfort it can cause, most instruments are affected by it. Sometimes the result is very noticeable as in the case of a high-class spirit level in which the bubble shrinks visibly and less obviously in the distortions of tubular instruments, stands and even iron surface tables.

6.4.1.2

An aid to right-angle measurement alone is a vee-block with its ends at 90° to the vee, positioned on an auto-collimator base. Cylindrical work can be held in the vee and examined to see how accurately an end is square to the axis. The straightness of the vee is a vital feature, but the orthogonality of the ends is not essential. If the block is known to be off-square, the rod under test has to be rotated by hand and the excursion of the return-image observed. The test is mainly used for laser rod working; other shapes can be treated in the same way provided that they are long enough to give adequate registration.

6.4.2 Spirit-levels and clinometers

6.4.2.1

The measurement of small angular departures from plane parallel, as in mirrors deliberately wedged in Q-spoiling cavities to induce mode-locking, can be achieved with spirit levels. The technique is mainly qualitative when normal levels are used in conjunction with precise packing pieces, but with the more advanced versions it is possible to work to better than ten seconds of arc.

Since the position of the bubble is an indication of the angle at which a precision level is tilted, the shorter the base the more sensitive the instrument. Where each scale graduation represents 10″, increasing (or decreasing) the height at one end by 2·5 μm on a base length of 5 cm produces a bubble movement of one division. Reducing the base length to 2·5 cm will result in a similar bubble movement (i.e. 10″) for 1·25 μm rise or fall at one end. Thus the sensitivity is inversely proportional to the length of the base. We have a block level made by Hilger and Watts which is adjustable over a range of a few degrees. Its tilt micrometer is calibrated in minutes and the adjustable vial divided into twenty seconds of arc.

6.4.2.2

Spirit levels are limited in range, though this can be extended either by using them on angle slip gauges or with a sine bar. However, a simpler solution is offered

6.35 Clinometer.

by the clinometer (figure 6.35) which is usually scaled over 90°. In consequence, it can be operated throughout 180° and although one 90° is often as good as the next, sometimes accessibility dictates a particular method of application. The instrument with which we have the most experience is the Hilger and Watts Field Clinometer, originally a piece of military equipment used in World War II. It has a quick-release worm drive giving one degree per revolution with single minute divisions which may be further subdivided by eye to give about 15 seconds of arc.

Apart from its value in the measurement field, the instrument is used for setting the orientation of faces of large quartz parallel to the sawing planes. In this method of setting it is an advantage to have the saw levelled. Tests are made on the orientation face of the crystal at approximately 90° rotational intervals; the clinometer should read 90° for the work and sawing plane to be parallel. The main advantage over dial-gauge adjustment techniques is that it is speedier since the reading is directly in units of angle and this requires no conversion.

There are similar benefits where parallel habit faces are being lapped. If the crystal is placed on a levelled surface table, a clinometer can be used to monitor angular differences on the uppermost face (always by swinging the instrument in order to obtain the maximum reading). Standard engineer's protractors (combination sets) are very often used for the purpose but in general they are an order of magnitude less sensitive then the clinometer.

6.4.3 Optical theodolites

Optical dividing tables are sometimes part of the equipment of an engineering metrology laboratory, but rarely found elsewhere. It was this belief which prompted the investigation of an optical theodolite (figure 6.36) for the purposes of angular measurement. As with gauge and polygon methods the work is rotated through a specified angle, but there is the advantage that any angle can be set and measured to within a few seconds of arc.

Two facilities are required in addition to the basic instrument. Firstly, a means of fixing a table to an elevation shaft on which the work to be tested can be placed; secondly, some means of tilting the work-table in two directions, either at 90° or less conveniently, two at 120°. If a special device has to be made to provide trimming of the work plane, a single tilt-table would suffice since the theodolite's own elevation control provides the second tilt at 90° to the first.

A method commonly used to square the worktable with the rotational axis of the theodolite is to place a spirit level on the adjustable mount and take readings rectangularly, correcting for level by means of the tilt controls. In a second method, the surface of the work-table is lightly polished and an auto-collimator erected above it. The return image can be examined and the table adjusted to give minimal movement.

For optical components which have to be orthogonal to the face that mates with the table, some means of setting the auto-collimator square with, or at some angle to, the worktable plane is needed. A 90° cube or polygon performs the function ideally. An acceptable approximation can be obtained, however by using a turned cylinder, constant in diameter with ends at 90° to the axis. A small slip gauge is attached to it

6.36 Theodolite used for accurate determination of angles.

with elastic bands or O rings and the image on the gauge face used to set an auto-collimator to some arbitrary zero. Cheap, precise cylinders are provided by hardened steel roller bearings. One of them is laid on its cylindrical surface and the reflection from its end face observed. Rollers taken at random from stock and lightly polished on their ends show typically 30–40 seconds of arc deviation from the expected ninety degrees.

A useful device for squaring purposes can be made from two 0·5 in diameter × 0·5 in long rollers. The ends are lightly polished, and, after degreasing them the rollers are laid on their sides on an optical flat or freshly lapped workplate. One roller has its end lightly pressed against the curve of the other and a spot of a rapid-cure epoxy resin used to cement them together. The device can be utilised either way up on any table requiring adjustment at 90° to an axis.

6.5 Measurement of Flatness

Once again techniques tend to divide themselves into those suited to the optical working environment and the more sophistacted ones used in the testing laboratory.

6.5.1 Contact methods

6.5.1.1 Straight-edge techniques
The conventional method of producing an engineer's surface table is to work it flat by scraping it after machining operations. A master surface plate is placed on the table with engineer's marking blue thinly applied, the high spots noted and progressively removed. The precision which results is achieved by by a combination of skill and patience on the part of the craftsman and can approach optical limits. Robust materials, both grey and specular, may be dealt with in a similar way where the use of a blue compound and a straight edge is principally an eye-fatigue saver. It is easier to see a dark blue mark on a lapped crystal surface than to observe a gap of say 10 µm beneath the edge of a flexible steel rule (one from a vernier protractor) that may be used in lieu of a straight-edge.

Some qualitative information about the departures from plane of a surface may be gained by holding the edge of the rule against the work and wagging the centre of the rule transversely. If it moves freely whilst the two ends contact the work then the figure is concave, and convex when the centre touches and the ends are free to wag.

Quantitatively, where fairly large errors are involved, an estimation may be made by inserting feeler gauges or polyester film between the straight-edge and work. In most cases, however, the technique is useful as a quick assessment of figure and usually gives way to optical methods as the work progresses.

6.5.1.2 Optical flats and monochromatic light sources
The phenomenon of the interference of light is seen everyday in the colours observed, for example, in soap bubbles and in changes in colour of thin oxide layers when tempering carbon steels. When two sheets of clean glass are placed in close proximity to one another and examined in strong daylight, faint irregular patterns may be seen, universally known as Newton's rings/fringes. The measurement of plane, or nearly plane, specular surfaces with an optical flat and monochromatic light source has adapted and refined these observations.

In the specific case where light is passing from a dense medium (glass) through a less dense one (air wedge) and subsequently being reflected from a material with refractive index greater than air, a few fundamental facts explain the phenomena of destructive and constructive interference (it is assumed that viewing is at or near normal incidence). Light may be considered in this instance to be purely wave motion, it undergoes no phase change on any reflection of the incident ray at the reference plane boundary (figure 6.37 (*a*)). The attenuated incident ray travels on through the air-wedge and is reflected at the work surface, it undergoes a phase change on reflection equal to $\lambda/2$ ($= \pi = 180°$). Therefore on its return to the reference plane it is either in phase with the incident ray and interferes constructively, resulting in a bright band (figure 6.37(*b*)) or it is out of phase, interferes destructively and produces a dark band (figure 6.37(*c*)). Mathematically expressed, constructive interference occurs when the thickness of the less dense air film d (that is, the path length of the ray on either its forward or return journey is:

$$d = \tfrac{1}{2}(n + \tfrac{1}{2})\lambda \qquad \text{when } n = 0, 1, 2, 3, 4 \text{ etc,}$$

that is at $\lambda/4$, $3\lambda/4$, $5\lambda/4$, $7\lambda/4$ etc, and destructive interference when the thickness is:

$$d = n\lambda/2 \qquad \text{when } n = 0, 1, 2, 3, 4, 5 \text{ etc,}$$

that is at 0, $\lambda/2$, λ, $3\lambda/2$, 2λ etc.
The change in separation of the two surfaces from the centre of a dark band to the adjacent centre of a bright band is thus $\lambda/4$, and from the centre of each dark band to the next (or bright to bright) $\lambda/2$.

Referring to the fringe pattern as a contour map is strictly inaccurate since it shows not only the topography of the work but the wedge-shaped air film between the reference plane and the surface undergoing test, too. As with the interpretation of contour lines, a steep gradient (equivalent to contact angle between work and flat) exists when the bands are close together and a shallow gradient when the bands are widely spaced. In any surface which is nearly a true plane the interpretation of curvature has to be made by observing the departure from straightness of the fringes and not their spacing. If there are changes from regular intervals between them, there must be some deformation of the work—local hollows or hills etc, figure 6.38 demonstrates this.

The air-wedge might be side-stepped by attempting to arrange the reference plane's surface parallel to that of the work. Consider what happens if the two surfaces are parallel. Since, at best, the work must be either purely concave or purely convex, one might expect to see concentric rings of interference fringes. However, if the work is flat to better than $\lambda/4$, one overall dark area will result and it will not be possible to estimate any departure from plane much smaller than this. Therefore it is neither easy nor desirable to arrange the flat parallel to the work and a wedge is invariably, deliberately introduced.

The wedge angle depends upon the air film and the residual particles of detritus between the two surfaces. Both the work undergoing measurement and the optical flat should be as clean as possible before contacting them. After the initial washing away of the polishing medium and drying, the work can be brushed with single strokes of a broad, fine-haired brush. The final contacting is done by interposing an optical tissue between specimen and flat and slowly withdrawing the tissue while holding the flat and preventing any lateral motion relative to the work (figure 6.39). Any temptation to move either of the components when they are in contact must be resisted since damage to either can easily result. If the fringe pattern is not satisfactory—bands too closely spaced or uneven balancing for example—it is necessary to restart the cleaning and contacting procedure, and repeat it until successful.

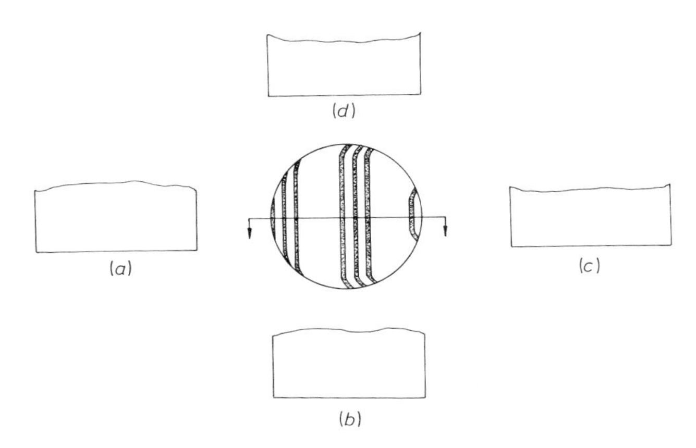

6.38 Possible interpretations (exaggerated) of simple fringe patterns without reference point: (*a*) convex with local ridge; (*b*) convex with local valley; (*c*) concave with local ridge; (*d*) concave with local valley.

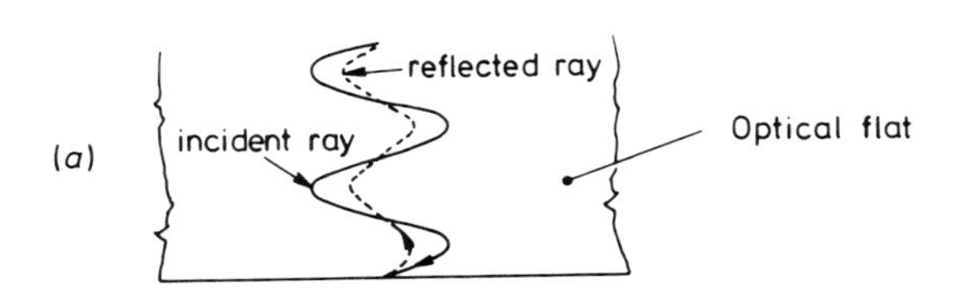

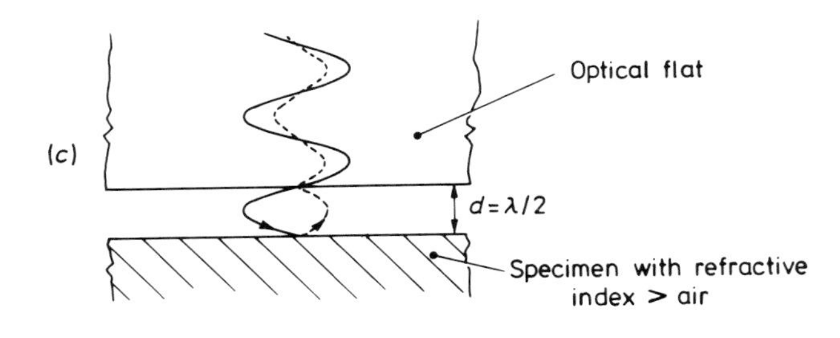

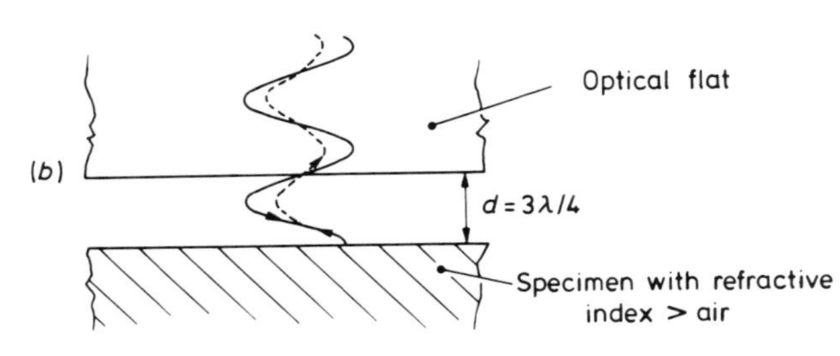

6.37 (*a*) No phase change at boundary of optical flat; (*b*) phase change at specimen surface when $d = 3\lambda/4$, bright band; (*c*) Phase change at specimen surface when $d = \lambda/2$, dark band.

6.39 Tissue withdrawal from between optical flat and jig-mounted specimen.

In the earlier stages of preparation for flatness testing it is important to clean the vertical sides of the work and the surrounding conditioning ring in order to remove any abrasive slurry in particular. Although the liquid may not be on the surfaces it is often drawn between by capillary flow when the optical flat is contacted. The effect is most noticeable and difficult to cure when working on specimens containing defects in the form of fissures and dislocations which are often present in research materials.

Transparent, distortion-free materials such as borosilicate glass and fused silica are used for the manufacture of optical flats. In use, the latter are capable of greater accuracies, not only because of their greater durability but for the reason that the coefficient of linear expansion of quartz is extremely low: $0{\cdot}42 \times 10^{-6}\,°C^{-1}$. Borosilicates—such as Pyrex—have expansion coefficients of about $3{\cdot}3 \times 10^{-6}\,°C^{-1}$. Since a temperature controlled environment is always advisable for polishing and measurement, errors due to thermal inequalities should be only those caused by handling. The diameter-to-thickness ratio recommended in order to minimise local deformation of the flat is $\leqslant 7{:}1$.

Both surfaces are prepared highly specular, usually with one definitively flat, designated as the reference plane. Some optical flats are worked both sides to the same standard, but the less expensive method is to prepare a more precise face and indicate it by means of an arrow on the side pointing towards the better surface. Flats in constant use deteriorate gradually and should be occasionally checked against a master flat, itself stored and handled with the greatest care.

White light is composed of all the colours of the visible spectrum from violet to red and is therefore a mixture of wavelengths ranging from about 4000 Å –7000 Å. In order to obtain precise measurement and contrasting fringe patterns, monochromatic light must be used—that is, light in which the rays have virtually the same wavelength. The interference bands produced when such a source illuminates an optical flat/work combination are termed isochromatics, but common usage of either bands or fringes is sufficiently explicit.

Helium (He) discharge tubes used as monochromatic light sources give more contrasting interference fringes than those commonly produced via open sodium or mercury lamps. The 'Lapmaster' version provides our standard He source for general use placed near to lapping and polishing machines. Although visually the lamp appears to be warm white its strongest line is at 5876 Å (therefore $\lambda/2 \approx 0{\cdot}3\ \mu m$). Any appreciable space between the work and the flat (caused by a scrap of tissue, for example) blurs the bands and makes accurate reading more difficult. In addition, the source should be near to the flat, but not so near that viewing at close to normal incidence is precluded. Errors of increasing magnitude are introduced by examining the fringe pattern more and more obliquely since refraction of the incident and reflected rays in the flat gives an apparent band shift. For the normal thickness of flat, viewing at 30° from the normal can give about 16% error (1 band is read as $\sim 0{\cdot}8$ band); at 45°, about 40% error (1 band is read as 0.6 band); at 60°, about 50% error (1 band reads as 0·5 band). In the last case a surface thought to be, say, flat to a quarter of a fringe in an inch would in reality be half a fringe.

In front of the discharge tubes there is an opalescent diffusing screen with a straight line engraved across its face. By suitable adjustment of the flat, work and eye position the reflection of the engraved line provides a reference of straightness with which the bands can be compared and their curvature, relative to their spacing, estimated. Thus if a pair of slightly curved dark fringes are joined by the line in three places (figure 6.40) (end, centre and end) the surface is said to be out of plane by one fringe— across the work diameter in the figure. To rationalise the measurement it should be quoted as bands/fractions of a band

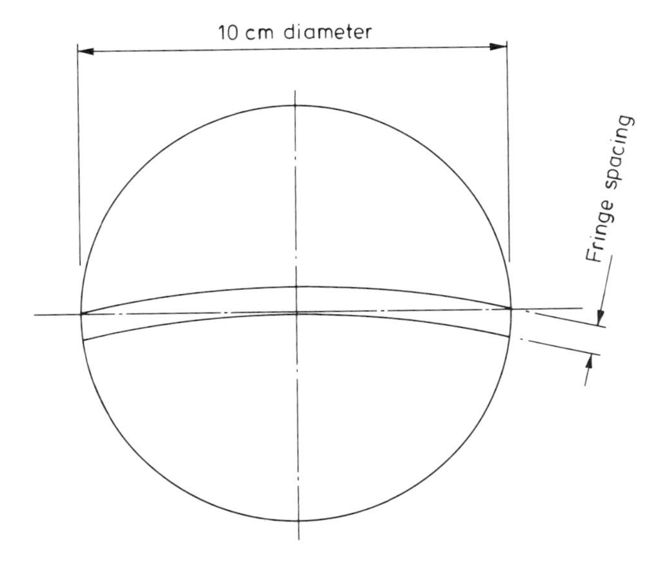

6.40 Interpretation of flatness: one band (in 10 cm)

or wavelengths/fractions of a wavelength per unit of length and the light source stated. In the example given, one band in the diameter of the work is one band in 10 cm, therefore it may be written as either one-tenth of a band in 1 cm or $\lambda/20_{He}\,cm^{-1}$. It follows that if a straight line 1 cm in length is drawn across the inside of the curve in figure 6.40, the distance between this line and the curve should be divisible into the inter-band width ten times. Unless a photograph is taken of the pattern, the measurement depends upon a mental subdivision exercise where the centre of a dark band is not easy to estimate. The flatter the surface, the more difficult precise measurement by the technique becomes, but through the test is not the most accurate it is eminently operable in the polishing environment.

Estimation of fringes gives the magnitude of curvature, but whether this be convex or concave can be determined by several methods the two commonest of which will be described. The first of these makes use of the apparent bandshift effect with increasingly oblique viewing. As the head is moved from the near normal to the oblique position—in practice, lowered—then the band pattern will appear to move away from or towards its centre of curvature. Thus the general trend of the fringes is observed and the direction in which their imagined centre lies is noted, then the head lowered. If the work is concave, the bands appear to contract towards their centre, if convex they expand away from their centre. A small mnemonic helps here: convex-expands. The effect is best seen on a lightly figured mirror with a convex or concave bullseye fringe pattern.

In the second method, the point of contact or near contact has to be known or ascertained. Usually, the gradual withdrawal of a tissue results in the greatest separation at the final removal point. The application of pressure at this spot should cause the fringe pattern to broaden out as the air wedge becomes thinner. Once the closest contact point has been found (opposite to the pressure point described) it is designated the reference point or line and the curvature of the bands relative to it determine the sign of the work. When they appear curved towards the reference, the surface is convex; when curved away, the surface is concave. Figures 6.41a, b and c illustrate this. On a bullseye pattern pressure applied at the centre which causes the pattern to fluff-out indicates a concave surface; whereas little or no obvious change in the pattern shows the figure to be convex.

In general, the interpretations of a fringe pattern need be quantitative only when it comes within obvious range of the accuracy of form required: usually flatness. Though other figures may be seen they are seldom measured but assessed qualitatively and corrective action taken to cure what are, after all, faults in polishing technique.

If the mounting, lapping and polishing procedures described in this book are followed carefully, the achievement of flatness on most crystals is not difficult.

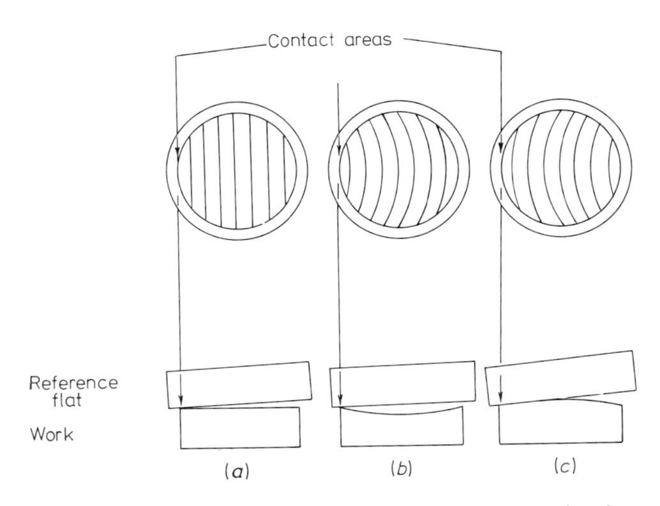

6.41 (*a*) Flat specimen, pressure applied at arrowed point; (*b*) Concave specimen, pressure applied at arrowed point; (*c*) Convex specimen, pressure applied at arrowed point.

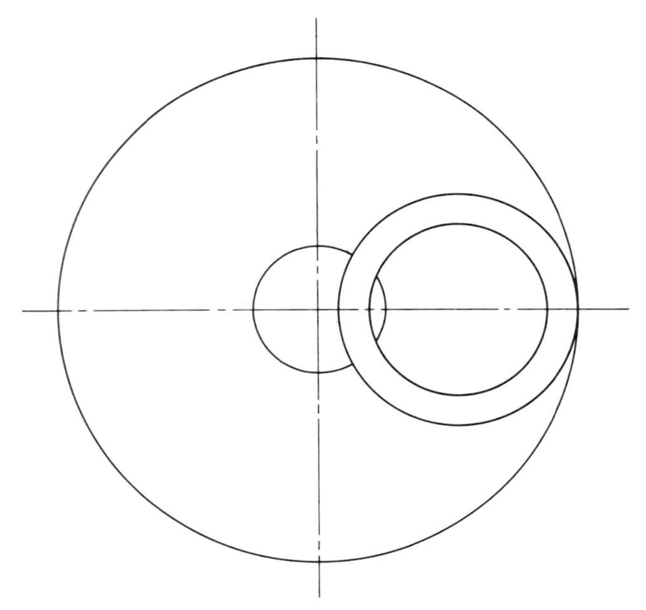

6.42 Convex work-surface and ring position.

The following illustrations outline the typical measurement/polishing process steps on a germanium specimen, 2 cm diameter, initially lapped flat on the 60 cm machine, then polished for 30 min. Figure 6.42 shows the slightly convex surface which resulted, curved by one band ($\lambda/4_{He}\ cm^{-1}$). The drawing (below, right) indicates the corrective measure taken to improve the figure, a slight outward repositioning of the roller bar and consequently the running centre of the work and ring. In figure 6.43 the surface has become nearly plane with both signs showing—flat to $\frac{1}{4}$ band across the diameter ($\lambda/16_{He}\ cm^{-1}$). Prolonged continuation of the polishing process would give a concave figure. This is shown in figure 6.44 curved by $2\frac{1}{4}$ bands ($9\lambda/16_{He}\ cm^{-1}$) as confirmation, not as recommendation! The compensating move is to re-position the running

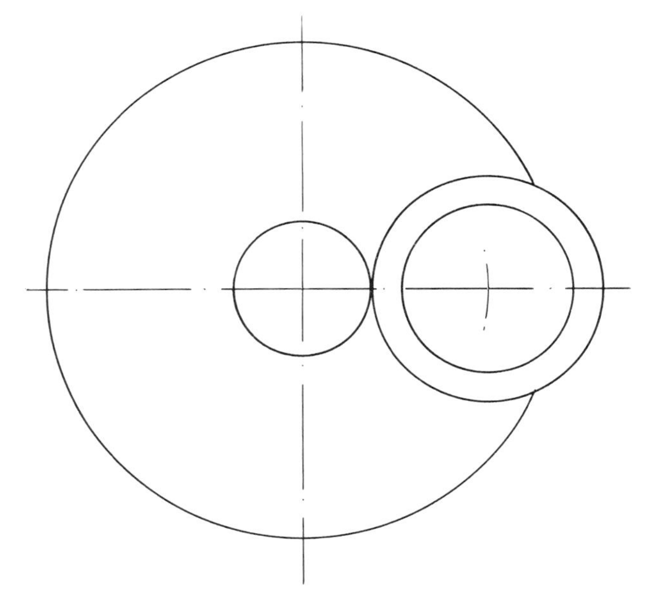

6.43 Plane surface and ring correction to illustrate sign change resulting. Transition stage:– mainly flat but showing some evidence of previous convexity and if continued future concavity.

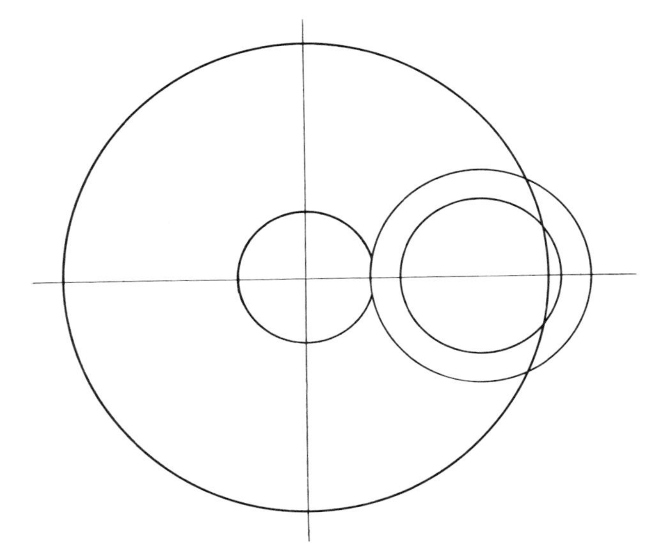

6.44 Concave surface, necessitating movement of roller-bar towards polisher centre.

centre of the work nearer to the polisher centre. It follows, of course, that the steps described could equally well be necessary in reverse order if the original figure were found to be concave.

6.5.2 Non-contact measurement

The non-contact technique makes use of an optical interferometer in which the work is independently mounted and adjustable in angle relative to the optical flat. An intense point source is provided by condensing light from a gas discharge tube on to a pinhole plate and thence via a part-silvered mirror and collimating lens through the reference flat to the specimen. In this way, not only is direct contact avoided but the techniques for adjusting the air wedge and aligning the fringe pattern are avoided.

There is quite a range of commercially available Fizeau interferometers, now often called interferoscopes. Some have very large diameter flats (up to 15 cm) which need considerable care in the arrangement of their support. The instrument that has proved sufficiently large for the type of work described herein is the 5 cm one made by Barr and Stroud. It uses a 5 cm diameter, $\lambda/50_{Hg}$ flat; a low pressure, filtered mercury source; and a large gap adjustable between the work alignment table and the face of the optical flat. The large gap is needed to accommodate the various adjustable-angle polishing jigs which are favoured by many crystal workers.

A coarse adjustment to the height of the instrument is provided by a clamp around the body. When this is locked, the fine control is brought into operation for reducing the work-to-standard-flat gap to an acceptable amount. The adjustment takes the form of a fast threaded column with a locking ring.

Finding fringes with some elementary interferoscopes can be a tedious process, however, with the model described there are two eyepieces provided: a clear one through which to see interference patterns and a second, acting as a 'finder', which carries angular scales to assist in aligning the two wanted images. A surface initially misaligned under test presents two images: one from the test surface and the second from the reference plane of the $\lambda/50_{Hg}$ flat. The angle of the work-face is adjusted until the two images are superimposed. A simple change is made by replacing the 'finder' with the clear eyepiece and the interference bands are revealed.

A modification to the commercial instrument has been made by substituting a standard two-axis mirror mount (PMM) for the three-axis table on the instrument. One reason for this change was the attraction of the finer micrometer-actuators; and secondly, adjustment in two planes provides a more sensible control on an instrument where the graticule has 90° scales. Figure 6.45 shows the complete test set-up.

6.45 Non-contact interferoscope with flat mounted on PMM.

6.47 Heat bump produced on concave surface (produced with a heated wire proving general work shape to be concave).

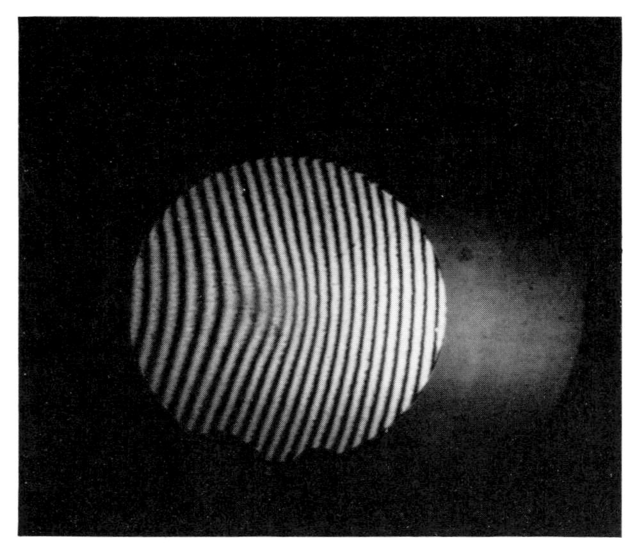

6.46 Heat bump produced on convex surface.

Discovering the sign of the curvature when the bands have been obtained interferometrically cannot be achieved by the head-moving method used with a monochromatic light source and optical flat. If turn-down (rounding of the edge of the work) is present, its pronounced convexity shows the general shape of the work very well, since a convex surface will be a gentler continuation of the fringe pattern seen at the edge, whereas a concave face reverses the curvature seen there. Where there is little or no apparent turn-down, localised thermal distortion can give the necessary clue to the sign of the surface. Figures 6.46 and 6.47 are interferograms showing indicative bumps in the work which have been raised by the application of a warm wire lightly contacting the polished face. In figure 6.46 the bump and the general curvature are the same, therefore the work is convex; in figure 6.4 the curves are of opposite sign therefore the general work-shape is concave.

In instruments such as the Twyman–Green interferometer, pressing the right-hand corner of the baseplate will cause the observed rings to expand on a convex surface. Another technique which may be used on the interferometers is to insert a very weak positive lens above the surface being measured. This has the effect of increasing the convexity of a weakly convex surface, or of reducing or cancelling the concavity of a slightly concave surface. The lens required need be convex only by a few fringes, with the opposed face worked plane.

6.6 Measurement of Concave Spherical Surfaces

In most of our work up to the present, plane and concave spherical surfaces have predominated with the greater emphasis on the former. Fortunately, the two traditional tests of Ronchi and Foucault are easy to interpret for the latter. Parabolic surfaces require a great deal of experience for their interpretation but they have not been, as yet, required. Knowing what to expect and what appearance well-known defects will present is always a difficulty when setting up tests for the first time. Mirrors of approximately correct spherical form but with a known defect such as a turned-down edge assist initial understanding.

6.6.1 The Foucault test

This is operated at the centre of curvature of the mirror with a brightly illuminated pinhole, about 50 μm diameter (see Chapter 5). In order to reduce parallax the pinhole is illuminated via a totally internally reflecting prism used at the extreme edge, allowing the hole and the knife to be as close together as possible (figures 6.48 and 6.49). It is difficult to obtain much light at the viewing point and for those setting up the test initially it is recommended that a silvered mirror be used which allows the worker to follow the return rays and adjust the position of the knife edge accordingly. Silvering or aluminising mirrors for test not only results in easier arrangement of the system but figure errors of the mirror appear in much greater relief, too. This is valuable to the inexperienced viewer who may not know exactly what he is looking for. Once accustomed to a Foucault view of the mirror, using the test on an unsilvered surface presents no problem.

When the eye is placed so as to receive the light returning from the entire concave surface it will appear uniformly bright. It is quite a feat of steadiness to maintain the eye position in order to achieve this. When the knife is moved across the beam at the centre of curvature—the focus of the mirror being examined—the surface appears to darken evenly if it is truly spherical (figure 6.50 (*a*)). When the knife is moved across the beam from a position a short distance inside the focal length, its darkening effect appears from the same direction as the knife (figure 6.50(*b*)). Conversely, when the knife is placed outside

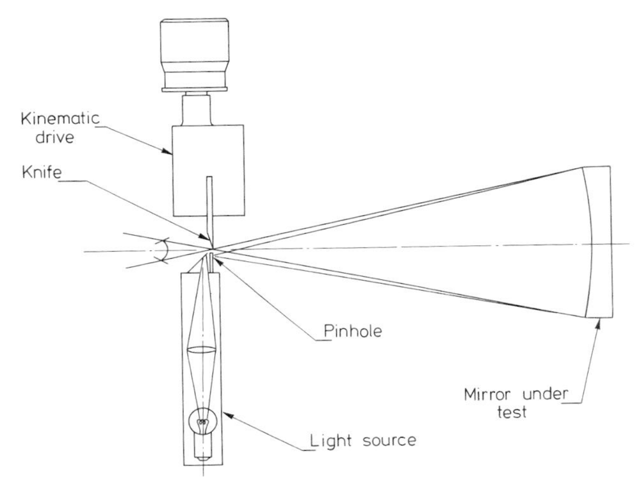

6.48 Foucault knife-edge test: diagrammatic.

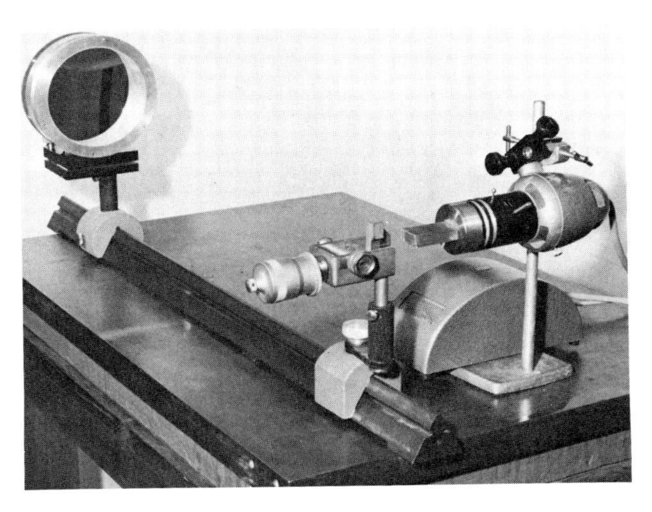

6.49 Apparatus set-up for a Foucault test.

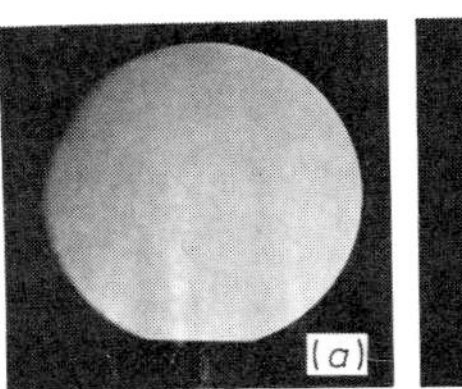

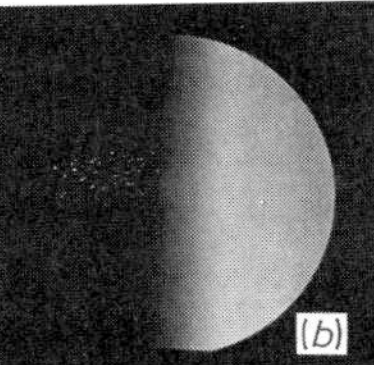

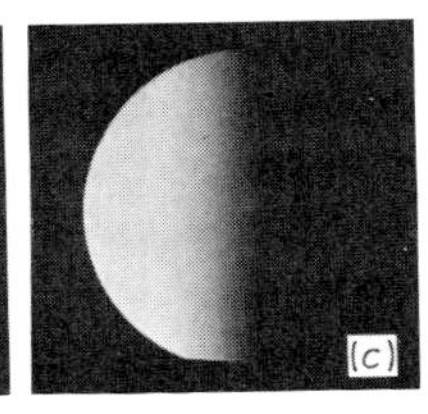

6.50 (*a*) Foucault test: knife at radius (focus); (*b*) Foucault test: knife slightly inside focus; (*c*) Foucault test: knife slightly outside focus.

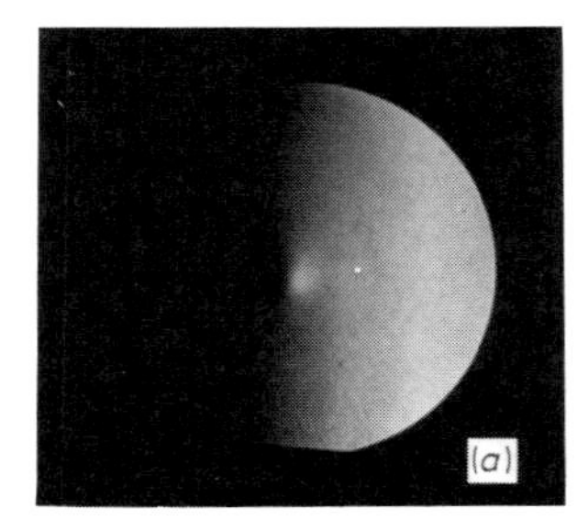

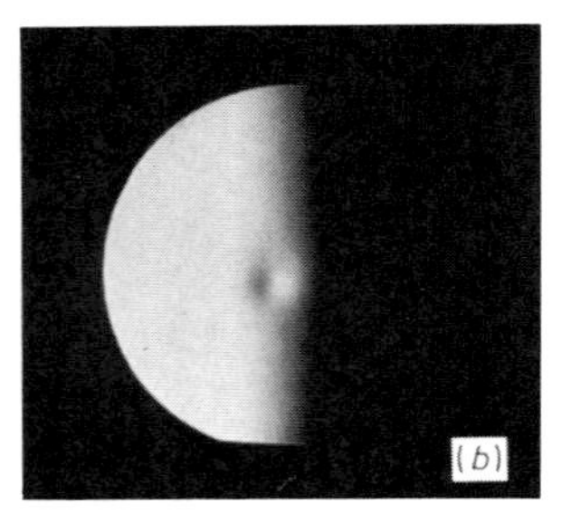

6.51 (*a*) A heat bump as seen on a Foucault test inside focus. (*b*) A heat bump as seen on a Foucault test outside focus.

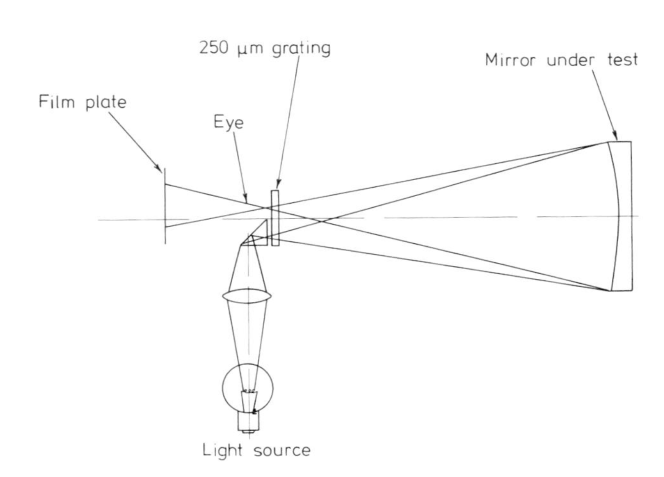

6.52 Diagram of Ronchi grating test.

the centre of curvature the shadow moves from the side opposite to it (figure 6.50 (*c*)). Once the test has been practised, a user will confirm that it is a sensitive radius testing method: both form and focal length are precisely obtainable. If any doubt exists about the sign of any observed defects, shown by contrasting darker or lighter rings or spots on an even background, the hot wire technique can be used to produce a bump. Figures 6.51(*a*) and 6.51(*b*) illustrate what is seen from positions inside and outside the radius when the hot wire has been applied.

6.6.2 The Ronchi test

The Ronchi grating test provides a simpler method for assessing the sphericity of a polished surface for those who are familiar with fringe patterns. In common with some other radius-measuring methods it can be used to establish flatness as well. A brightly lit portion of a grating, comprised of alternately opaque and transparent bars, is set at the approximate centre of curvature of the surface being tested. The return beam from the mirror is passed through an adjacent part of the grating and thence to the observer's eye, or at a suitable distance, to a screen or film plane. Since the grating need not be too fine it may be made by winding small gauge wire between two cylindrical posts thread-cut 100 TPI (i.e. 0·010 in pitch) and spaced a suitable distance apart. The wire is cemented to the rods and the winding on one side cut away subsequently (Kirkham, 1946). Alternatively, photolithographic mask making is a commonplace technique in many laboratories and it facilitates grating manufacture.

A typical apparatus is shown in figures 6.52 and 6.53, the lamphouse and prism arrangement taken from the standard Foucault test equipment with the pinhole graticule replaced with a grating. Some parallax occurs, but a semi-reflecting pellicle of polyester film can be used to minimise the effect.

6.53 Apparatus set-up for Ronchi test.

Straight fringes appear from the curved mirror surface when the grating is just inside or outside the focal length (figures 6.54 (*a*), (*c*)). The bands become broader at the precise radius and appear very diffuse at this point (figure 6.54(*b*)). Thus accurate determinations of the focal length can be made by the Ronchi test and these can be made more precise by arranging a controlled axial movement to either the mirror or the grating. This adjustment is used to

7 Application

There's a divinity that shapes our ends,
Rough hew them how we will.' (Shakespeare)

The foregoing six chapters have described a variety of equipments and techniques with occasional illustrative references to specific materials. All of the following sections amplify the tables—to be found at the end of this chapter—which summarise many polishing procedures. A comprehensive account of each of the substances worked would be repetitive, so it is intended to concentrate on several main classes of polishing indicative of techniques that may be more generally applied. It is expected that the reader will refer to the previous chapters for more explicit instructions about, for example, machines, jigs, fixtures, mounting and measurement.

Some of the techniques described are expanded notes made during the past fifteen years and often underline the fact that achieving a satisfactorily specular surface is not always the major problem. A number of methods can achieve this on many materials, but the shape and size of the components can be very demanding—for instance: small diameter laser rods and miniature glass prisms.

7.1 Laser Rods

Adjustable-angle jigs (§2.3) have made the production of single laser rods (e.g. Forrester, 1974) from both hard and soft crystals a relatively routine process. With them, parallelism of the ends to better than six seconds of arc can be achieved (by optical measuring techniques), flat to about $\lambda/10_{(He)}$. However, polishing a single rod or several simultaneously whilst achieving parallelism routinely to better than 2″—monitored by electromechanical measurement and corrected by push-rod bias polishing methods—is the forte of the block technique. Whichever method is used, the initial cutting and shaping of the cylindrical rods is achieved by the processes described in chapters 1 and 5.

7.1.1 Block methods

In order to give stability to a single rod during the optical surfacing of end faces by conventional hand techniques, some form of holder is used, either with a packing washer, or with several feet or wasters, (figure 7.1). The rod is reversed in the holder for working the second end, and an auto-collimator used to indicate parallelism. However, for mechanical lapping and polishing, and electromechanical measurement, a precise double-ended block is desirable (Bond, 1962; Fynn and Powell, 1969a).

7.1.1.1 Hard crystals: plane parallel

For a single ruby rod, typically 6·35 mm, the block is a mild steel cylinder of 32 mm diameter, faced at each end with either a ruby or a sapphire ring, or three separate wasters as feet. To achieve a uniform rate of wear during lapping and polishing, it is important that these stabilising facings be of the same crystal cut as the central laser rod. They can be cemented in place with a thermoplastic resin such as Tan wax and thus recovered easily if required. Epoxides may be used, too, with an attendant increase in permanence. A cast-iron lap coated with 25 μm diamond is used to lap the facings parallel to about 2·5 μm (measured near the periphery) and square with the axial hole. The overall length of the block should be made about 0·25 mm shorter than that of the sawn laser rod.

Both block and rod are heated to about 100 °C and a coating of optical stick wax worked into the extremely small gap around the rod when it is in place. A small, even fillet of wax is left at each end. The assembly is allowed to cool in air and, when the cement reaches the plastic state, the rod is positioned axially so as to protrude approximately 0·125 mm at each end. Finally, the block is left to cool to room temperature in a horizontal position.

A flat cast-iron lap, of 15 cm diameter, is screwed to the lapping-machine spindle and the sweep of the conditioning ring in its roller-bar adjusted to give about 1·5 cm sweep driving as close to the lap surface as possible. The flat face of the lap is coated with a suitable abrasive which in this case would by Hyprez OS 25 μm diamond diluted with Hyprez fluid. With the block positioned in the conditioning ring, a Tufnol protection cap (figure 7.2) is mounted on the exposed end of the block and pressure applied to the workface by means of the spring acting through the vertical push rod. On a general purpose polishing machine (see §2.2.1), the position of the leaf spring must be adjusted so that when the wanted load—say 100 g—is applied by sliding the counterweight, the workarm is clearing the end stop by 0·25–0·5 mm. A fixed workarm machine, such as the 'Multipol' (§2.2.4) does not allow this, but the pressure is varied by screwing the micrometer push-rod spring unit. The applied load is easily measured by hooking a spring balance under the tip of the leaf spring and lifting vertically until the push rod moves freely in its locations; a micrometer spring unit can, of course, be calibrated and marked linearly by acting against a pan-balance.

The machine is run at 100 rpm with replenishments of diamond and lubricant at about two minute intervals. Treatment is continued until an even grey surface is produced on the entire washer and rod surface. This process is repeated using another cast-iron lap with 14 μm diamond abrasive; then individual solder-faced or tin-plated polishers with 8 and 3 μm diamond. The surface finally obtained is highly specular, but wood-faced cast-iron (§3.2.5) and 1 μm diamond abrasive can give an improved finish.

Depending on the nature and axis of the crystal being worked, a moderate polish will appear after a few minutes running with 3 μm diamond. Tests for flatness can be made at this stage, preferably with an interferoscope using a non-contacting optical flat. A simple monochromatic light source and a contacting optical flat can be used with the attendant risk of damage to the flat itself. If this method is adopted, the optical flat should be first spaced from the work by a sheet of tissue which is slowly withdrawn (§6.5.1). There must be no sliding movement between the components: a short wooden rod should be used to manipulate the flat for the desired fringe pattern, but only by very carefully applying pressure to influence the air-wedge.

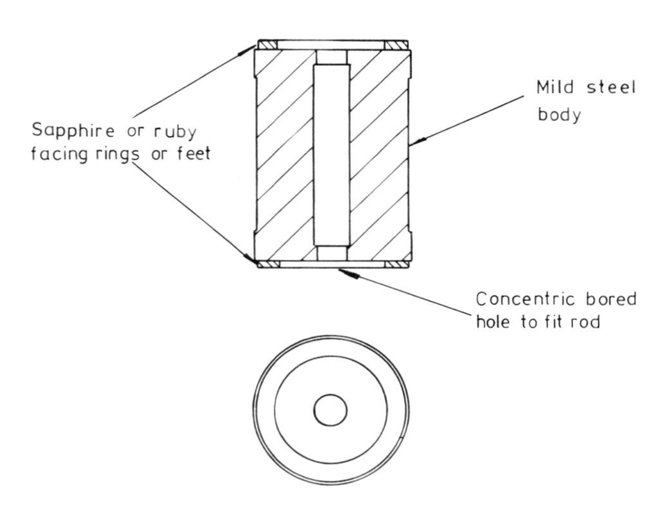

7.1 Lapping and polishing block for a single, hard laser rod.

7.2 Tufnol dimple cap for application of bias polishing loads.

Parallelism measurement and correction is made as described in §6.3.1, and the necessary adjustment of the push rod to effect bias polishing carried out from the data obtained. Thus, if the electromechanical readings indicate a higher point, this is marked with a crayon spot on the block rim, and the off-centre dimple in the Tufnol cap positioned in line with the mark. The process is repeated as necessary, until the ends are closely parallel: for 2″ departure, for example, the difference in linear length measured on a 3 cm diameter is 0·3 μm.

Small concave or convex errors on the work, also shown by similar errors on the conditioning-ring's face, may be corrected by adjusting the running centre of the ring. As detailed and illustrated in §6.5.1, a concave error on the work is corrected by moving the roller bar slightly towards the centre of the polisher and vice versa for a convex error. When 8 and 3 μm diamond are being used on soft-metal polishers, they maintain their form well as long as the conditioning ring supplies the dominant load. A single rod, typically 0·3 cm^2 in area, being lapped within an open packing ring is initially the most unfavourable combination, since the mass is large and acts over a small area. While the rod is still protruding, it should be closely monitored to ensure that minimal channelling of the lap surface is taking place. As it wears down and the packing washer makes contact with the lap and its area caters for a larger load. Thus the same rod within a washer, 3·17 cm outside diameter and 1·9 cm inside diameter, would have an effective area of about 2·5 cm^2 and would require about 250 g lapping and polishing load. Even though the polishers subsequently used may retain their flatness, occasional resurfacing on the surfacing and scrolling machine (§3.3) generally reduces the sleeks on the rod end.

During the finer lapping stages, either a polished radius or, somewhat easier, a rounded chamfer on the exposed edges of the ruby rod and packing must be maintained. This reduces the scratching of the optical surface which results from the presence of breakaway particles of crystal material. The chamfering is done by lapping with a hollow, conical iron or copper tool, fed with the grade of diamond in use at the time on the faces.

Finally, optimum specularity can be achieved by polishing for a short period on polyurethane foam with an alkaline silica sol. All traces of diamond abrasive must be scrupulously removed from the block before any change to a smaller grade of abrasive, and this is equally important when employing the alkaline sol.

When several uniform rods are needed, a technique is used which again gives no close support to the crystal edges. A mild steel cylinder is jig-bored with one central and six surrounding holes (figure 7.3) (then an outer ring of twelve if a larger quantity should be required) and is axially shorter than the rods by 6 mm.

The rods are a close sliding fit in their holes, protruding an equal amount from each end of the block and held in place with a thermosetting cement such as optical stick wax, glycol phthalate or Tan wax.

The configuration results in fairly even wear on cast-iron laps with diamond abrasives and with the soft-metal polishers. Since several rods have a correspondingly increased surface area, they present less problem in non-uniform lap wear and so do not need additional wasters. Seven 6·35 mm diameter rods, for example, require a total load (that is, block dead weight plus rods plus spring pressure) of about 220 g. Although hard crystals will withstand pressures up to 200 g cm^{-2}, conditioning-ring masses and areas are standardised at 100 g cm^{-2} and polishing loads are, of course, adjusted to conform with these.

Measurement of parallelism is made electromechanically on the outer ring of rods—made possible by their being harder than the stylus and thus

7.3 Lapping and polishing block for seven rods.

7.4 Fringe pattern obtained on seven rods polished simultaneously.

scratch-proof. Flatness is achieved as before by adjusting the running centre of the conditioning ring. Figure 7.4 shows a typical fringe pattern on seven rod-ends.

7.1.1.2 Glasses and soft crystals: plane parallel

A softer crystal (Mohs's value ~7) requires close support to its periphery and its block is, therefore, more difficult to manufacture. The internal diameter of the packing washer must be worked to fairly close limits to fit the rod if their support is to be of any real value in maintaining flatness to the extreme edge of the crystal—particularly during the later polishing stages. For the same reason, the rod must be ground and lapped to be as cylindrical as possible. It is assumed in the following description that no precise glass-boring facilities are available.

A stainless steel or aluminium alloy block is turned truly cylindrical, bored concentrically and has, for convenience, its ends faced closely square to the axis of the hole. Two glass washers are prepared by trepanning their central holes 125–250 μm below the rod size at this stage. A special lapping jig (figure 7.5) is required to make the washer hole 25 μm below nominal rod size while keeping it precisely at right angles to the face. The lapping operation consists of slowly rotating the brass tapered plunger while lightly holding the face of the glass washer against the end of the jig. The plunger is coated with BAO 303 grade abrasive (or 25 μm diamond) and fed through the hole until the parallel section emerges at the far side of the washer.

Another brass plug is used to register the first washer centrally on one end of the block, with the register hole and plug coated with a release agent such as wax polish. A slightly flexible epoxy resin is recommended for cementing the washer in place because of the differing coefficients of expansion of the components: CIBA AY/HY 111, for example, allowed to gel at room temperature, then post-cured at 60 °C for three hours. The register plug is withdrawn and a tapered lapping plug used with 600 carborundum to size the hole finally to about 8 μm above the laser rod's diameter. The second washer is registered, cemented and lapped in the same way. Figure 7.6 shows a typical finished block.

A simpler technique which copes with a widening range of crystal diameters uses stainless steel blocks bored to a standard diameter of 10 mm for all rods up to that size. Only the washer-hole has the same diameter as the rod and the only remaining variable (usually not critical) is the block length. Registration of the first washer is by means of a two-diameter brass plug; the second is registered for epoxy-cementing by means of a two-diameter plug machined from solid polystyrene, since it will be trapped in the block between the washers and require dissolving away—e.g. by toluene. Dissolving is slow if done statically; however, either stirring or, better still, pumping speeds it up. Finally, the last few micrometres are lapped from the washer-holes with a very slightly tapered lap, using 600 carborundum.

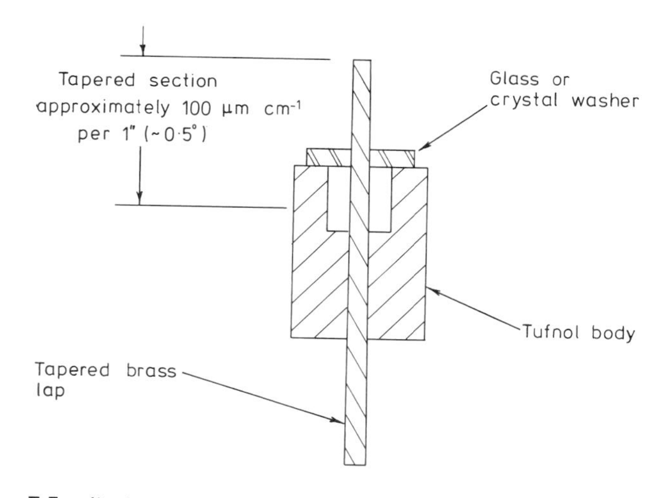

7.5 Jig for size-lapping a single, soft rod block (for precise packing washers).

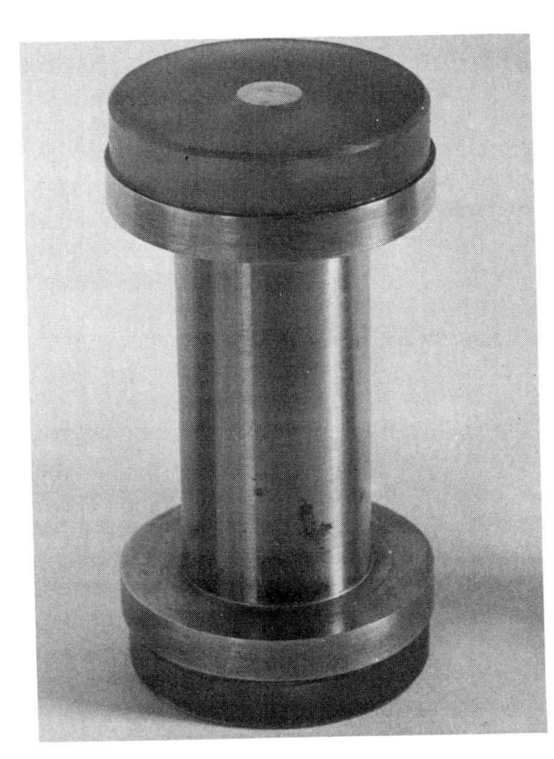

7.6 Lapping and polishing block for a single, soft laser rod.

During the lapping process, extreme care must be taken to avoid chipping or bell-mouthing the washer-hole edges but, if it should happen, it can be remedied by prolonged treatment of the end faces. For this, the body of the block is fitted into a cylindrical housing with squarely faced ends (figure 7.7) and the washer and housing are lapped simultaneously, thus maintaining approximate squareness.

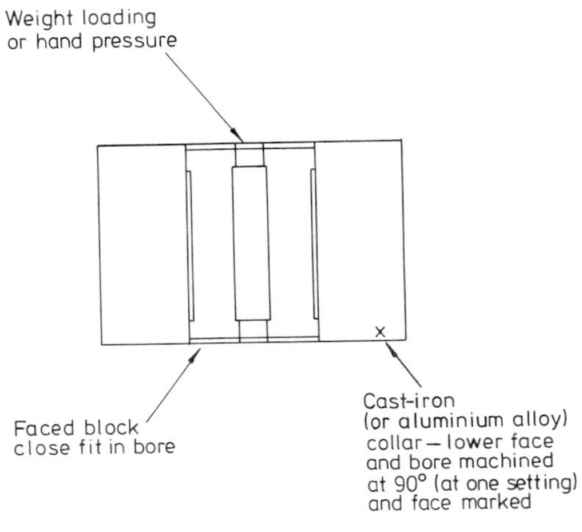

7.7 Housing for squaring laser rod block ends.

Where the thermal expansion of the laser material is sufficiently close to that of glass, all-glass or silica-faced glass blocks are very satisfactory. When glass trepanning facilities are available, the process is an obvious choice. In order to obtain the effect of relief in the bore of the block, each assembly consists of a series of collars with two facing washers at the ends. Again, the end washers have bores to suit the laser rod while the spacing coller or collars are bored to a generous clearance on the rod's diameter.

A typical block of this type, at present in use on 3 mm diameter, yttrium aluminium garnet (YAG) crystals, consists of a 3 mm thick silica disc on each end bored to fit the crystal with spacing collars trepanned from 6·35 mm plate glass with 6 mm diameter holes. The assembly is cemented together with CIBA AY/HY 111 resin. Concentricity is obtained by resting the assembly in a vee-block during cure; or, if the operation is to be done separately, a sleeve bored in PTFE as a close fit on the outside diameters of the collars can be used as an assembly jig and left in place during the curing cycle. The cemented blocks are subsequently treated for final hole sizing and end lapping in exactly the same way as the metal ones.

It will be noted that glass washers are normally used to face all the blocks for soft crystals and that, although their erosion rate matches laser rods such as neodymium-doped glass ones, softer crystals like calcium fluoride would be mismatched. Relatively rare and expensive crystals do not always allow the luxury of matching wasters, therefore a process is needed which increases the erosion rate of glass washers (Fynn and Powell, 1969b). The general technique for calcium fluoride is described.

In principle, the process relies upon the fact that during pitch polishing, the erosion of a lapped (grey) surface is affected by the surface roughness. In general optical work the grey surface is lapped as smooth as possible in order to reduce the polishing time. Conversely, a rough grey surface will erode rapidly during pitch polishing, but will be very slow to attain a highly specular finish due to the continued presence of deep pits. It follows that if differential grey surfaces can be prepared in the same plane, different natural erosion rates can be matched.

A block is prepared with the end faces in the plane parallel state and with a fairly rough surface resulting from lapping with 220 carborundum. The laser crystal can now be cemented in, the only difference being that some delicate crystals will require careful warming and cooling, to avoid cleaving. A satisfactory method in common use is to buffer the rod and block assembly by wrapping it in several layers of paper tissue and applying the cement—usually beeswax/resin—in the oven. The assembly is left to cool whilst remaining rolled in the tissue.

When cool, it is lapped in the housing (figure 7.6) until the crystal is within 50 μm of the washers, and the remaining length then removed by hand or machine lapping. In either case the lap is cast iron and the abrasive Aloxite 50 or, in some cases, finer abrasives such as BAO 304 may be used. The applied load should not result in a total pressure greater than $100\,g\,cm^{-2}$. Whilst the rod is still proud, the assembly pivots freely on the face of the rod. It is most important that all lapping is stopped when the rod is reduced to the level of the peaks on the washer's surface. This is essential in order to preserve the surface roughness of the glass. If there are no scratches on the rod face, the block is ready for polishing, preferably on a 2·5 mm pitch polisher with 0·3 μm alumina (Linde A) and water. The block loading is about $30\,g\,cm^{-2}$ and an average time for achieving the necessary finish and a flatness of $\lambda/10_{(He)}$ is forty minutes. However, should scratching occur on the first end, the block can be warmed and the rod slid through far enough to allow for a second attempt. Scratches on the second end, which may be still in evidence at the co-planar stage, can be remedied only by removing the rod, shortening the block by about 100 μm then starting again. The first end, when polished, is especially vulnerable to accidental damage while the second end is being lapped because of the environment of relatively coarse abrasives.

In cases of lapping difficult crystals, the method of spreading and controlling the working thickness of very fine abrasives is important. When using BAO 304, liquid detergent is used to reduce agglomeration of the particles and it is advisable to use a crusher when working by hand. This is a 6 cm diameter, 2·5 cm thick block of glass ground flat on one side, slid over the freshly applied abrasive to reduce it to an even, thin coating on the cast-iron lap. The condition obtained, though ideal, is very short-lived in terms of lapping duration because of its thinness and, consequently, needs constantly reworking with the crusher. The water content is also critical—too much can encourage scratches and too little allows the accumulation of a thick blanket of abrasive and detritus which will cause the edges of the work to be turned down. A locally damp atmosphere is desirable, both in the final stages of lapping and during the entire pitch polishing process.

To illustrate the factors involved in the differential erosion rate techniques, a series of surface profile charts of adjacent washer/rod regions are shown in figures 7.8–7.11. Figure 7.8 shows nearly identical materials lapped to the grey state—a glass washer in conjunction with a neodymium-doped glass rod. With this combination, the surface roughnesses with any given abrasive are very similar and, consequently, both acquire a polish at the same rate. Figure 7.9 shows the same treatment given to a calcium tungstate

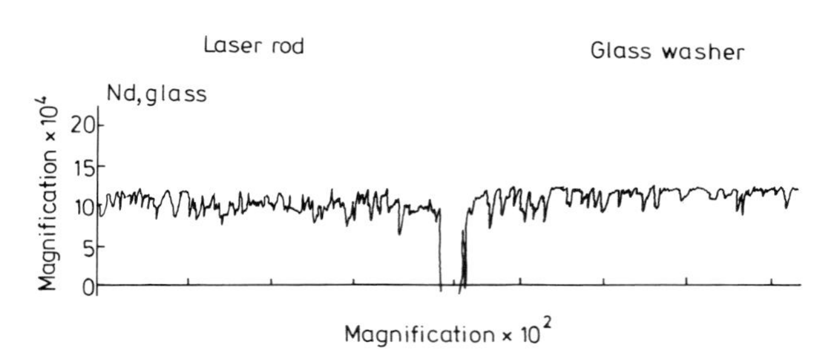

7.8 Talysurf record of nearly identical rod and washer materials, lapped to the grey state (abrasive BAO 304).

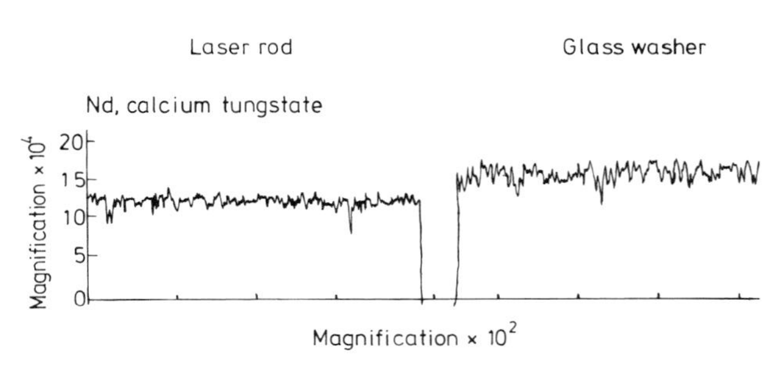

7.9 Talysurf record of calcium tungstate rod, glass washer, treated as in figure 7.8 (abrasive BAO 304).

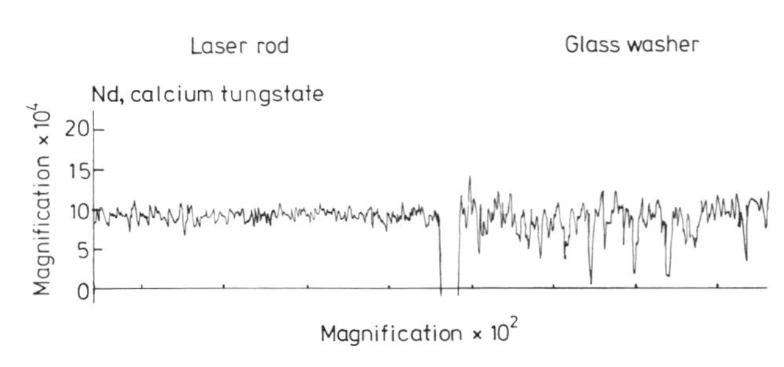

7.10 Talysurf record of calcium tungstate rod (abrasive BAO 304), glass washer (abrasive BAO 303), differentially prepared.

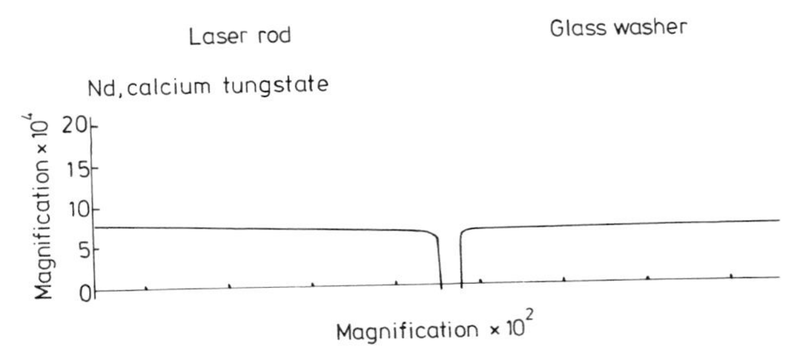

7.11 Talysurf record of calcium tungstate rod, glass washer, final polished condition

laser crystal within a glass washer. Due to the relatively softer nature of the crystal, it has eroded to a level lower than that of the glass. If this were subjected to pitch polishing, the rod would not contact the polisher for a long time; the polishing would be slowed down due to the absence of pits in the face. Figure 7.10 is a surface-profile chart of the same combination of rod and washer, as in Figure 7.9, but the finishes are prepared by the techniques described above—that is, with a rough grey surface (from BAO 303) on the washers prior to mounting the rod. When it is mounted, allowing about 50 μm to protrude, the crystal ends are lapped with BAO 304 to bring them just co-planar with the washers. The result is a very fine crystal surface in line with the peaks of a coarse washer face. When this composite surface is subjected to pitch or wax polishing, the rapid erosion of the coarser surface provides an erosion rate matching that of the laser rod.

Figure 7.11 is the surface of rod and washer when fully polished, at the same magnification as figures 7.8, 7.9 and 7.10, traced with a stylus 2·5 μm radius. Stylus profile measurements of polished surfaces can be made down to less than 100 Å and, in fact, the chart shown here is of a finish commensurate with the optical quality of the material. No defects are visible in the finish, but turndown on the edge of the crystal can be seen clearly. This is, at least in part, due to too slack a fit originally between rod and washer. When the latter is viewed microscopically, deep but widely spaced pits can be seen. These are not indicated on the surface-profile chart because the wide spacing between the pits allowed the stylus a clear track across the polished glass. Figure 7.12 is an interferogram of a similar optically polished rod.

Quantities of identical soft crystal and glass rods have been polished in multi-hole blocks (figure 7.13). On the occasions when these have been used, matching washers have always been available and so the differential erosion rate technique has not been applied. The cementing operation must preserve the alignment of the washers which have been initially waxed together and trepanned as a pair; therefore, one is cemented on the block using a plug to preserve concentricity and the second aligned by a short rod fitting the centre holes with two ground-steel rods fitted to a pair of diametrical holes, preventing relative rotation. The assembly is supported between two ground-steel blocks on a surface plate while the epoxide cures. Approximate squareness is once again obtained by hand-lapping or 'Lapmastering' the block in a housing (figure 7.7).

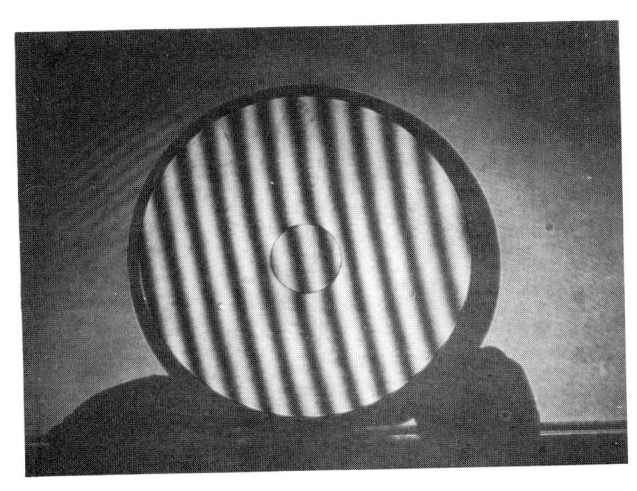

7.12 Interferogram of rod and washer produced by differential erosion rate polishing.

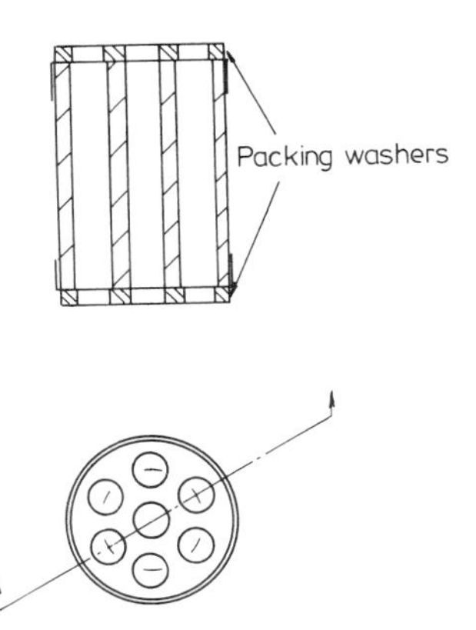

7.13 Block for lapping and polishing seven soft rods (for glass or soft crystal).

Both the hand-working and Lapmaster processes require that the relative rotational positions of block and housing be changed at approximately five second (time) intervals to give good results: about one minute of arc is not difficult to obtain. Accurate parallelism of the block facings is achieved by bias-lapping on a Draper-type machine with electromechanical measurement on the washer. The block is ready for the rods to be cemented in place when the washers are parallel to better than four seconds of arc (i.e. 0·6 μm total length variation measured on a 3 cm pitch-circle diameter). Once the rods are in place, they are worked with fine abrasive and 0·3 μm alumina as detailed previously.

It should be emphasised that these operations in highly skilled hands can be performed as quickly as the equivalent machine operations. However, unlike the hand, the machine is freer from temperature excursions, while such error patterns as are produced by worn laps and polishers are symmetrical and, therefore, predictable. They are easily removed at the later stages and do not lead to erroneous parallelism readings, as do the facet errors produced by some handworkers.

7.1.1.3 Confocal crystals

With the exception of the fit between the washer hole and the laser rod, the construction of a confocal block (figure 7.14) is very much less critical than its plane-parallel counterpart, since errors of a few minutes of arc in alignment of the ends can be tolerated. Confocal washers are available from the trade (Stanleys) in the diamond-ground state, with or without holes; if the former, the hole dimensions are specified well below the rod size and then subjected to the jigged-hole lapping processes described in §7.1.1.2. The hand preparation of confocal washers may be done with male and female cast-iron tools of the correct radius which have been purchased or prepared by milling or turning (§3.1.2). Plane-parallel washers are cemented to the block and the exposed faces roughed to the wanted curvature in cast-iron tools, typically with 220 carborundum and water. This stage which is intended only to give an approximately correct form, may be carried out on a flat lapping plate by working a family of facets on the end of the block, checking the shape with a metal template. Final trueing is carried out in an accurately turned spherical lap.

Mounting crystals in confocal blocks is basically similar to the plane-parallel preparation. Depending on the required curvature of the ends, varying amounts of protrusion are needed. In the case of a 5 cm radius

7.14 Lapping and polishing block for working confocal ends on rods.

of curvature, there is a minimum calculated protrusion of a 6·35 mm rod of 100 μm; but that supposes no contact at the rod's centre until its edges are lapped to the correct radius. In practice, at least twice the calculated protrusion is arranged and, since this results in a dangerously unsupported edge which receives preferential erosion, great care is needed initially to avoid chipping. In cases where the material is known to be brittle or friable, a chamfer should be carefully worked on the rod, before mounting it, by one of the methods described in §5.4.2.

Brass roughing tools which are easy to turn or mill have proved more economical than precise cast-iron ones. The latter can be kept for fine finishing operations, used only with BAO 303 and $303\frac{1}{2}$ abrasive. Under these conditions, they maintain their form well.

Pitch polishers, for working surfaces with 25 cm radius of curvature or greater, can be made on small cast-iron discs: about 6 mm thickness and 2·54 cm diameter for blocks between 2·85 and 3·125 cm diameter. Confocal polishing is carried out by reversing the polisher and work positions (see figure 2.35). Thus the block is held in a housing screwed to the machine's spindle and the polisher is swept through a small arc, held in contact with the work by a ball-point attachment.

The preparation of pitch facings has been outlined in §3.2.3. The problem lies in forming, netting and finishing convex tools. It is difficult to obtain the correct radius on the pitch surface by simple hand pressing. However, a correct curvature can be made by using a small press, or drilling-machine spindle, to apply pressure to the polisher for forming and netting. In order to remove any suggestion of tilt in this operation, a ball bearing is registered in the central hole at the rear of the polisher and pressure applied on the ball.

No attempt has been made to polish blocks containing numbers of confocal rods. Clearly, it is not possible to expose both ends of groups of rods since they would intersect at the block's centre. A precise radially bored block could be made to hold several rods, with only one end exposed for polishing at a time. The parallelism of the ends would depend, apart from the initial boring, on the accuracy of registration of the rods in their respective bores. In laser work there is generally no surplus material on the diameter of the cylinders, so the normal edging technique applied to lenses cannot be used to centre the body of the rods.

7.1.2 Adjustable-angle jig methods

Blocks are a common form of specimen holder in which the control of angle over a small range is effected by adjusting the point of application of lapping and polishing pressure. Though the cylinder with ends slightly off-square to its axis can have them corrected by biasing of the applied load, there is an obvious practical limit to the angular range of any fixed block for related-angle working. Wedges greater than a few minutes are not normally deliberately worked by this method. Moreover, because the attainment of a precise angle is by a dynamic process, varying continuously from too great to too little and vice versa, the change must be closely followed and held at the null point. Once orthogonality has been achieved on a block, on-axis pressure via the central dimple in the Tufnol cap should maintain it.

Adjustable-angle jigs, as described in §2.3, are continuously variable within a range of $\pm 3°$ and their method is to present the component at a required angle on the end of a sliding shaft rather than by altering the load application point. Usually, as with a laser rod, the angle is perpendicular to the conditioning-ring's face—the polishing plane—but even at this stage it should be remembered that optical measurement and setting of components worked in jigs can give precise small-angle wedges. Thus with a jig, three variables can be independently controlled: shape of the polisher via the jig's conditioning ring and thus the flatness of the work, the angle of the specimen and the pressure applied to it.

With suitable tailoring of the jig's workplate, laser crystals ranging in hardness from α-Al_2O_3 (9) to calcium fluoride (4) can be produced. Hard crystals need cementing in a workplate but do not need additional stabilising feet. For glasses and softer crystals, it is advisable to use the workplate as a single-ended block, complete with washer to minimise turndown and amenable to differential erosion rate polishing. There is the obvious point that the load control nut allows the polishing load on a single rod to be suitably low and, in practice, an unsupported 6 mm diameter soft crystal can be worked in this way; but the risk of edge-damage and increase in the area of turndown should be avoided. Jig-working of confocal crystals, though possible, has not been carried out.

The techniques as they are applied to the softer laser crystals are outlined in the following description of the making of a calcium tungstate rod:

(1) low-speed diamond-saw the crystal into square bars (figure 7.15) larger than the required diameter by about 1 mm (chapter 1).
(2) Either: (a) resin-encapsulate the bars in cylindrical PTFE moulds (figure 5.19) for centreless-grinding, or, (b) attach extension centres (figure 7.16) for centre-grinding, or, (c) hand-lap the crystal to rod form (§5.3).
(3) Centreless grind, centre grind or hand lap to the required diameter; then half-lap if necessary to minimise lobing errors.

7.15 Sawing laser material into square bars prior to grinding.

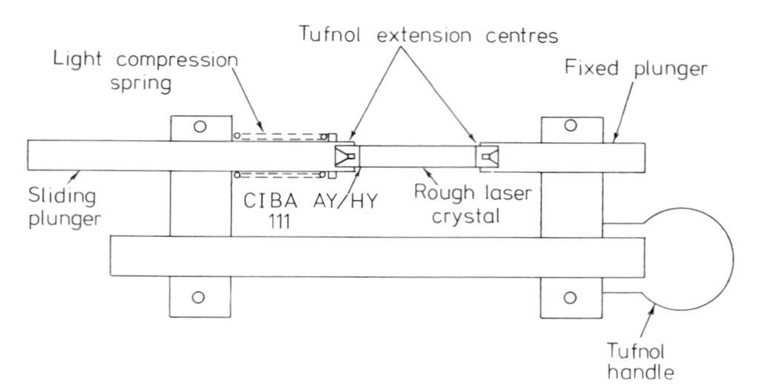

7.16 Extension centres and arrangement for centre-grinding bars.

7.17 Vee-block guide for temporarily polishing rod-ends.

7.18 Cementing a rod into an adjustable-angle jig-workplate.

(4) To align the first end, a temporary polish is required at 90° to the cylindrical axis of the rod; one face is lapped and polished, using an accurate vee-block to guide the rod during hand-working (figure 7.17). A barely specular finish is all that is required.

(5) Prepare a washer in calcium tungstate; if available, a calcium fluoride or soft glass washer can perform equally well. The washer-hole must be a close fit on the laser rod's outside diameter.

(6) Cement both washer and rod to a standard jig workplate with dental or brazil wax (with due regard to minimising thermal shock); the temporarily polished end facing upwards (figure 7.18).

(7) Fit the workplate assembly to the polishing jig, (figure 7.19).

(8) Check that the end of the rod and surrounding washer are level with the conditioning ring at the middle of the jig shaft's travel; lock the conditioning ring tightly.

(9) Retract the front face of the rod from the plane of the ring by screwing down the load control nut, thus removing it as far as possible from the lapping plate (figure 7.20).

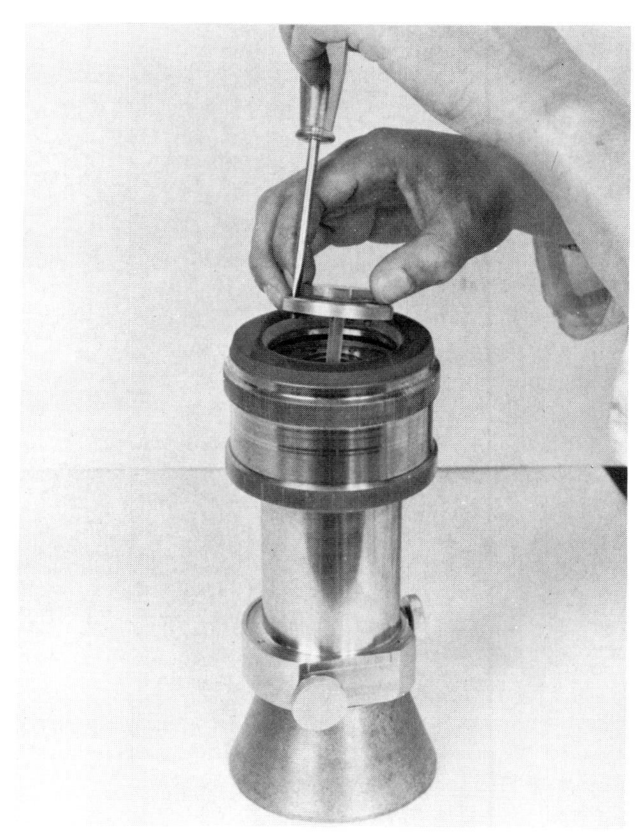

7.19 Fitting the workplate and rod in the jig.

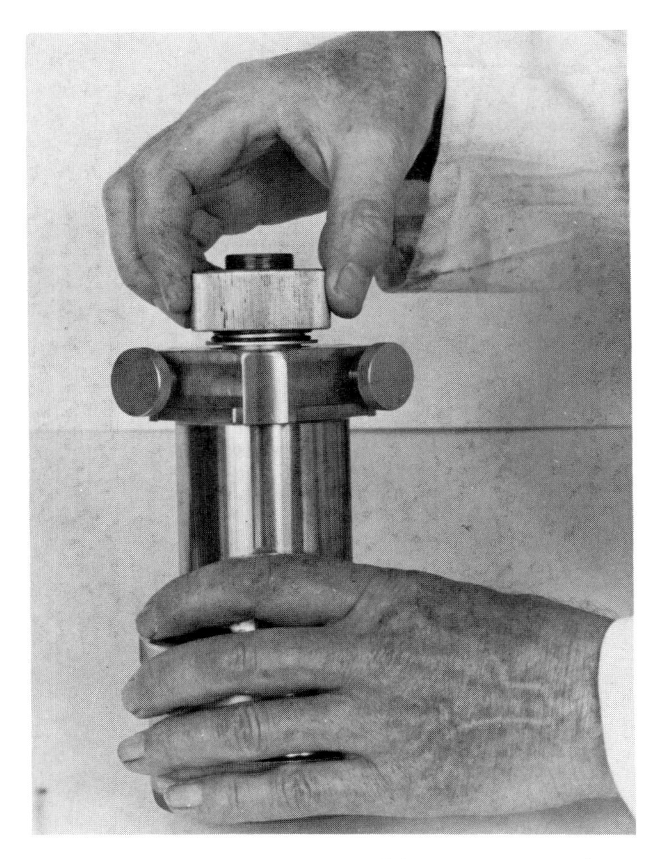

7.20 Load control nut of jig screwed down to retract specimen.

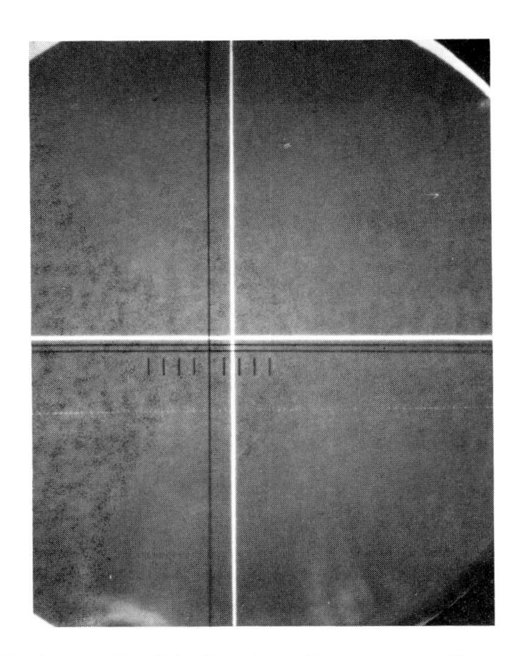

7.21 Image of polished work-end in an auto-collimator.

(10) Lap the conditioning-ring's face to about $\lambda/5_{(He)}$ and clean it carefully. *Note*: This stage should not be disregarded, however recently the jig has been used. Any adjustment of the conditioning ring and its consequent re-clamping usually alters its form and misleading readings can result when it is set up on a kinematic plane for auto-collimation.

(11) Stand the jig on a previously aligned kinematic plane, positioned so that an auto-collimator looks vertically down the hollow jig-shaft. The image of the auto-collimator's graticule (figure 7.21) reflected from the polished upper-end of the laser rod is then adjusted to replicate the arbitrarily set reading arranged initially on the master stainless steel flat.

(12) Return the jig to the lapping plate, with the work lowered into contact by unscrewing the load control nut. Lap-work the washer to an even grey surface, using 600 carborundum and oil. The pressure is arranged to be about $100\,g\,cm^{-2}$, measured with the jig vertical by a tension gauge applied to the workplate and pressed upwards until the shaft lifts (figure 7.22).

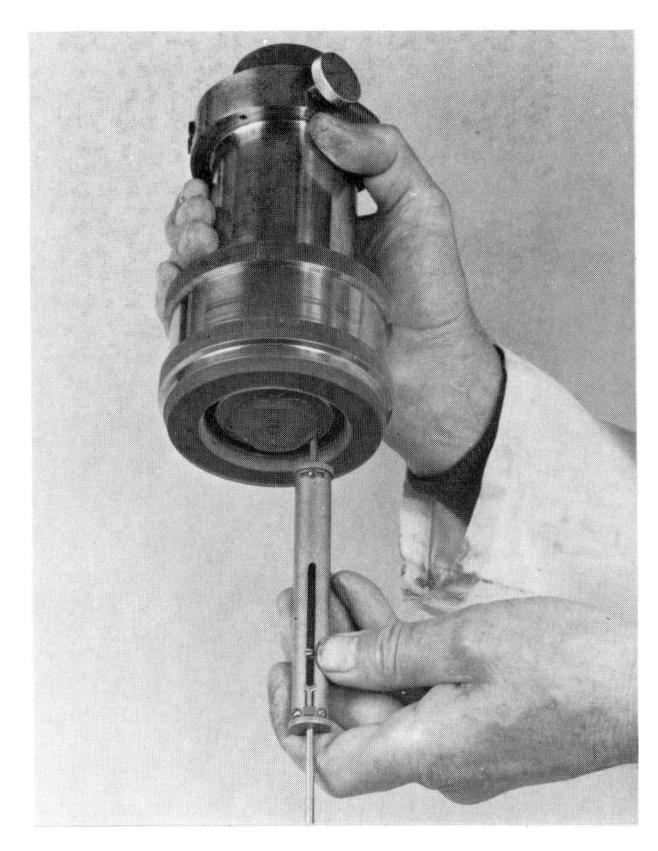

7.22 A method for measuring jig-shaft load.

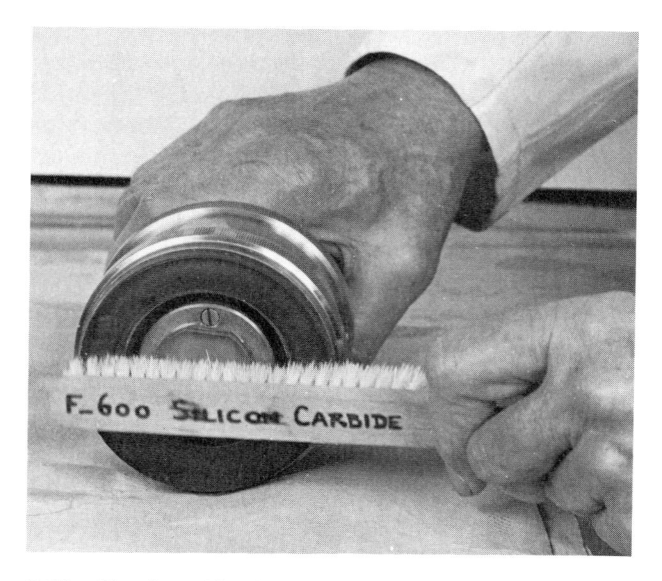

7.23 Cleaning with a labelled brush.

7.24 Tubular workplate.

(13) If a second lapping machine is available, clean the jig carefully (degrease and scrub with detergent/water and labelled brush—(figure 7.23)) then continue with 800 carborundum or 1900 alumina. This has the effect of reducing the subsequent polishing duration.

(14) Clean with detergent/water and labelled brush; check collimation then polish. If a closely matching washer/rod combination has been used, wax or pitch polishers with Linde A in water or ethane diol

7.25 Interferogram of a soft (CaF_2) rod in a jig.

are satisfactory. The standard spiral grooving is used on both types of polishing matrix and a pumped, recirculating slurry feed is an advantage. If a non-matching glass washer has been cemented on the workplate, the differential erosion rate technique may be adopted (see §7.1.1.2). A tubular form of workplate is best (figure 7.24) so that the rod can be slid forward slightly through the washer without losing alignment. Polishing can be carried out initially on a soft-metal matrix such as solder or tin with 3 μm diamond; then a change to alkaline silica sol (Rabinowicz, 1968) recirculated at about $2\,l\,min^{-1}$, on polyurethane foam. The load is $100\,g\,cm^{-2}$ for about three hours for the complete process.

(15) The polished face is inspected for defects and flatness (figure 7.25) and, if acceptable:

(16) the rod is reversed in the workplate and the processes repeated from (7), omitting (8) if the ring is not moved. Slight adjustment to the angular setting of the jig may be needed during the later stages of polishing; drifts of one minute of arc on soft crystals are easily removed by polishing.

(17) Remove the finished rod, clean and chamfer ends with a steel or glass hollow cone and 600/800 carborundum.

(18) Box, label and record.

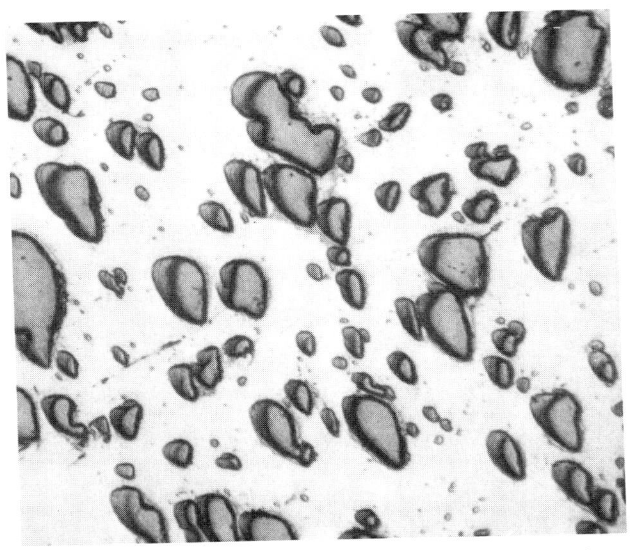

7.26 Pitting of stainless steel conditioning-ring's surface: yttrium lithium fluoride rod and Linde A/water slurry.

Jig conditioning rings are usually faced with replaceable glass washers, though for many applications stainless steel facings are equally serviceable. However, yttrium lithium fluoride (YLF) rods, worked in stainless steel-faced rings react in polishing slurries such as Linde A in water and attack the metal, producing deep pits (figure 7.26). The surface oxides which make chrome/nickel steels stainless are constantly eroded away during polishing and become vulnerable. Two remedies are apparent:

(i) discover the eroding mechanism (intergalvanic corrosion, acid, halide, etc) and circumvent it;
(ii) use glass rather than stainless steel for ring-facings.

The first is, of course, scientifically satisfying and may yet be carried out; but the second, lateral thinking, solution has been preferred.

7.2 Miniature Glass Prisms

Some of the optical systems designed for laser instrumentations have required small prisms with unusual angles, though with loose tolerances allowed on the linear and angular dimensions. The prism used as the example (figure 7.27) has six faces: (*a*) two parallel to each other, finish as sawn; (*b*) two plane-parallel and polished; (*c*) one at 45° to (*b*), polished; and, opposing it, one at 67·5° and polished. The only modification needed for their production on the equipment used for working electronic material involves machining an angled-flat on a polishing jig's work-plate—in this case, one at 22·5° caters for both of the off-square ends.

7.2.1 Sawing

A small block of crown glass, large enough to provide four or more prisms, is cemented to a glass platform and this assembly then mounted on a worm driven rotary dividing table (see § 1.6.3) which forms part of the standard equipment of a low-speed diamond saw. The long axis of the block is arranged to be vertical, thus these first saw cuts will define (*b*) and (*c*) faces that are large enough to provide at least four prisms after the (*a*) cuts will have been made. Dimensional stability is important, therefore both mounting operations require a strong cement such as shellac based Tan wax. The table is clamped in position so that its rotational axis is at 90° to the axis of the saw shaft, and its friction-adjustment tightened to prevent the alignment changing during sawing.

An electrometallic disc, 100 mm diameter ×0·5 mm thick, with 220 grit diamond abrasive held in place by a nickel-bond, is used to saw slits parallel to each other and 5 mm apart, preferably with the degree scale set at zero so that the other angles may be read-off directly.

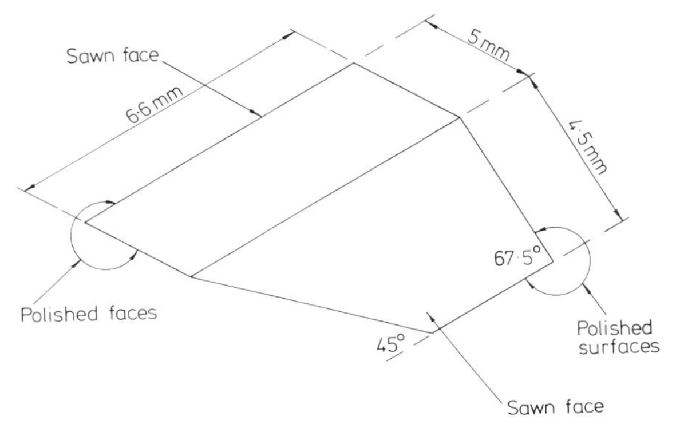

7.27 Diagram of a small glass prism.

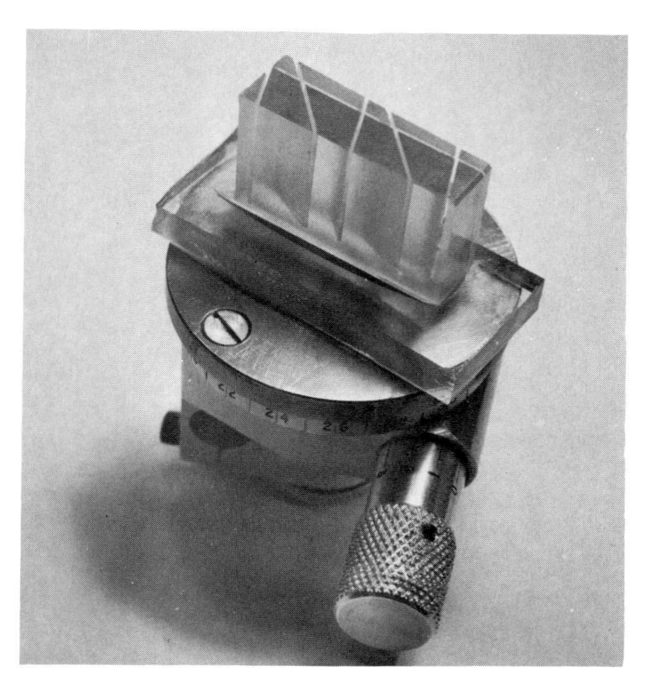

7.28 Sticks of prisms.

These two cuts must be continued into the glass mounting platform to isolate the glass plate. The table is indexed to give 45° and a saw cut made which stops short of entering the platform by some few millimetres; the operation is repeated at 67·5° to the parallel sides, again without sawing through into the glass mounting platform. As many angled cuts may be made as there is material available, and the result will be several columns (or sticks) of potential prisms, the relationship between their parallel sides solidly retained by the backbone of glass uncut on the platform (figure 7.28). This enables all the plane-parallel faces to be polished in two operations by using an adjustable-angle jig (§2.3). Finally, the work is removed from the sawing platform and cleaned in a solvent such as warm methylated spirits.

7.2.2 Lapping and polishing: parallel faces

In order to polish the first (*b*) surface, the array of prisms is mounted on a flat jig-workplate using a soft paraffin wax, and filling the sawn slots in the process. This is contrary to the usual practice of cleaning out potential traps for abrasive particles, but the depth of slot precludes complete wax-removal. A DTI and ring base mount (see figure 6.8) is used in conjunction with the jig's tilt controls to adjust the exposed sawn faces of the prisms so that they become co-planar with the conditioning-ring face.

Lapping is ideally carried out in two stages: initially on a machine with 600 carborundum in oil, then transferred to a machine using 1900 grade alumina powder. All the work can be carried out manually on two separate cast iron or glass plates, suitably lapped flat, and fed with the same abrasives in water. However, small surfaces already sawn flat to within 1 μm cm^{-1} often make it hardly worthwhile either to use a machine or to set up lapping plates for a process lasting about two minutes. There is the added advantage that no abrasives can be trapped in the wax prior to polishing. Since small surfaces can tolerate only the very smallest amount of turndown, a fairly hard polisher is needed Several polishing matrices/abrasives are indicated, such as soft metals (§3.2.2) with 3 μm diamond, polyurethane rigid foam (§3.2.6) with alkaline silica sols, or a conventional mixture of pitch, resin, beeswax and woodflour (§3.2.3), cast on a stainless steel plate and used with cerium oxide. The latter combination has been preferred, the process carried out on a 'Multipol' polishing machine (§2.2.4), modified to provide a pump recirculated slurry of cerium oxide, water and YMS (Youngers Miracle Suspender—Autoflow, Rugby) fluid.

It is advisable to use a Perspex cover over the machine in order to maintain a humid environment. Spiral grooving of the polisher (whichever type might be used) must be both shallow and closely spaced: the surfacing and scrolling machine (§3.3) can be adjusted to give a very fine scroll. The machine is run in the ring-polishing mode (i.e. the jig is free to rotate within the roller-bar) with one centimetre of sweep.

The polisher is run-in to an acceptable state of flatness—for example, with a Minijig (§2.3.4)—then the adjustable-angle jig and work is thoroughly cleaned and placed on it. Rotational speed is initially 50–70 rpm to polish most of the greyed surface away, then the speed reduced to 5–10 rpm to give the maximum edge definition. When the surface is flat and specular enough, the workplate is removed from the jig and the work demounted and replaced, polished face downwards, on a waxed tissue. A central portion must be clean and visible through a hole in the tissue and one in the workplate, for alignment purposes. The assembly is remounted in the jig and, assuming its conditioning ring to be flat enough, the tilt controls are used with an auto-collimator and kinematic plane to bring the rear face parallel to the polishing plane of the ring (see §§6.1.3.1 and 6.1.4). Polishing the second (*b*) face of the array of prisms follows the same procedure as before.

7.2.3 Sawing prisms to length

The array of prisms is removed from the workplate by warming gently to melt the wax and transferred while still warm to a saw table. A piece of tissue is interposed between the prisms and a glass platform which is sufficiently thick to allow the saw disc to plunge cut completely through the work and into the plate without entering the worktable. Additional wax is added to the array so that the cuts made previously are filled up, thus minimising chipping which will occur at the intersections. The reference ends of the prisms are aligned with the sawing plane, either by dial gauging from the saw shaft (figure 1.14), or more simply, by pressing the reference face firmly against the saw cheek while the wax is warm and flexible. Cuts can now be made to divide the sticks of prism-sections into 5 mm lengths—the indexing of the slide will be 5 mm plus the kerf loss. Each prism should be separated, cleaned and dried.

7.2.4 Lapping and polishing the 45° and 67·5° faces

Although general optical techniques cater routinely for polishing prism faces, it is convenient to use standard crystal working equipment and practices for small numbers of components. Since the range of tilt on a jig's central shaft is about 3°, a special plate has to be made with an angled-flat machined upon it. An existing brass workplate, 8 mm in thickness and polished on the upwards-facing side—for auto-collimation purposes—has a flat platform machined on it 22·5° to the polished face (figure 7.29) large enough for a small group of prisms to be cemented to it. In order to preserve reasonable orthogonality, the sawn edges of the prisms must be aligned with the machined straight-edge on the flat. The platform is coated with a thin layer of beeswax/resin (§4.11) in which the prisms are embedded so that the faces presented for lapping and polishing are parallel to the rear side of the workplate. At a temperature at which the wax is still flexible, the assembly is placed under a low-power binocular microscope.

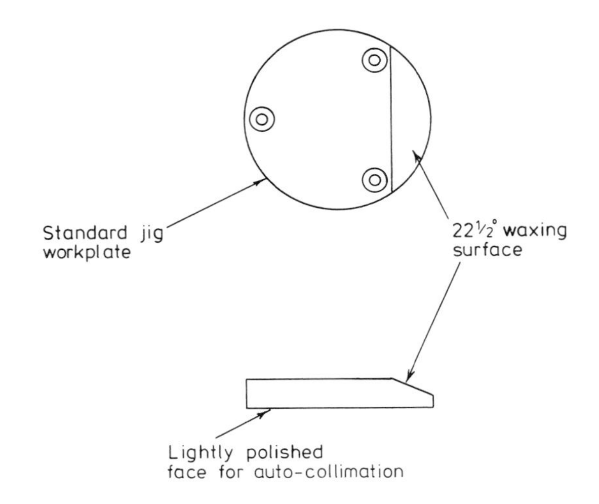

7.29 Adjustable-angle jig-workplate for $22\frac{1}{2}°$ prism.

Surplus wax is brushed away from the alignment edge with a stiff brush and methylated spirits and the prisms eased into position with a wooden rod. If the angled-platform is arranged to be horizontal at the time of this manipulation, the work should not drift in position as it cools. The processes of alignment in the jig, lapping and polishing are repeats of those outlined for the plane-parallel faces.

Mounting, lapping and polishing the second non-parallel face again makes use of the angled platform. By cementing down the recently polished surfaces of the prism and suitably orienting and aligning their edges with the reference straight-edge as before, the second (*c*) faces become parallel to the rear of the jig-workplate. Thus, when the workplate is assembled in the jig and the polished rear-face auto-collimated, the sawn faces of the group of prisms will be closely co-planar with the polishing plane of the ring. After a light lapping operation and the full polishing process, the prisms are finally demounted and cleaned.

Chamfering the edges of the prisms helps to reduce accidental damage, and the least demanding method is that of using 600 carborundum in detergent/water on microscope slides (see § 5.4.1). Rubbing small work on a lapping plate at an inclined angle is a fairly skilful operation which is preferably reserved for larger, less sensitive components.

7.3 Surface Acoustic Wave (SAW) Plates

The preparation of flat, polished, damage-free surfaces on SAW plates is necessary both for device performance and for very exacting photolithographic processing (Smith, 1974; Rich, 1976). Crystal quartz, usually hydrothermally grown, is the material which currently provides the greater number of plates. Though lithium niobate is worked in fairly large quantities, its processing is similar to quartz and the techniques for polishing the latter may be taken to be equally suitable unless otherwise noted.

7.3.1 Cutting

Figures 7.30 and 7.31 identify the faces and planes that are referred to below. On most of the large quartz crystals so far obtained, the major R and minor r (rhombohedral) faces have been very flat and closely angularly related as grown. A precise Z face is established from these by lightly lapping it to coincide with the major R faces at the theoretical angle of 51° 47″, measured by clinometer (see §6.4.2). A temperature invariant face—the ST (stable temperature) face—occurs at 42° 45′ to the Z, and the cutting of ST plates is achieved by setting Z at this angle to the sawing plane. This setting is made simpler by initially adjusting the sawing plane of the machine to be vertical: by testing a sawn plane surface with a clinometer, for example, whilst packing-up the machine. None of our existing standard equipment is large enough to cut a 7 kg crystal, therefore a modification to a reciprocating-blade saw has been made and allows an area of about 225 cm^2 to be produced at one setting. The reciprocator motion is utilised to drive a large arm carrying one or more tensioned blades, while the crystal, cemented to a large plate and suitably bolstered with glass packing, can be tilted and rotated for alignment purposes, (figure 7.32).

Worn power-hacksaw blades provide a cheap source of high speed steel strip possessing sufficient rigidity to make the usual tensioning procedure unnecessary. They can be used without additional treatment, but slots ground in them at intervals (figure 7.33) serve to retain abrasive particles and thus make them faster in cutting. The abrasive slurry pumped to the cutting region is 120 carborundum in oil.

A slab is taken from the crystal to give an ST face, then remounted and sawn into bars, all with Z edges and ST faces. These bars are again mounted on glass bolsters and cut into slices, usually on the modified reciprocating-blade Norton 262 saw itself (see § 1.4). A DTI is mounted across the slideways and the crystal face aligned with the blade pack (figure 7.34). It will be noted that the worktable is pivoted above the blade pack (in the manner of the Abraslice saw) so that the specimens descend through the blades. This reduction of mass is very important with the more delicate materials (see §1.4.1) but its advantage on quartz and lithium niobate is in the ease of setting-up and

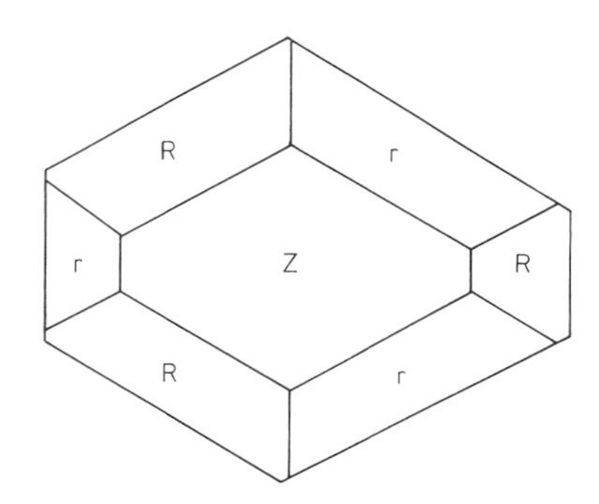

7.30 Surface acoustic wave: quartz crystal with R (major rhombohedral faces) and r (minor rhombohedral faces).

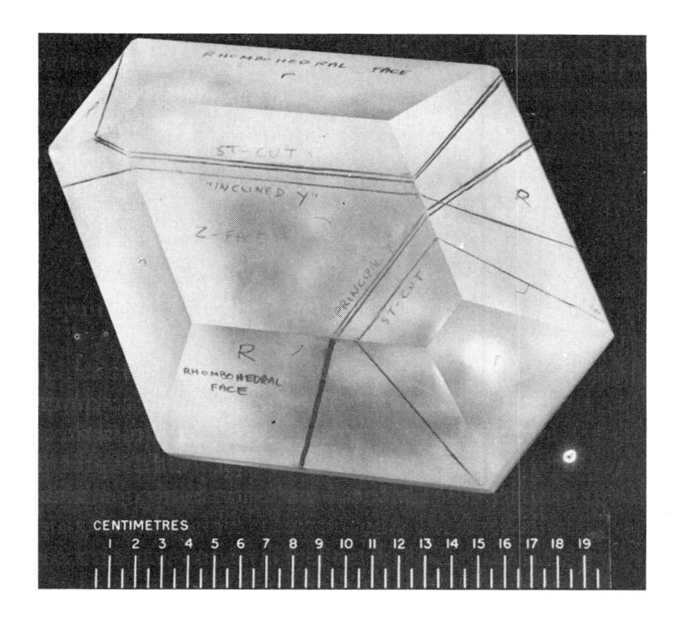

7.31 Quartz crystal with ST and Z drawn on prior to sawing.

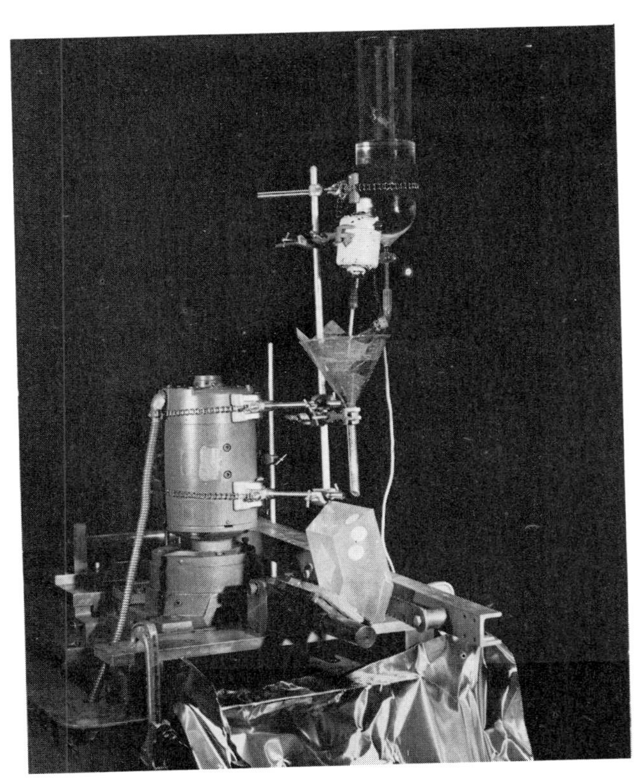

7.32 Sawing SAW plates on modified Norton 262 Saw.

general accessibility, since both materials saw readily on the unmodified machine. A copious supply of abrasive slurry ($\sim 10\,\mathrm{l\,min^{-1}}$) is desirable and in our experience greatly increases the life of a blade pack.

Sawn plates have to be trimmed to the required length and width: the reference Z face must be retained, thus each strip cut from a plate must carry two accurate Z edges. Low-speed diamond saws (§1.2) are used for trimming, since the number and variety of shapes and sizes required in the experimental and pilot production programmes have been about a thousand plates per annum.

Many electrometallic annular blades have been very short-lived when used for cutting crystal quartz; however, peripheral cutting is the most convenient method for SAW plate trimming and electrometallic peripheral wheels are extremely durable. The commonest size of disc in use is 100 mm diameter × 0·5 mm thick, with a 12 mm diameter hole; made more rigid with cheeks usually 75 mm outside diameter. Discs used for sawing between finished devices are 75 mm diameter × 0·225 mm thick, with 50 mm diameter cheeks. All sawing operations require some lubricant —for example, any standard lapping oil such as Paynes No. 3 or Abralap No. 2A from Payne Products and Lawrence Chemical Ltd respectively. Water-soluble lubricants, like those available in machine shops, are satisfactory. Speeds and loads are detailed in table 1.2.

The sawn pieces are mounted, singly or in packs, on a glass platform which is in turn cemented to a saw-table. When packs of plates are cut, they are mounted with their Z edges in contact with the vertical face of a glass angle-plate, comprised of two strips of 6 mm glass cemented together with an epoxy resin to form a right angle. The use of a thermosetting

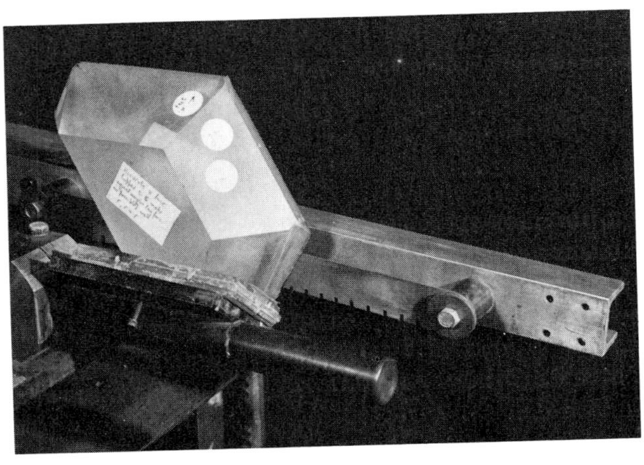

7.33 Large slotted blade used on reciprocating-blade saw.

7.34 Dial test indicator used for realigning crystal on 262 saw.

7.35 Dial test indicator used for re-establishing Z or ST edges on sawn plates prior to trimming the ends.

7.36 Rotable twin-toom worktable on low-speed saw for cutting angled ends.

resin for bonding the glass-angle enables the rest of the mounting to be carried out with a thermoplastic wax, leaving the glass-angle joint unaffected. Alignment of the Z edge, or edges, for trimming is carried out by attaching a dial gauge to the saw shaft, its stylus contacting an edge (figure 7.35). The worktable adjustments (see §1.6.3) are used to position the plate accurately. Plates can be cut to length using a similar procedure. Any specified angle for the ends can be obtained by rotating a worktable of the twin-boom type (figure 7.36). Alternatively, an angle can be set up by using a template or a vernier protractor with the saw-supporting cheek as the reference for the the sawing plane.

7.3.2 Lapping

Profiled SAW plates are mounted with beeswax/resin on stainless steel workplates (blocks) previously lapped flat on the 60 cm Lapmaster machine (§2.1). For easy handling, these blocks should be as thin as possible, consistent with stability, and their engagement with the roller-bar on the recirculating chemech polishing machines (§2.2.2). The weight of a solid stainless block, 9 cm diameter and 1·5 cm thick (0·75 kg), will be insufficient to load a well covered surface with plates; therefore additional thicker blocks of the same diameter are added to give about 100 g cm^{-2} on the work. Double-sided pressure-sensitive adhesive tape can be used to hold the extra block(s) in position. Alternatively, a hole can be drilled centrally in each block and a stub used to couple them together. However, any hole, whether plain or threaded, is a trap for abrasive particles and must be cleaned out meticulously before the transfer to a polishing machine.

The plates must be distributed to present an even, simulated ring or disc, lapping and polishing situation (e.g. figure 2.41, and figure 4.6). Glass wasters, if needed to fill up the interspaces when only a small area of SAW plate is available, can be old photographic plates or microscope slides. They should be as thick as, or slightly thicker than, the SAW plates and checked with a dial gauge after mounting, as shown in figure 6.5. Any appreciable variations in the thicknesses could affect the orientation of the plates if they erode unevenly. The standard conditions on either a 30 cm or 60 cm Lapmaster are adhered to: pressure about 100 g cm^{-2} and 600 carborundum suspended in Lapmaster oil No. 3. When large quantities of plates are being lapped, the workplates should be equivalent to, or slightly larger in area than, the conditioning rings which they replace. The SAW plates can then be arranged to equal a broad ring face or a complete disc with small gaps between the work. In this way, more plates can be worked than would be possible if the blocks ran within the conditioning rings.

Accurately sawn plates need lapping for a few minutes only to make them co-planar and suitable for polishing. It is difficult sometimes to assess the overall uniformity of the greyed surface since abrasive-slurry sawn plates and lapped finishes vary little in appearance. A short duration run on a polishing machine will result in sufficient specularity to produce fringes with an optical flat and monochromatic light source (§6.5.1); if these continue to the extremities of all the plates, the lapping has been adequate. The disadvantage of this method is the cleaning needed before polishing, which may have to be repeated if re-lapping is required. Sometimes, with high-quality lapped surfaces, fringes can be seen directly by holding an optical flat in place and viewing obliquely, and the efficiency of the lapping process established. Oblique incidence interferometry provides an excellent instrumental method for determining grey surfaces (Birch, 1973).

Either before lapping or, more usually, before polishing, all surplus wax must be cleaned away from the mounting block surface and the crystal-edges with a plastic or Tufnol scraper (figure 7.37). The wax becomes contaminated with abrasive particles which would be almost certainly released into the polishing slurry. Most polishing fluids are able to erode or dissolve cements used for mounting, affecting the figure of many less robust specimens (see figure 4.12). Those polishing processes where the slurry is pump-recirculated seem to be more tolerant of contamination from earlier working, and even appear to be less sensitive to ragged edges on plates. It may be that large abrasive particles and agglomerations fall or settle in unstirred places in the slurry tank, possibly retained by an abrasive like cerium oxide or α-alumina which has sedimented—which, in its turn, might be a favourable result of a pumping system not sufficiently powerful to keep all abrasive particles in suspension. With alkaline silica sols, particles of abrasives may settle but are the least likely to be entrapped since there is little or no sediment. Nevertheless, the process seems to be the least sensitive to contamination.

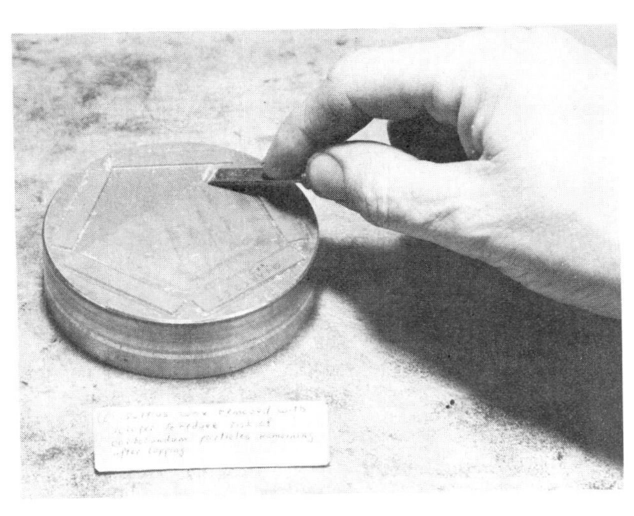

7.37 Tufnol scraper in use for removing superfluous wax on lapping and polishing blocks.

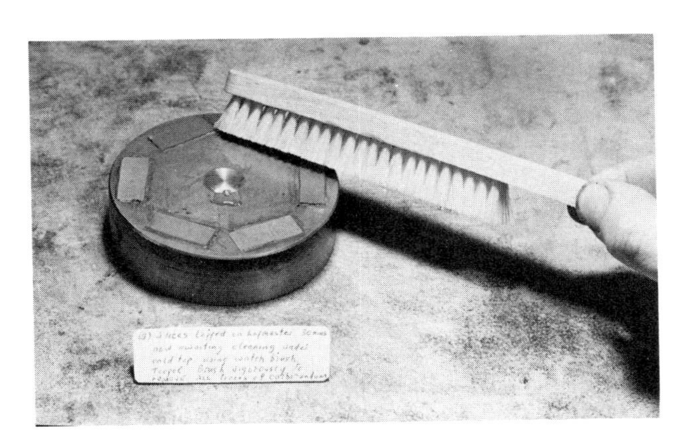

7.38 Cleaning block face with a labelled brush.

All cleaning operations before polishing must be carried out as nearly as possible at room temperature. The hand-worker traditionally cleans with a small sponge and a bowl of water at ambient temperature—any abrasive particles fall to the bottom and stay there. Liquid detergent and water applied with a soft brush to all faces of the block and work (figure 7.38), followed by thorough rinsing, is a fairly satisfactory pre-polishing clean-up. The process has been made easier by designing the work blocks to be of the simplest possible form: that is, flat-ended solid cylinders of stainless steel and borosilicate glass. Relatively inaccessible contamination traps, such as threaded holes, should be avoided if possible.

7.3.3 Polishing

The simplest, but somewhat slow, polishing technique is to use a spirally grooved polyurethane pad cemented to a stainless steel backing plate (see §3.2.6) mounted on a recirculating chemech polishing machine (§2.2.2).

An alkaline silica sol slurry—a colloidal suspension of silicon dioxide in a basic electrolyte, pH $\sim 10{\cdot}5$—is recirculated at about $2\,\mathrm{l\,min^{-1}}$: $600\,\mathrm{cm^3}$ of the fluid lasts for about 150 h, running continually. Any pause in the process requires that the polishing slurry be pumped out of the machine and put into a closed container, while large quantities of water are circulated through the pump and over the polisher. Failure to do this results in a hard white silicate coating, forming over the machine's parts and the polisher, which is very difficult to remove.

Cleaned, suitably loaded blocks of work are placed on the polyurethane rigid foam pad and slid against the rollers which are set so that the work overlaps both the central recess and the periphery of the polisher. Three similar sets of workblocks or two blocks and a jig can be run on the one machine (figure 7.39). Starting from a grey, 600 carborundum surface, 24 h of polishing produces a very deeply specular finish. If the aim is to produce optical surfaces as distinct from electronic perfection, a much shorter time will suffice. In a few hours all visible evidence of the grey surface (examined under a microscope with oblique incident illumination, §6.7) will have been removed.

One of the snags encountered with alkaline silica sol polishing is the glassy deposit that forms on and in the polishing pad with an attendant reduction in material removal rate. Scraping the polisher every hour with a razor blade while the machine is running (figure 7.40) removes the glaze and restores the polishing rate. At much longer intervals the spiral groove needs to be cleaned with a pointed instrument held in the groove while the polisher is rotating (figure 7.41). In this way pads have lasted for more than twelve months, running continuously.

Special running-down procedures are needed to remove traces of the sol from SAW plates, somewhat

7.40 Scraping the rigid polyurethane foam (polyfoam) polisher with a razor blade to promote speedier polishing.

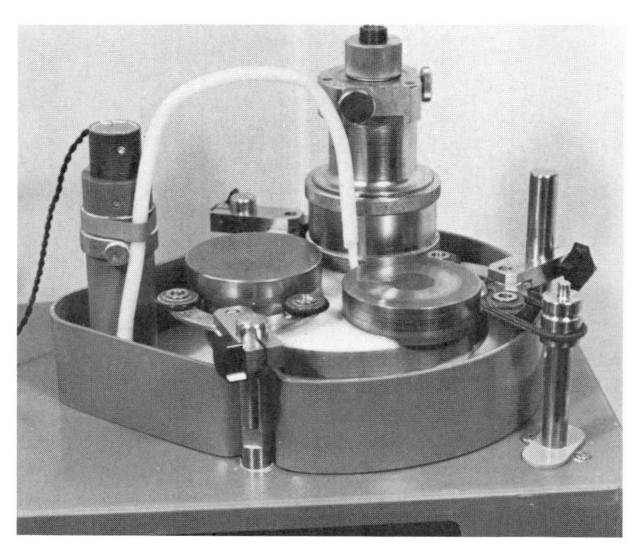

7.39 Three blocks on recirculating chem-mech polishing machine (RCMP)

7.41 Using a scriber to clean-out the groove on a polyfoam polisher.

similar to those used for silicon. A Microcloth polisher (§3.2.7) glued to a stainless steel plate is used during the first 20 min while continuing to circulate the same slurry. Owing to the large increase in torque (drag) associated with this type of polishing cloth, it is necessary to reduce the polishing load on the work to about 40 g cm^{-2}. This is an additional reason for having relatively lightweight workblocks, since the additional blocks required for the earlier weighting can be removed for the run-down process. During the following 40 min, the sol is pumped from the machine and stored. Six or seven complete changes of distilled water are fed through and run to waste. The appearance of the last one should be quite clear. The blocks are finally removed from the machine, rinsed with distilled water and blown dry with nitrogen at about 25 psi.

All the usual flatness tests are made on the polisher shape during the run by observing the work figure at intervals, with an optical flat and monochromatic light source, §6.5.1. Checks every 24 h are all that are needed when conditions of reasonable stability prevail. It is a fortunate characteristic of rigid polyurethane foam pads that they retain their shape for long periods. Of course, it is equally true that if they are allowed to drift far in one direction, either concave or convex, it takes many hours of compensatory running to establish flat working again. Convex work can always be corrected more readily than concave, since a concave polisher is remedied by repositioning the running centre nearer to the outer edge of the pad where the peripheral speed is greatest. Though flatness is not usually very important on SAW plates generally (since 0·1 cm thick material several square centimetres in area cannot maintain precise flatness when demounted) both laps and polishers should be kept true in form. This is because polishing should occur evenly over the entire surface on transfer from the lapping machine, thus ensuring that the depth of polishing, below the last of the coarse abrasive damage, is uniform. A great amount of time is wasted in putting concave or convex work on a flat polisher, or vice versa.

A further reason for keeping the polisher reasonably flat, beyond the requirements of the devices, is that work differing in figure from the pad is often reluctant to rotate. Positive drive to a roller has been incorporated on later machines (see §7.9) and can remedy this situation.

If work suddenly shows unexpected convexity and non-spherical features, totally at variance with previous performance, it can be due to the slurry's forcing its way under the edge of the pad, or the local production of a bubble. The loose polyurethane foam which is difficult to spot, flaps up and down slightly as the work rides over it, producing the unusual figure. Stripping the pad away, re-lapping the base and recementing the pad (or its complete renewal) are necessary. It must be emphasised that overall adhesive contact, by the method detailed in §3.2.6, is essential.

Although the greater part of the surface acoustic wave plate-polishing programme so far has been carried out by the methods described, pitch polishing—with a final stage of alkaline silica sol and Microcloth run-down—is faster. The pumped slurry pitch polisher (PSPP: see §2.2.2) with a spiral-grooved matrix and a recirculated feed of cerium oxide has reduced the overall time to approximately one quarter. Figure 7.42 shows the effect of some loads and slurry/polisher combinations. The only

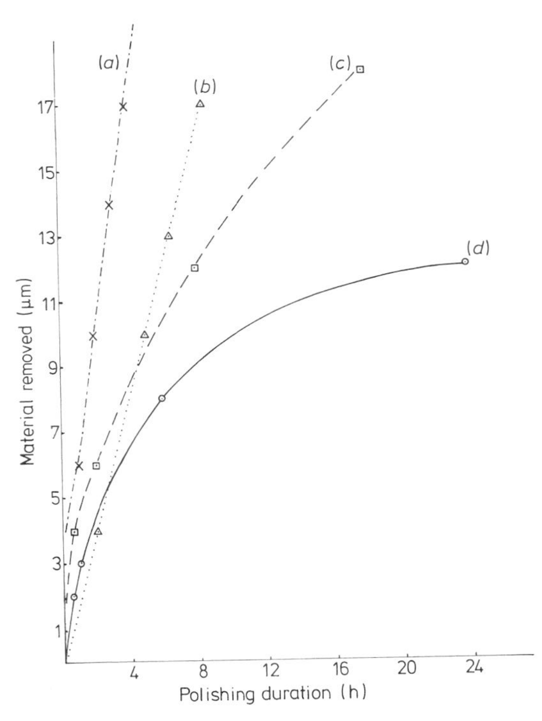

7.42 Graph of material removal rate against time for different pressures and polisher/abrasive slurry combinations. (*a*) Cerium oxide slurry/pitch (200 g cm^{-2}). (*b*) Cerium oxide slurry/pitch (100 g cm^{-2}). (*c*) Alkaline slilica sol/polyutethane foam pad (200 g cm^{-2}). (*d*) Alkaline silica sol/polyurethane foam pad (100 g cm^{-2}). General data: three station polishing machine; polishers, 20 cm spiral grooved; rotational speed, 60 rpm; slurry constantly recirculated at 2 l min^{-1}.

snag with pitch polishers is that they require skill for their preparation and maintenance. However, a polisher surfacing machine (§3.3) or precision lathe can de-skill the process, although not to the same extent as cementing rigid polyurethane foam to a stainless steel plate.

A warning should be given that polishing on soft cloths (such as Microcloth) directly from the grey surface results in turned-down edges and a rippled texture on the work—known as an orange peel finish. Though wax or pitch impregnated fabrics may be used to correct these faults, they have to be worn into shape rather than being formed or turned.

Working lithium niobate is similar to working quartz: it is slightly softer and works faster, therefore a block should not hold mixed plates, but the materials are compatible on the same machine. Plates produced both by reciprocating-blade sawing and on the automated low-speed annular Microslice III have been polished from the as-sawn surfaces without intermediate lapping. It is more susceptible to thermal shock, therefore mounting and demounting operations are carried out with gradual changes in applied heating and cooling.

The advantage of alkaline silica sol slurries, typified by Syton W.30 fluid (John Morrison Instruments Ltd) are its longevity, comparatively mild chemical properties and suitability with many electronic materials. Its chief disadvantage is its persistence, and the run-down method described is successful—on the harder materials, at least. Stabilisation of colloidal particles of silicon dioxide is maintained by a basic electrolyte. If the pH value is reduced below about 7, or the solution dries in air, the resulting solid is extremely durable. One of the reasons for a PVC casing for the polishing machine is that it allows slight flexing to crack-off poorly adhering silicates. Less obvious is the sol that becomes attached to the surface if normal tap water (usually slightly acidic) is used as a rinse: though the solid is probably only physically absorbed it is hard to remove it without damage to the specimen. One usually successful method for its removal from delicate electronic materials (§7.8) is equally applicable to SAW plates. The blocks are removed from the polyurethane foam pad and immediately dipped in a solution of 0·01 M NaOH, carried *in situ* to a fume cupboard then sprayed-off with the same solution for one minute. This is followed by an intense spray with distilled water, then a nitrogen blow-dry. If traces still persist, the block must be repolished for a minute or so and the cleaning process repeated.

7.4 Lithium Fluoride

The transmission of lithium fluoride (LiF) in the ultraviolet region of the electromagnetic spectrum makes it a suitable window material for space research and other UV applications. Surfaces obtained by cleaving thick plates from bulk crystals can result in an optimum transmission of about 62% (*Lα* 121·6 nm), but it is difficult to produce closely toleranced components by this method. LiF is a soft material, Mohs's value about 2·5, easily damaged during polishing processes, and the following technique has been successful in producing optical-quality windows with about 57% transmission (Fynn and Powell, 1970). It has long been recognised that wax polishers (§3.2.4) give excellent surfaces on glass and many crystals but that the polishing process is appreciably lengthened. Their relatively narrow-band melting range compared with that of pitch does not allow them to conform readily to a heated flat surface as pitch does (§3.2.3.1); painting pitch polishers with wax is a technique aimed at circumventing this difficulty. The surfacing and scrolling machine (§3.3) has, however, made the preparation of accurately flat wax polishers a rapid, routine process. In the techniques described here, pitch polishers are used to speed-up the intial removal of material, and pure wax matrices are used for final polishing. This extended polishing process is applicable to a wide range of soft optical materials.

For this work, the sweep on the polishing machine is restricted to about 7 mm. Sweep prolongs the life of a polisher even when the favourable effect of a conditioning ring has been utilised. For specimens mounted on a block of 75 mm diameter, a 120 mm diameter polisher is used and polishing carried out within a conditioning ring of 76 mm internal diameter. Block and ring are run continuously on one side of the polisher with a typical mean-spacing of 35 mm between work and polisher centres.

Initial flatness testing on the machined polishers can be done by using a lapped (grey) glass-faced conditioning ring. The grey surface is given a polish sufficiently specular to show fringes by running it on the pitch or wax matrix using a polishing grade of alumina abrasive or cerium oxide. It will be immediately evident if it is either concave or convex, and the compensating adjustment may be made (see §6.5) using the general rule that if the conditioning ring shows convexity its running centre must be moved out. Conversely, if it is concave, it must be moved in. On pitch polishers, rapid corrections can be made, but on wax correction is slow since the polishing rate itself is slower.

The general technique for polishing LiF is to avoid, if possible, greying-off the surfaces since the micro-cracks are thought to reduce the ultimate transmission of the polished specimen (Makarov and Novikov, 1967). Because the material can be cleaved readily near to the require thickness, the final surfaces can be polished directly from the cleaved faces.

Since wax polishing is slow, all the irregularities of a cleaved face are removed by some four hours of pitch polishing with 0·3 μm alumina abrasive and water (5% Na_2CO_3 added) as a lubricant. This leaves the surface covered with fine sleeks which are then removed by polishing on a wax polisher for about 3 h, with 0·05 μm γ-alumina (Linde B). Latterly, a drip-feed of water plus 5% sodium carbonate without an abrasive is used, since lithium fluoride is sparingly soluble in water 0·27 g LiF/100 g H_2O) giving a slightly acidic solution.

A single specimen is normally polished in an adjustable-angle jig (§2.3) with a typical polishing pressure of about 100 g cm^{-2}. The loading maintained on the specimen relative to the jig's conditioning ring should not exceed the latter since a track gradually develops in the polisher and thus degrades its form. If such a state should occur, resurfacing and scrolling is the best remedy. Small errors may be corrected by running the polisher against a freshly greyed conditioning ring alone. The specimen is replaced after flatness has been regained.

When a number of small discs or windows have to be polished, it is the practice to mount them with dental wax spaced out on the surface of a 75 mm diameter solid workplate, its mass calculated to give 100 g cm^{-2} on the work. Conventionally, the lowest number that can be worked is three around the centre; then seven (i.e. one on centre surrounded by six) etc. However, these arrangements are intended for the standard poker-driven polishing machines (see §2.2.1 and §2.2.4) and are designed to minimise the formation of high or low zones. With conditioning rings—or plate working where bias polishing could be involved—the specimens can be arranged in rings without the necessity of filling the central area.

With a 20 mm diameter LiF window a typical polisher speed of 120 rpm is used, both for pitch polishing and in the early stages with wax. This is reduced to about 40 rpm for the final wax stages in order to reduce the heat generated during polishing and consequent reduction in time required for the specimen to attain room temperature before flatness testing. The best edge definition has been obtained as a result of machining the wax polisher again during the final polishing stage. This improves the cutting characteristic of the wax matrix/abrasive with an attendant reduction in edge turndown. The effect is also noticed by experienced hand optical workers, who periodically scrape hard pitch polishers and thus renew their polishing qualities.

The largest single specimen polished has been 56 mm diameter by 10 mm thick: the polishing of larger specimens might well require either a relaxation in flatness specification or a larger polisher. Flatnesses obtained are estimated to be $\lambda/20_{(He)}$ cm^{-1} because this was the limit of the testing technique, carried out by placing an optical flat on the conditioning ring of the jig and raising the specimen on the sliding shaft carefully into close contact (cf §7.5). It should be possible to obtain surfaces flatter than this, if required, since conditioning-ring techniques may permit a slow drift from a state of concavity to one of convexity and vice versa to be used to advantage. Thus, if this transition were made sufficiently slowly, the polishing process could be stopped at the critical time for the desired flatness. Polishing on proprietary cloths gives surfaces which result in comparable transmission (~57%) but their flatnesses are inferior.

Although the process described here is specifically related to cleaved LiF surfaces, it is applicable to low-damage sawn or lapped specimens. In these instances, the need for pitch polishing or some extended low-damage technique is even more important, since the surface layers must be progressively removed without propagating fissures. Polishing times and loads are, up to a point, interdependent. Excellent surfaces have been prepared from pressures lower than 40 g cm^{-2} with an attendant decrease in material removal rate. Though pressures in excess of 100 g cm^{-2} might well give results faster, wax polishers begin to 'pick-up' (the wax smears the surface) as temperature and pressure increases much above 20 °C and 100 g cm^{-2} respectively. Measurements made by surface profile and interferometric techniques (figure 7.43 and 7.44) gave figures of the order of 2 nm peak-to-valley for residual scratches with agreement of 20% for the methods used. Deep polishing is an improvement over most conventional techniques which give figures, typically, of the order of 40% transmission.

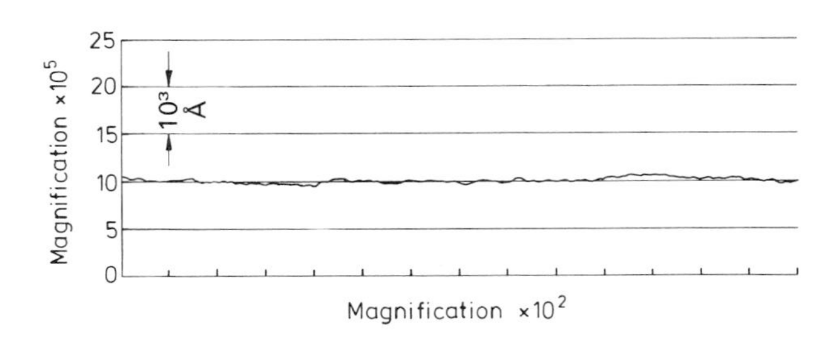

7.43 Talysurf of polished lithium fluoride surface.

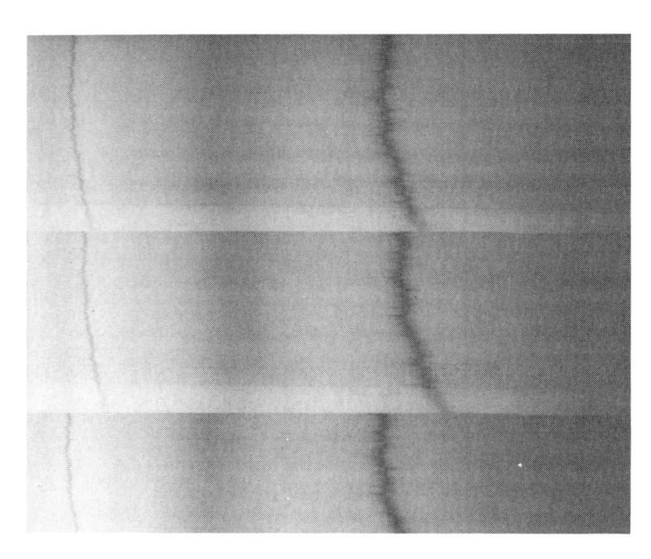

7.44 Interferogram of polished lithium fluoride. Bands: left, $N = 2$, 6100 Å; right $N = 3$, 4550 Å.

7.5 Proustite

Large single crystals of synthetic proustite (silver orthothioarsenite, Ag_3AsS_3) have been grown for optical mixing and electro-optic applications (Hume *et al*, 1967; Bardsley *et al*, 1969). They have a scratch hardness range of 2·0–2·5 (Mohs's scale) and thus proustite can be classified as a difficult material in the cutting and polishing field. This does not imply that is can be neither sawn nor polished by conventional optical processes, but that the quality of such cutting and polishing would be relatively poor when compared with that of harder materials. The attainment of some specular quality on most crystal surfaces is not difficult; but the achievement of a surface flat to better than $\lambda/8_{(\mathrm{He})}$ cm^{-1} with a finish satisfactory to the discerning eye necessitates both technique and understanding (Fynn *et al*, 1970). Low-speed sawing processes yield surface finishes directly comparable with those of the more robust crystals; polishing, however, provides surfaces which, though adequate for present-day quality material, are barely comparable with the optimum surfaces available on glass or quartz. Residual scratches of about 5 nm peak-to-valley commonly occur on the final surface.

7.5.1 Cutting

Contemporary electro-optic devices often require proustite to be in the form of a cube of 1 cm side, allowing some relaxation in the necessarily strict waxing techniques in use for repetitive thin slicing of similar friable materials. Thus the glass-angle/epoxy resin/Tan wax encapsulation method (§1.3.3) can be dispensed with. It is mentioned here because any requirement for smaller, thinner devices would almost inevitably require its use. Proustite is susceptible to thermal shock, especially when in volumes of greater than 1 cm^3. In order to minimise this, the specimen is placed on a glass substrate with a small quantity of low melting point cement such as dental wax placed under and around it; and the glass on a turntable used with an extension finger, (figure 1.48). The assembly is placed in an oven at room temperature and allowed to warm-up slowly to 60 °C, when a fillet of cement forms round the bottom section of the crystal. The cooling process follows the same leisurely pattern.

An end sawn relative to the crystallographic axis of the material is usually available from the x-ray

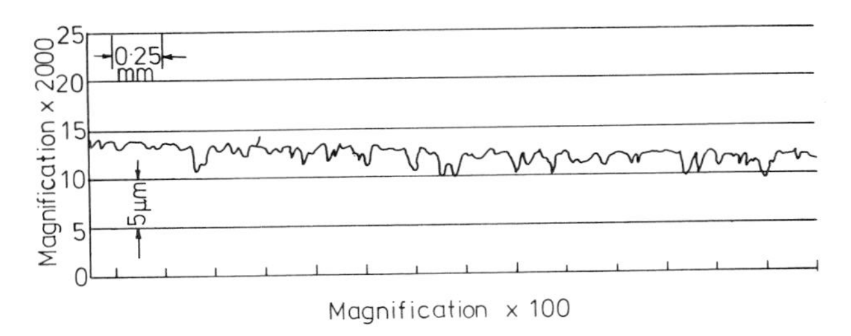

7.45 Talysurf of sawn proustite surfaces. (Meter cut-off: 750 μm).

room, and this axis can then be re-established as described in §1.3.3. The cutting method used is sawing with an electrometallic annular diamond blade operated at about 180 rpm. The kerf loss with an Impregnated Diamond Products disc, Type FX, on a 50 μm centre, is about 150 μm; the surface finish obtained has been measured with a surface profile indicator and a figure of 0·5 μm centre-line-average surface roughness is typical (figure 7.45).

The first contact with the blade should be made at a speed of less than 20 rpm and a counterweight over-balance of about 20 g. These conditions are slowly changed to 150 rpm and 50 g cm^{-1} cut-length respectively. Ideally, the speed and weight should be reduced as the end of the cut is approached, but the cutting rate of the wax and glass finger can have an overriding effect and make the adjustments unnecessary. Additional related cuts (e.g. for a cube) are made similarly after the specimen table has been rotated and aligned.

Finally, the sawn specimens are removed from their wax, either by immersion in cold solvent such as trichloroethane or by warming in the oven and being allowed to fall on to tissue. In the latter case, a solvent clean-up is still necessary.

7.5.2 Polishing

Angle-control jigs (§2.3) are used for polishing cubes of proustite, running within the roller bar on a machine such as the GPPM (§2.2.1). There is no reason why much of the work should not be done with its vertical axis freely floating. For this, the specimen is held loosely in an aperture in a Tufnol ring, surrounded in turn by a conditioning ring in the form of a Minijig (§2.3.4). The pressure can then be applied by a push rod via a dimple cap mounted on the upper face of the specimen. Various advantages are gained from a free system: the crystal needs no waxing down and it can be turned over at intervals during polishing with, possibly, less strain and thus better overall flatness. However, parallelism correction is similar to that used with block-polished laser rods: essentially a continuous feedback process that must be held at the null when reached. Adjustable-angle jigs provide an easier though more expensive method and the technique as applied to them will be described.

The material which holds the embedded abrasive particles (i.e. the matrix) may be soft metal or wax for polishing proustite. The selection of which of these is used is determined by the specification for the specimen. The polisher criteria may be summarised as:

(*a*) soft-metal polisher only: adequate polish, extreme flatness and edge definition;
(*b*) (i) soft-metal polishers; then (ii) wax polishers: improved polish, while maintaining flatness and slight degradation of edge definition;
(*c*) (i) sawn or grey surface; then (ii) wax polisher: best polish at expense of edge definition.

The soft metals that we use in a two-stage polishing process are either cast solder (§3.2.2.3) followed by cast indium (§3.2.2.4) or electrodeposited tin (§3.2.2.2) followed by electrodeposited indium (§3.2.2.4). The results from both are directly comparable, but the high price of indium favours the latter process since only a small quantity of the metal need be electrodeposited. The wax polishers used are made by pouring molten 'Okerin 100' (Aster Petrochemicals Ltd) on to 120 mm diameter × 18 mm thick cast bronze or stainless steel plates using a wall of masking tape to contain the wax (see §3.2.4). When cool, the front surface is machined on the surfacing and scrolling machine (§3.3).

The sawn cube—or other shape—requires its edges chamfering before either lapping or polishing is attempted. Chamfers typically 0·5–2·0 mm wide are produced on all edges either by rubbing them against 600 'Bramit' paper affixed to a microscope slide (see §5.4.1), or by drawing them lightly across a lapping plate using 600 carborundum and Hyprez fluid. The latter technique can produce a keener and less-chipped chamfer.

The x-rayed face is then given a light polish for auto-collimation purposes. Again, there are two techniques for achieving this:

(i) the face can be held against a slowly rotating

solder or tin faced polisher, fed with 3 μm diamond abrasive and Hyprez fluid;

(ii) the specimen is cemented to a jig-workplate (see below) and the x-rayed face aligned with the plane of the conditioning ring using either a DTI (figure 6.9) or a 12 mm parallel optical flat (§6.3.2).

The jig's tilt controls are adjusted to give direct alignment both where the DTI is used and where, for the optical flat, the image from the specimen surface is superimposed on the image returning from the conditioning ring. Though (i) needs far more skill to carry out than (ii), it is a very quick process; both techniques should result in greater accuracies than can be indicated by x-ray orientation.

A piece of washer-shaped optical tissue is impregnated with paraffin wax at about 60 °C and placed on a cleaned worktable surface; the proustite specimen is placed in position with its lightly polished face in contact with the tissue (see figure 4.16) and the assembly transferred to an oven. By this means, a minimal quantity of wax is used, and the hole through the tissue ensures that the wax does not obscure that specular area of the specimen which will be viewed in the auto-collimator. The oven temperature is slowly raised to 60 °C and as slowly lowered.

When it has cooled to room temperature, the workplate is fixed to the sliding shaft of the jig with three lightly tightened screws. The conditioning ring is screwed in the direction which aligns it with the crystal-face at the mid-point of the jig-shaft travel and the crystal then retracted by screwing down (i.e. clockwise, from above) the load control nut. Thus the proustite is protected from accidental contact damage during collimation.

The rear, temporarily polished surface of the specimen is brought parallel with the face of the conditioning ring by using an auto-collimator (see §6.3.2). For moderate accuracies, for example about 30 seconds of arc, the graticule of the autocollimator is set at some arbitrary reading (say 20–20 on the respective scales) whilst viewing the cleaned baseplate of the instrument, then the jig registered on this baseplate and the returned image from the crystal can be brought into 20–20 alignment by using the tilt controls. However, for work that must be accurately parallel within a few seconds of arc, straightforward location on the baseplate is no longer satisfactory, and a kinematic plane (§6.1.3) is used as an aid to precise location. The three balls define a plane which is nearly parallel to the baseplate (it should include an angle of 1′). Thus by placing a stainless steel master flat on the kinematic plane, alignment is obtained by rotating both together until coincident at 90° to the auto-collimator graticule axis. Adjustment to the graticule itself is used to achieve final alignment in the second direction. The jig is substituted for the master flat and the specimen aligned by adjusting the tilt controls. When the alignment is exact, the image will remain stationary as the jig is rotated carefully on the balls. A loose fitting brass collar is slipped over the periphery of the kinematic plane to locate the jig during this rotation.

In some cases where the sawn finish is adequate and the length dimension correct, polishing can be carried out immediately on the sawn face. Frequently, however, some lapping will be needed to make the second face parallel to the auto-collimated one and, at the same time, align it with the plane of the conditioning ring. A cast-iron ungrooved lap, 20 cm diameter and flat to $\lambda/2_{(\mathrm{He})}$ is used for greying-off. A better 'feel' for this state is obtained initially by hand working, since the constantly changing lapping conditions affect the drag on the work and are interpreted by the experienced operator. Once a fairly stable and successful set of conditions has been recognised, the work can be transferred to a machine with confidence, but constant monitoring is still advisable since conditions change more quickly on a machine than manually. Moreover, the first few strokes must always be made on an ungrooved section of the lap to avoid an edge being chipped on engaging with a slot.

A heavy slurry of carborundum and Hyprez fluid is kept ready mixed. When used, it is spooned on to the plate and either spread with a glass disc (a 'crusher') or with the conditioning ring of the jig. It is advisable to keep the specimen in the retracted position for the latter operation, ensuring that its delicate edges miss all of the initial bumping during the period of abrasive spreading. When the jig slides smoothly over the lap's surface, the load control nut is screwed anticlockwise, gradually introducing some load to the crystal. The maximum load of about 100 g cm^{-2} can be applied when the whole surface is seen to be in contact with the lap. The process is continued until the required dimension is obtained.

Cleaning the proustite and jig front prior to polishing is the next important stage. Low-power ultrasonic vibration in a solvent such as stabilised trichloroethane is often used, but a safer, though more tedious, process is to brush the abrasive-contaminated areas lightly with liquid detergent and water. At all times the direct application of an aerosol spray must be avoided, since the sudden fall in temperature may strain or shatter delicate crystals.

The chosen polishing process depends upon the specification of surface-quality requested. If extreme edge definition (freedom from turndown) is needed, a somewhat inferior surface finish may have to be accepted. This could be achieved by jig-polishing on indium with 3 μm then 1 μm diamond abrasive and Hyprez fluid. A typical programme is five minutes on 3 μm diamond; clean-up, then two minutes on 1 μm—both at pressures of about 100 g cm^{-2}.

On work where the edge definition is not too critical, pure wax polishers with 0·3 μm alumina and ethane diol lubricant may be used. This combination polishes proustite directly from the sawn or the greyed-off state and the process should take about thirty minutes again at about 100 g cm^{-2}. Though appreciably longer than the soft metal plus diamond process, it is not as time consuming as one might expect. In fact, many soft crystals are characterised by fast removal rates on scrolled wax polishers.

Both of the preceding processes can be combined to produce surfaces which are needed with a highly specular finish and a minimum of turndown. The grey surface is polished on solder, tin or, best of all, indium—using 3 μm diamond and continued until the pits from the greying process have been removed. This leaves the proustite covered with a haze of fine sleeks, usually defined as very shallow smooth-sided scratches. After cleaning, the second stage is a very short-duration polish (about two minutes) on a wax matrix. Two factors are important in the second stage: that the polisher be new and of the correct form and that the duration should be as short as possible, consistent with the removal of the sleeks. The result of prolonging the process is to produce a highly specular surface with increasingly rounded edges. If the trend has gone too far, polishing should be carried out again on the soft-metal/diamond polisher, then a return to the wax.

Surface flatness (figure 7.46) can be inspected by interferometric techniques varying from contacting optical flats (§6.2.1) to a non-contacting interferoscope (§6.2.2). A favourite method for quick testing is to place an optical flat on the conditioning ring of the jig with the work retracted and then raise the specimen carefully into close contact with the flat. This arrangement results in clear fringes when placed under a monochromatic light source and viewed as nearly normal to the surface as possible For work which has to be tested after it has been removed from

7.46 Fringe pattern obtained from polished proustite.

the jig, a true non-contact method should be used. The Barr and Stroud two inch interferoscope (figure 6.45) used with a $\lambda/50$ flat is adequate for most specimens. When the built-in mercury source is used, the face undergoing test should be kept within 1 cm of the standard flat. If a gas laser (e.g.: helium/neon) can be used as an illuminant, this distance is no longer critical.

Once the specified figure and surface finish have been achieved, the workplate is removed from the jig. The method used for demounting the specimen depends on its volume. For less than 1 cm^3, the table is placed on a cool hotplate and the temperature slowly raised to about 60 °C. The specimen can be carefully slid-off and its temperature allowed to fall gently. Specimens larger than this (typically cubes 1 cm side) need even more considerate treatment. The workplate is put into an oven at such an angle that the specimen will slide into a prepared bed of tissue when the melting point of the wax is reached. At this point, the oven is immediately switched off and the work left to cool inside it. The specimen and tissue are placed in a dish of warm solvent (see table 4.2) and, in a fume cupboard after standing for a while, the remains of the wax can be eased away with a very soft brush. It is given a final rinse in clean solvent then allowed to drain dry.

Collimator techniques for the second face are the same as those used for the first. For very accurate work it is advisable to make periodic checks of the alignment on a Mark I jig so that any slight drift which may occur is corrected. It should be mentioned that quite large angular readjustments—up to several minutes of arc—can be made by polishing on solder polishers with 3 μm diamond abrasive: thus a 10 μm wedge on a 1 cm side may be corrected. The advantage of this is that there is no need to re-lap with 600 carborundum which avoids tedious cleaning before a return to polishing.

Final packing of the polished components requires special care since proustite is easily scratched by the standard optical tissue wrapping. Ideally, an individualised non-contact system should be devised: possibly one which holds the specimen by its chamfered edges, or corners; but the present compromise is to line a plastic specimen box with chamois leather. This does not scratch the proustite and enough is used to prevent it from moving about. One final word of warning: proustite contains arsenic and should, therefore, be handled minimally—preferably with plastic tweezers, rubber gloves or some other disposable buffer material. For the same reason, all tissues, sawing lubricants and polishing slurries contaminated with its residues should be kept, contained and labelled for disposal.

7.6 Lead Germanate

Lead germanate is one of the materials that can be used for the manufacture of pyroelectric detectors (Jones *et al*, 1972). The crystals, as grown, often taper in a series of steps each flat, specular and parallel to its neighbours (figure 7.47) so that there is little effect on an auto-collimated image.

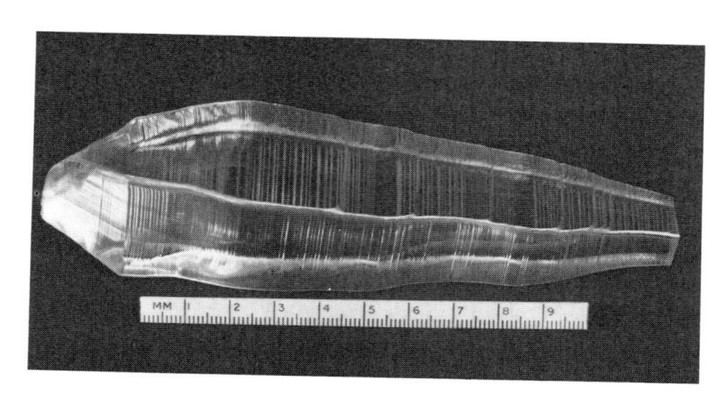

7.47 Crystal of lead germanate.

7.6.1 Trepanning and polishing cylinders

Where circular slices are required, parallel to the optical axis, cylinders typically 20 mm in diameter are trepanned from the crystal. It is mounted on a glass platform 6 mm thick and packed-up with glass to bring the optical faces approximately parallel to the platform. The complete assembly is cemented together with dental wax and, while at the flexible stage, positioned beneath an auto-collimator and aligned accurately. Some pressure is needed to maintain the crystal's position until the wax has thoroughly cooled, and a few layers of tissue may be used to buffer the fingertips against an uncomfortable heat input.

The biscuit cutting technique (§5.2.1) is used with free-abrasive slurry: a plasticene moat is moulded around the coring region and contains 600 carborundum in water. In many trepanning operations the more usual grade of abrasive is 220, but the finer one has been chosen because of the material's fragility. Relatively thick-walled cutters (1·5–2 mm) are safest for lead germanate because, with the larger effective contact area, lapping pressures are easier to control and optimise: they reduce the reliance on the sensitivities of the operator and of the drilling machine in use. With the finer abrasives, too, it is an advantage to use a tool with a large number of saw-cuts or milled slots in the tube (figure 5.7) to provide better access for the abrasive. Since thick-walled cutters are themselves abraded and acquire a radiused trepanning face, the cuts must be continued deeply into the glass mounting plate. This avoids a rim being left at the lower edge of the cylindrical core and makes subsequent centreless grinding or circular lapping easier to carry out. The cores and crystal-remnants are demounted in an oven, by sliding them into a bed of tissues.

All traces of wax are carefully removed from the cylindrical surface of the core, then it is waxed by its one end to a brass rod, so that it can be rotated slowly in a lathe while being lightly lapped and polished. Half-laps (§5.3.3) of brass, glass or steel are used to smooth the core with successive grades of abrasives: 600 carborundum and 303½ BAO alumina, each mixed with water. The surface roughness left after the final lapping stage is reduced by two half-polishers, 0·8–1·0 mm internal radius larger than the work. They are lined with self-adhesive Microcloth or some other similar pad. While the core is rotated slowly in a lathe (held by its brass shaft) the half-polishers are held on by hand (cf figure 5.26) and supplied either with 0·3 μm polishing alumina—Linde A— and water or, for a speedier polish, 3 μm diamond abrasive and Hyprez OS fluid. As with all cylindrical polishing, the rotational speed of the work must not greatly exceed the longitudinal stroking rate. Complete freedom from pits in the cylindrical surface is not necessary but a moderate polish helps to reduce edge-breaking on finished, thin (25 μm) slices.

7.6.2 Sawing

Slicing the cylinders into discs has been carried out by low-speed annular blade cutting (§1.3) with very fine fixed-diamond abrasive (IDP Ltd 320 mesh) and Lapmaster oil as the lubricant. Some of the first crystals of lead germanate were satisfactorily sawn on an Abraslice reciprocating-work saw (§1.4) but the slices were less uniform in thickness than those produced on the stretched-blade saw; consequently, they required slightly longer times for their polishing. Much of the routine sawing is done on the modified Capco Q.35 (§1.3.1), its spindle speeds decreased to be variable between 0–300 rpm; but the Microslices are equally satisfactory. Peripheral wheel low-speed saws can be used too, but the slice thicknesses have to be increased to minimise the risk of breakage. A typical specimen thickness cut by both annular and reciprocating-work saws is 500 μm and, occasionally, 250 μm when required; for peripheral sawing, 1 mm is advisable.

The cement widely used for mounting lead germanate cylinders is Tan wax (§4.1.1), though in the case of annular-blade sawn work which involves a rigid epoxy-resin bonded glass angle-plate (§1.3.3), softer waxes like beeswax/resin mixtures are satisfactory. The counterweight load applied to cut a 20 mm diameter slice when using a 125 μm thick annular blade is about 75 g—assuming that some hand control will be used for the first few sawing revolutions on a small surface area. Lightening the load by hand near the cut exit point is unnecessary because the saw by that time will be well immersed in the glass mounting block. Automated saws such as the Microslice III have a resilient constant feed mechanism which helps to protect the material from changing pressure stresses. Section 1.6.2 describes several techniques that can be applied to minimise the effect of changing cut-length and thus applied load, on a cylindrical slice.

7.6.3 Lapping

As the first faces of the slices to be worked do not need to maintain parallelism at this stage with what will be the second faces, large numbers can be mounted with dental wax on a workblock and processed together (figure 7.48). Because of the high quality of most contemporary sawing, it should be unnecessary to lap lead germanate purely for flatness reasons before polishing it. However, should it prove necessary in order to remove unwanted material, it can be carried out on a ring-lapping machine (§2.1) either with 600 silicon carbide and oil; or with 303½ BAO alumina powder and oil/Hyprez OS fluid. Hand-lapping a few blocks covered with lead germanate slices is a rapid process and, if the necessary skill has been acquired, a machine may not be regarded as worthwhile. However, if much material has to be removed the machine is preferred, since it will do with one fine abrasive what is normally done by hand

7.48 Lead germanate slices mounted on polishing block.

with a range of particle sizes and the necessarily tedious cleaning processes between successive changes of powder. It is advisable to reduce the thickness of the slice by lapping until at least 150 μm is left for polishing away from each side to reach the specified final thinness.

7.6.4 Polishing

Polishing many hard crystals is a process intended to result in a correctly figured, specular surface; with many fragile electronic materials the problem is rather one of reducing damage to a minimum (commensurate with thickness) whilst maintaining flatness, often prior to a short chemical etch. Lead germanate targets present yet another problem: that of producing thin, parallel, specular, relatively large diameter discs which must maintain their flatness, finally, unsupported. Thus their polishing is often a lengthier process than usual, designed to remove material and, at the same time, leave the both sides of the disc in a state of strain that is small and equal.

One method of improving the figure of the targets is to retain them on the blocks, which are warmed at intervals during polishing, thus allowing the slices to change shape and partly relieve the strain. After the surfaces on one side are satisfactory in finish and form, the work is demounted and reversed onto polyester (Melinex) film, or tissue impregnated with wax, to protect the polished face from damage. These protective pads should be cut slightly smaller than the slice diameter. A non-eroding wax, such as brazil wax, is an advantage since polishing slurries can attack others in time, causing them to swell and thus exaggerating the turndown of the edges of the work, (figure 4.12).

Though the previous method has been reasonably successful and simple to carry out, it has been largely superseded by a two-stage process in which several of the slices are firstly pre-polished, unmounted, in a Tufnol carrier, itself within a conditioning ring (cf the Minijig, §2.3.4); then finally polished, mounted on porous substrates, in adjustable-angle jigs. For the pre-polishing 'free' process, the load can be applied by a block with a lapped lower face, via a precise neoprene pad inserted between the weight and the slices (figure 2.67). By reversing them and changing their positions relative to one another, very flat and parallel discs can be obtained.

The Tufnol (or Melinex) carrier has its apertures machined accurately to fit the work and its external diameter fits the conditioning ring. If the carrier is thin—to work thin slices—the internal lower edge of the ring must be sharply defined—that is, with little or no internal chamfer. This pre-polishing stage produces slices polished on both sides and in a reasonable state of balanced strain. When the final polishing is done to thin the discs (again from both sides) very little warping takes place after demounting. The use of porous substrates in jigs: full waxing procedures and alignment techniques are fully covered in §4.1.6 and §6.1.4 respectively.

The polishers used for all processes are standard spiral-grooved wax matrices (§3.2.4) with 0·3 μm particle size α-Al_2O_3 (Linde A) for the block and pre-polishing processes, and 0·05 μm γ-Al_2O_3 (Linde B) for final polishing, each suspended in ethane diol. Pump-recirculated slurries are almost a necessity and the machines that are suitable for small quantities of work are the general purpose polishing machine (GPPM, §2.2.1) and the Multipol, and for larger numbers of slices, the three-station recirculating chemical–mechanical polisher (RCMP, §2.2.2). The small, high-density abrasive particles readily settle out of solution and are kept in suspension by a stirring paddle such as a neoprene blade attached underneath the polisher plate.

When unlapped work is being polished, it is advisable to use a worn polisher at first, since slight unevennesses of the work and cementing errors can produce very high local loading which will score the wax matrix. Lapped work will be free from these defects and, provided that the loading has been

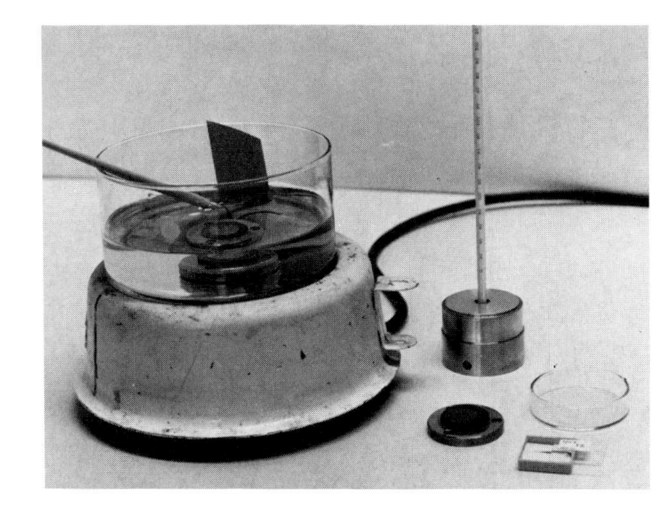

7.49 Demounting a finished thin slice in bulk-solvent.

correctly calculated and applied, it will not disturb the surface of the polisher. The softness of lead germanate allows relatively large volumes to be speedily polished away at loads of about 100 g cm^{-2}. Although the finish obtained with wax polishers and Linde B/ethane diol is adequate, it is optimised by refacing and regrooving the wax surface (§3.3) towards the end of the process. As with thinning operations in general, it is advisable to complete the final stages without undue delay, so that the mounting wax does not undergo sufficient attack to cause breakages at the slices' edges.

Where dental wax has been used for mounting the slice on a porous substrate for the final process, demounting is carried out in a bath of warm solvent (figure 7.49) such as trichloroethane or toluene, in a fume cupboard. A piece of folded card can be arranged to receive the disc as it slides off the substrate whilst submerged in the solvent. Thin slices are held in folded paper and cotton wool and boxed individually as shown in figure 7.50.

7.50 Boxing a single slice of lead germanate.

7.7 Water-Soluble Materials

Many well established infrared and ultraviolet transmitting materials are etched by atmospheric moisture and therefore undergo accelerated attack in aqueous solutions; in consequence, they present problems not only in polishing but in storage and use too. Whenever possible, it is suggested that hydroscopic components should be replaced with less exacting alternatives. The following notes extend the techniques (described in Twyman, 1957) in which minimal quantities of lubricant are advised, or the addition of sodium thiosulphate suggested.

7.7.1 Mounting and sawing

Since many hygroscopic compounds are susceptible to damage by thermal shock, the mounting of large pieces is a critical operation. Tri-glycine sulphate (TGS—a pyroelectric material (Putley, 1970; Watton *et al*, 1977)) provides an example of the general technique used. A glass angle-plate made from 6 mm thick plate glass is required for cradling the crystal. It is formed by cementing two smaller plates together at 90° with an epoxy resin; a third plate, on which the crystal will rest, is bonded at a suitable angle, like a one-sided vee block (figure 7.51). Since little or no heat can be safely applied to the TGS crystal, a cold-mounting technique is used: the glass cradle is cemented to the saw's worktable with Tan wax, and, when cool, the horizontal and angle faces of the cradle are coated with Sira wax (table 4.1). In order to make the wax pliable it is kneaded like putty then spread on the glass vee-support. The vertical face of the cradle is coated with a cold-curing expoxy resin, the crystal pressed into the wax and against the resin and this assembly left to cure at room temperature.

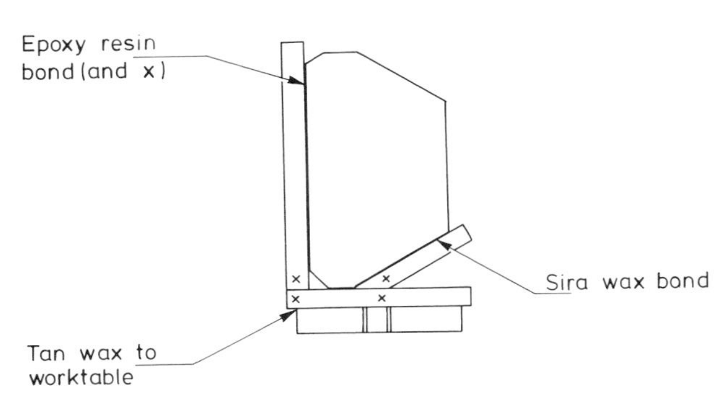

7.51 Mounting tri-glycine sulphate with Sira wax (for sawing).

When cured, the crystal on its worktable is mounted and aligned—by the DTI method, §1.1.3—on a Microslice III low-speed saw and cut into slices 0·75–1·0 mm thick. The work in-feed is controlled by the Delicut mechanism. Electrometallic 280 grit diamond, 30 cm annular blades are used with an oil lubricant such as Paynes No. 3. Although the crystal is epoxy resin bonded to the mounting cradle by one end, each slice as it is cut-off is held by a strip of very soft, weak Sira wax. The auto-hold position allows time for the slice to be removed before starting the next cutting cycle. Crystals up to 65 mm square have been mounted and sawn in this way. Some softening of the wax by the cutting oil takes place and, in order to retain the slice against the oil drag induced by the blade's rotation, extra wax (or plasticene) should be pressed onto the exposed crystal face and into the cradle before starting a cut, thus providing additional support finally for the sawn-off slice. The technique is akin to that used with a waxed angle support (§1.3.3)—which cannot be directly adopted for large slices of materials like TGS because of their thermal frailty.

By tradition, thin discs have been prepared by slicing the crystal intially then trepanning the required diameters from the slices. The techniques developed for coring lead germanate (see §§ 7.6 and 5.2.1) make it possible that trepanning cylinders first and slicing afterwards could be used for materials as delicate as TGS, KDP and ADP. A broad-rimmed trepanning tool is needed—to reduce the dangers of high-pressure cutting regions—with carborundum/oil as the abrasive slurry.

7.7.2 Lapping

It is unnecessary to lap the first surface of an annular sawn disc before polishing it, but where a quantity of bulk material has to be removed to obtain a definitive thickness, lapping it away has obvious advantages. Components which require minimal surface and sub-surface damage should be polished on both sides without recourse to lapping (cf §§7.4, 7.6 and 7.9). A Lapmaster (§2.1) supplied with 600 silicon carbide in oil provides a slurry sufficiently water-free for the majority of greying-off operations, though lapping soft materials is such a speedy operation that is is often done by hand. Where a material tends to become loaded with free abrasive particles, bonded silicon carbide paper, like the self-adhesive Bramits used in metallographic work, can be used with oil lubricants. They have a limited life, due not only to rapid diminution in cutting effectiveness but also to the rapid attack of the adhesive by the oily lubricant. The latter results in swelled-edge effects and produces uneven work (cf figure 4.12).

7.7.3 Polishing

Because of the sensitivity of TGS and similar components to sudden changes in temperature, a low melting point paraffin wax such as 'Paraplast' is used with leisurely warming-up and cooling-down of the work. As with lead germanate, (§7.6) or cadmium telluride (§7.9) groups of slices are worked preferably on a lightweight stainless steel block (about 50 g cm^{-2}). Adjustable-angle jigs are, as usual, suitable for precise working either of single specimens or of valuable materials that are better worked singly.

Many water-soluble crystals may be polished on pitch or wax matrices with 0·3 μm α-alumina (Linde A) abrasive and ethane diol as the lubricant. Cloths, too, are satisfactory but can give the usual slight degradation in micro-finish and form. For the more hygroscopic materials such as TGS, very much reduced quantities of lubricant are essential: a drip-feed which just keeps the work sliding smoothly seems to be the optimum condition. In the hands of an experienced operator, excellent work has been done on pitch suitably grooved or reticulated with the lubrication reduced to a breath-film. However, this practice is more suitable for the production of simple optical components rather than the repetitious manufacture of electronic devices.

Soft-metal polishers, too, have proved their value for obtaining very flat surfaces rather than optimum specularity (often all that is required on transmitting materials for long wavelength uses). Diamond abrasive, 3 μm particle size, is used with Hyprez fluid on solder, tin or indium matrices (see §3.2.2).

Cleaning the slices for inspection is done during the polishing process by repeated wiping with clean chamois leather. The final clean-up can be carried out by wiping with chamois leather and a degreasing solvent with little or no affinity for water, such as dichloromethane. All specimens are then very susceptible to moisture etching and should either be kept in a desiccator, or stored under dry nitrogen conditions. Alternatively, but more hazardously, oil can be left on the specimens for later removal.

Perhaps the only long term durability can be given by an evaporated or sprayed inert coat which does not degrade the desired properties of the component.

7.8 Zinc Sulphide and Zinc Selenide

Because they transmit well in the 8–14 μm window, these zinc compounds are made into optical components in infrared systems. They are particularly attractive as window materials since they transmit visible light, too. Both materials can be cut readily, either by peripheral wheel (§1.2) or by annular blade (§1.3) low-speed diamond sawing methods and should do so equally well on the reciprocating-work saw (§1.4). However, since blocks and windows, rather than thin slices, have been requested, the Abraslice remains untried. The manufacture of zinc sulphide prisms for refractive index determinations illustrates the general techniques involved.

7.8.1 Sawing

A plate of zinc sulphide of the required thickness is cemented with dental wax onto a glass plate which has been previously mounted with a higher melting point cement on a rotary worktable such as a worm-driven, calibrated type (figure 1.52). The wax is placed on the glass, the block added and the complete assembly put in an oven, cycled to 70 °C and returned to room temperature. For pieces of greater than 1 cm^3 a hotplate, gently heated and cooled, is safer with a beaker or dish placed over the assembly helping to exclude cold draughts.

Three cuts are made on the saw: two at 90° and the third, the hypotenuse, at 18° 45′ in this instance. The cut is arranged to leave a small flat parallel to the 90° base (figure 7.52), since at this acute angle a knife-edge would be too fragile to remain intact. Suitable lubricants are Payne's No. 3 oil or Lawrence Chemical Ltd's Abralap 2A, used, for example, with the disc and conditions noted in table 1.2. The work and its off-cuts are demounted in an oven and allowed to slide into a bed of tissues, then the parts separated in a bath of warm solvent. The latter process should be carried out in a fume cupboard or with some vapour extraction system in operation when either trichloroethane or toluene is being used.

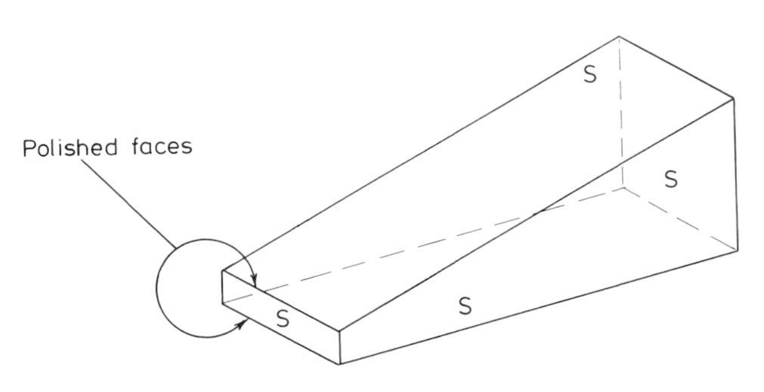

7.52 Zinc sulphide prism dimensions (S; sawn faces).

7.8.2 Chamfering

The cleaned work requires chamfering on a glass or cast-iron lapping plate with 600 silicon carbide abrasive. More than the usual care is needed to ensure the correct spreading of abrasive and the gradual increase in pressure applied to the chamfer. It is very easy to chip delicate materials when the chamfering operation is started. If the design of the components allows, wide chamfers are the safest to produce although they are not so attractive in appearance as narrow ones. The light-duty coverglass filing technique detailed in §5.4.1 is also effective.

7.8.3 Polishing

Before mounting the prism for polishing, a check is made on the angular accuracy and orthogonality of its faces by using angle slip gauges (§6.4.1). These are normally used by engineers in mounting work for machining operations and they are accurate to better than one minute of arc. A short period of bias lapping will bring them to within a few seconds of arc and accurately flat.

All traces of the 600 carborundum from the chamfering process must be removed and temporary polishes worked on the two faces of the prism that include the angle of 18° 45′. The gauges and the prism being tested are slid against an angle-plate mounted on the base of an auto-collimator (figure 7.53). Once the angle-plate's position has been adjusted, the prism and slip gauge resume the same attitude to the auto-collimator on repeated tests. Orthogonality measurement makes use of the 90° angle-plate too, with the prism held by a selected side face and swung through the optical axis of the auto-collimator, previously set at right angles to a slip gauge held against the same angle plate.

If the surface sawn and lightly hand-polished for testing is not sufficiently accurate, then one surface and the reference side is bias-lapped to the correct angle. Temporarily polishing surfaces at each testing stage can be dispensed with if a small optical flat is used to provide the specularity necessary for auto-collimation (§6.1.4).

Like the angles on the miniature prisms described in §7.2, 18° 45′ is too large to be directly covered by the shaft tilt-adjusters on a Mark II polishing jig (§2.3.8) which was designed originally for plane-parallel work. However, a thick workplate can be modified to bring the prism within the range of the adjusters by milling an 18° slot in it, wide enough to be a generous clearance on the work (figure 7.54). For the maximum convenience in setting the angles, the longest dimension of the prism (i.e. the wedge) should be aligned with one of the jig's tilt controls. The benefit of an oriented mounting is that one angle-control knob effects change to the 18° axis (and trims it to 18° 45′) while the other maintains the orthogonality at 90°.

The prism is waxed (as before) in its slot in the work-plate and bolted in position in the jig. A DTI mounted on a ring base which slides on the inverted jig's conditioning ring (figure 6.8) may be used in conjunction with the tilt controls to bring the prism face co-planar with the polishing plane. However, zinc sulphide is easily damaged and a smooth, precisely parallel thin plate must be placed between the stylus and the prism's face. This is a purely mechanical alignment technique and it may be replaced by optical methods where an auto-collimator is arranged to overlap both a portion of the conditioning ring and part of the prism, showing two images. The prism's image is superimposed on that of the ring by adjusting the jig's tilt controls; although the method is not very accurate, it brings the faces close enough to co-planarity to make a second method possible. An optical flat is placed on the work and ring, illuminated by monochromatic light and the controls adjusted until the fringe patterns agree (figure 7.55).

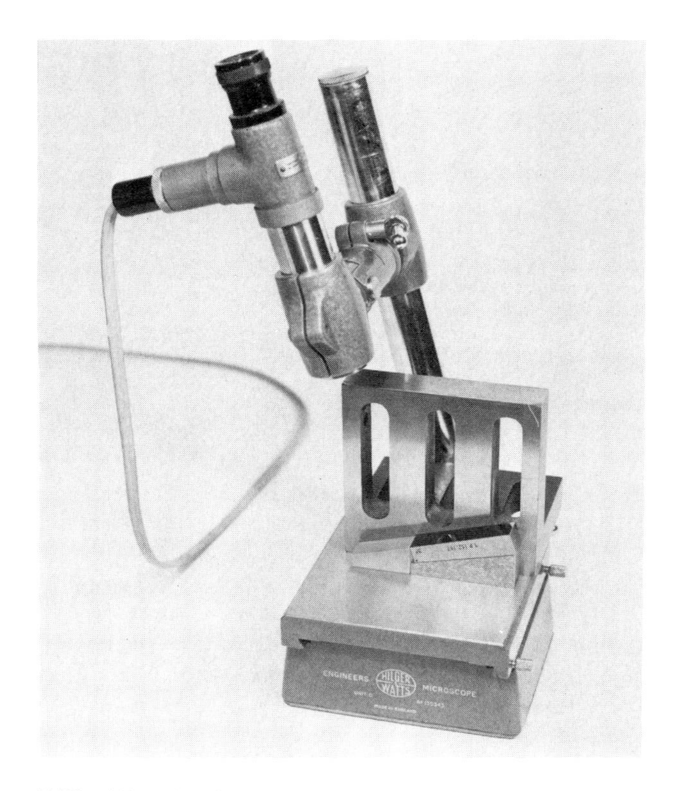

7.53 Workplate for prism with 18° sloping section.

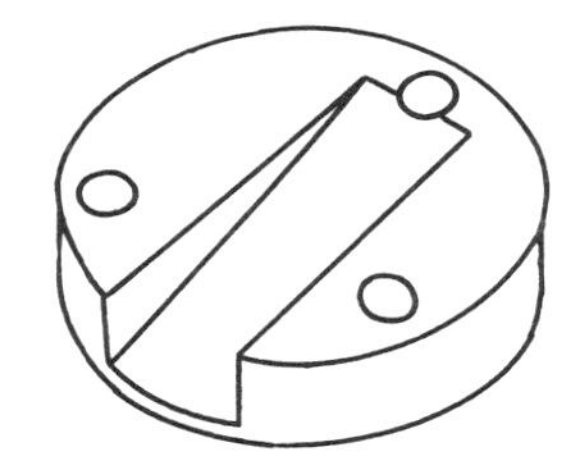

7.54 Angle-plate used for holding prism beneath the auto-collimator.

7.55 Ring and specimen patterns coinciding for accurate alignment of polishing planes.

A large capacity Fizeau interferometer offers a very convenient arrangement since the proximity of work-and-ring to the reference flat is not then critical.

Polishing is carried out on a Draper-type machine such as the GPPM (§2.2.1) or Multipol I (§2.2.4) with a few centimetres of sweep applied to the roller-bar within which the jig runs. Soft-metal matrices (§3.2.2) are satisfactory, if the need is for extreme flatness (the case with these prisms) rather than for the best obtainable surface finish. Since solder and tin are too hard, indium polishers supplied with 3 μm diamond abrasive and Hyprez OS fluid are used. If a slightly greater turndown is permissible, wax polishers (§3.2.4) with 0·3 μm α-alumina (Linde A) and ethane diol produce a highly specular surface. Cleaning the face for periodic inspection and testing can be done with tissues; but they should not be used after the final polish since they produce detectable marks on the surface. A soft brush and alcohol solvent are suggested for removing the ethane diol/alumina slurry; diamond abrasive/Hyprez OS fluid can be brushed gently away with detergent and water.

The second side is polished by reversing the prism in its special workplate and, since the side polished firstly is not going to be used for alignment purposes, the prism can be laid on an unperforated sheet of waxed tissue. When the assembly has cooled, it is returned to the jig and the prism aligned, as before, from the temporarily polished face.

The usual dicta apply for packaging cleaned, easily damaged materials. If possible, support the component outside the required area; alternatively, retain it by its chamfered edges. If the recipient is able to clean the component safely, probably the kindest method is to cocoon it in paraffin wax and, where its mass is fairly large, wrap it in tissue. Smaller work can be mounted by an unpolished face on a carrier-substrate, and the substrate lightly held by double-sided pressure-sensitive adhesive tape in a small box. It is advisable to stick, say, half of the carrier on the tape so that removal may be helped by inserting a thin probe under the unstuck edge and prising gently until the bond creeps apart.

7.9 Low-Damage Polishing of Soft Semiconducting Compounds

Much of the polishing described so far has been aimed at obtaining surfaces that are not only flat but as little perturbed as possible crystallographically. Thus high transmission of lithium fluoride (§7.4) is obtained via an extended polishing process, and of course, thin, whole slices of relatively soft pyroelectric materials (§7.6) would not withstand rough processing.

When overall technologies are considered for many semiconducting devices made from soft materials, there is often a need for parallel, homogeneous slices with minimal surface and sub-surface damage. Irrespective of the quality of the bulk material, the intention must be to preserve its structure as grown and finally pass it on to whatever photolithographic fate awaits it, scrupulously prepared and uncontaminated. Chemical and electrochemical etches, with and without physical polishing on cloths, are the compromises usually made when preparing surfaces with minimal damage, though often at the expense of the macro-finish—that is, brilliant polishes may be obtained (the micro-finish) with attendant loss of form. For many purposes (e.g. light emitting diodes) the results are satisfactory enough to warrant no further trouble being taken; but there is an increasing case for less chemically active processes than, for example, halide/alcohol etchants (typified by the reaction of bromine in methanol (Strehlow, 1969)) used for III–V and II–VI compounds. Comparatively benign standard polishing processes like the one to be described are less demanding on the health of personnel and the durability of equipment, and a final short-duration etch is all that may be needed to remove any residual damage. Perhaps even more fundamental is the limitation on device-performance that can result from unwanted absorbed species: if the process can be carried out successfully without even a final etch, it circumvents their problematical removal.

The low-damage polishing process refined here caters for these objectives but in so doing it becomes greatly extended in duration. This is largely balanced by the fact that it can be carried out unattended on an RCMP (§2.2.2) and yields results which when measured both by Scanning Electron Microscopy and Rutherford back-scattering (Morgan and Bøgh, 1972), are at best, as good as etched samples—that is, a faithful crystallographic termination to the bulk material.

7.9.1 Mounting and measuring

Although several methods of slicing-up friable materials may be used successfully, the standard machine now employed is the reciprocating-work saw (§1.4). Its philosophy, mechanism and techniques have been already fully described as they relate to cadmium telluride, therefore they will not be repeated here but those slices will be taken as an example of the polishing processes, too.

The conditions are determined initially by polishing one (or more) 1·2 cm diameter slice(s) mounted on a porous substrate (§4.1.6) and bolted to an adjustable-angle polishing jig (§2.3), where load control is available. Thus the pressure applied can be pre-set at a minimum to cater at first for an inevitably small contact area, then gradually increased as the slice makes uniform contact with the polisher. Mounting a single slice at one side of the substrate follows the method described in §4.1.6, except that dental wax is used (at $<70\,^{\circ}C$) to load the oilstone. Since precise parallelism is not usually needed—that is, to better than 0·5 μm cm^{-1}—no pressure-applying jig is involved. However, if more than one whole slice, or several pieces are mounted and worked together so that the surface area of the work is increased, the small gaps adjacent to touching, or between barely separated, specimens can be the indirect cause of devastating scratches later. They become packed with alumina particles from the first polishing process, and, after the transfer to the extended-run polishing machine, release abrasive into the alkaline silica sol. Figure 7.56 shows slices mounted together on a porous substrate: their arrangement may be aesthetically pleasing but it is demonstrably hazardous in our process. A long, gentle brushing carried out on the work immersed in alcohol/water, while viewing through a binocular microscope, is needed to remove the particles and even this may not be wholly successful. Avoid the close-packed configuration!

With the data which accumulates from jig-polishing—rate of material removed for a measured load—it is possible to mount a given number of slices on stainless steel workblocks, calculated to have a deadweight suitable for the final polish. Figure 7.57 shows the dimensions of a block designed for twelve 1·2 cm diameter specimens ($\sim 12\,cm^2$), its rim wide enough to engage with the wheels of a roller bar and its rear-side relieved to reduce its mass to about 300 g. There are obvious risks with multi-slice polishing since one break-away chip can cause general disaster and larger quantities of wax need cleaning away meticulously if they are not to trap abrasive particles. Less apparent is the unfavourably high load which must be applied to some three highest points at first contact, resulting in damage being caused to the bulk material. The demands on uniformity of sawing and mounting are impossibly stringent if this snag were to be countered by trying to achieve co-planarity of all the slices on the block before polishing. It is advisable to incorporate three thicker wasters, spaced at 120°: either of the same material as the slices if cross contamination is thought to be potentially a problem; or, preferably, with a slightly harder material the erosion rate of which will dominate the abrasion process.

The first side mounted (as sawn) needs no protective film between it and the substrate, but since approximately equal amounts of material should be removed from each side, on reversal the polished face can be buffered by polyester film. Waxing methods relevant to this technique may be found in §4.1.3.

7.56 Three slices of soft semiconducting compound incorrectly mounted on porous substrate.

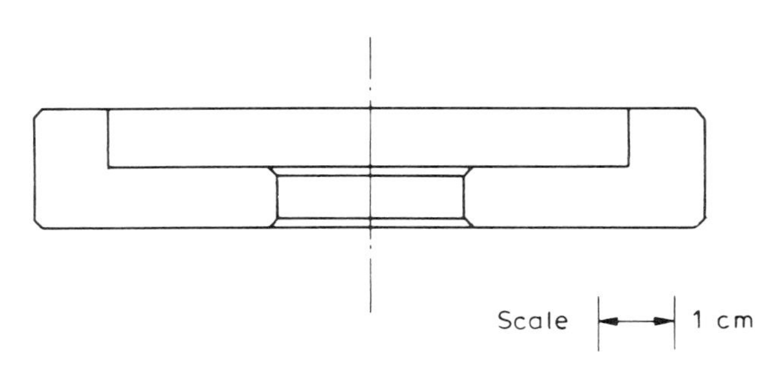

7.57 Dimensions of light weight stainless steel block for multi-polishing easily damaged materials.

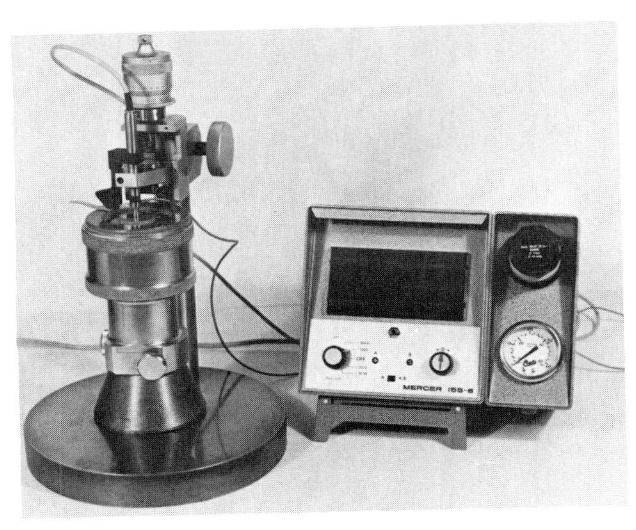

7.58 Measurement of specimens *in situ* in jig, using a pneumatic gauge.

Each mounted slice is measured under a pneumatic gauge (§6.2.2) to establish the overall difference in height between it and the block: when polyester film is used for protecting the first, polished side, its known thickness plus the less certain ones of the two wax layers may be averaged from a marked, previously gauged specimen. A slice mounted in a jig can be measured *in situ* as shown in figure 7.58, the tilt controls used initially to bring the porous substrate's upper surface parallel to the flat baseplate of the air-gauge. Additional advantages of porous substrates are their insignificantly thin yet uniform cement layer and the ease with which parallelism can be attained by non-contact auto-collimation, (§6.3.2). Blocked slices must be either bias polished to correct any wedging indicated by pneumatic gauging, or special blocks used with limiting shoulders such as that illustrated in figure 2.51(*c*).

7.9.2 Polishing

In a polishing process for soft materials which is predominantly mechanical the major problem is not one of abrading away material, but one of slowing down the erosion rate. Therefore no greying (lapping) process is necessary. There is a useful (but potentially dangerous) rule-of-thumb relationship which is thought superficially to exist between abrasive-particle size and depth of damage on any specific class of materials. Unfortunately, it is often quoted without reference to effective surface areas and load applied as if it were sublimely independent of pressure. Figure 7.43 (§7.3) shows the effects on a hard material of applying 100–200 g cm^{-2}, both with cerium oxide slurry. An obvious conclusion is that if the surface is acceptably specular and flat, a process which removes a certain amount of material faster merely as a result of increasing the load must be economically sounder. But, of course, the depth of damage increases even in the hard material, and in a softer one the increase may be catastrophic.

Consider the various polishing parameters in this more critical situation.

(1) The abrasive used:

(*a*) the material e.g. γ-alumina is friable and breaks down easily, then can agglomerate to result in gross surface-damage.

(*b*) Particle size: if carefully graded by elutriation, as with 0·3 µm alumina, its uniformity is not generally a problem, but any tendency to agglomerate, or form a mortar with polishing debris, must be minimised.

(*c*) Shape: not expected to be a problem with sub-micrometre particles of polishing abrasive, though in diamond polishing compounds it profoundly affects rate of material removal.

(*d*) Density: the denser the abrasive, the more difficult it is to keep it in suspension; increasing the viscosity of the liquid to counteract this can bring its own problems (see 2(*a*) and (*c*) below).

(2) Liquid used (the vehicle):

(*a*) its viscosity: if too viscous, it may promote aquaplaning of the work and thus prevent fine abrasive particles from cutting.

(*b*) Stability: that is, its ability as an electrolyte under some circumstances to keep colloidal particles dispersed.

(*c*) Solvent power: not only on specimens, but on mounting waxes, too. When taken in conjunction with 2(*a*), less chemically active liquids such as undiluted ethane diol can fail to remove, for example, calcium fluoride. Water is apparently necessary, not only to reduce aquaplaning but to achieve solvation.

(*d*) Liquid/abrasive ratio: the greater the concentration of abrasive in the liquid (the stiffer the mix) the drier it becomes and the resulting drag can induce damage. If the concentration, is too low, however, either aquaplaning or wringing may result.

(3) The matrix

(*a*) material used: generally the softer it is, the less the damage caused; but it must be firm enough to stay in shape and stable enough not to be attacked by polishing slurries. Its ability to bury large-sized abrasive particles (see wax, §3.2.4) is important.

(*b*) Surface interruptions such as grooving effects uniform slurry supply at the specimen/polisher interface. Antithetically, a continuous surface can slow down material removal rates, and may be useful if wringing, or near wringing, conditions are not allowed to arise.

(4) Pressure applied: (weight/unit area of work) increased load, increased removal rate and therefore damage, until abrasive grains are excluded and wringing takes place. In addition to this direct effect on the work, the matrix may become distorted and degrade the figure of the surface produced.

(5) Rotational speed of polisher: though low rotational speeds are used to give very flat work, there is a probable compromise to be made between high speed and consequent aquaplaning and low speeds with possible near wringing conditions.

(6) Work-rotation and sweep: rotation of the work within a roller bar (precession) either free running or deliberately driven, helps to maintain flatness and prevents patterning on the specimen's surface, Sweep is important, especially for a single slice mounted close to the rotational centre of a jig, since it prevents a witness forming on the work.

From all these considerations, several fundamental factors appear to be most favourable: small, discrete abrasive particles contained in a plentiful supply of slightly viscous fluid, and used on a soft polishing matrix with the work lightly loaded. As with sawing, the problem is better examined in terms of rate of removal of material to establish the load for a specific system which abrades away, say, 1 $\mu m\,cm^{-1}$ as a first polishing stage, then 1–2 $\mu m\,h^{-1}$ (a monolayer or so a second) for the second stage. The initial polishing must be continued until damage resulting from cutting has been removed; the second, until damage left from the first has completely disappeared without being propagated further.

A spiral-grooved wax 15 cm polisher is used for the first process, copiously supplied with Linde A polishing alumina suspended in water which has been made more viscous by the addition of polypropylene glycol. The 40 $g\,cm^{-2}$ pressure applied to the work deserves a little more discussion. A spiral-groove machined in the polisher surface results in a significant increase in load since it decreases the contacting areas: load calculations usually assume fully mating flat surfaces, but this cannot be the case where the groove is, and its land provides the contact. For example, where the ratio is 1:2, there is a fifty percent increase in effective polishing pressure, and, of course, for small specimens equal in diameter to the width of two lands or two grooves and one land, at the worst the mean effective pressure can be doubled. Hence fine grooves should always be used with soft materials.

However, polishing in a jig, measuring and calculating material removal rate and adjusting the load accordingly, automatically caters for the effective load. Once the rate has reached 1 $\mu m\,min^{-1}$, a gauge is used to measure the resultant weight of the sliding shaft assembly as it is off-loaded by the spring, and a block tailor-made for multi-polishing if required.

At least 100 μm per side is removed by this process, the exact amount depending upon the sawn thickness and that of the final devices. Either a Draper-type machine (§2.2.1) with roller-bar and sweep may be used; or a three-station recirculating chemical–mechanical polishing machine (RCMP, §2.2.2). Polisher rotation is 45–60 rpm. The surface obtained is specular but covered in fine scratches. Cleaning is carried out by rinsing the work thoroughly with running water, inspecting it for traces of abrasive and brushing them away if necessary. The bulk of the slurry contains potentially toxic debris and, when discarded, must be stored to await safe disposal.

The logic of the second process's being a more delicate version of the first does not hold true in practice. Even finer particles seem to be needed and less load applied, but 0·05 μm γ-alumina (Linde B) is not wholly satisfactory. Since some of the smallest particles exist in colloidal form, such as silicon dioxide in an alkaline solution, one has been tried and gives an improved surface with slight orange-peel and sleeks. To redress the former, there are several possible oxidising agents that have proved to be efficacious in alkaline silica sols and the most universally successful of these is hydrogen peroxide. Although it is not as powerful an oxidant in basic solutions as it is in acidic ones—its E° value is greater in the latter—it is more rapid in action and is not discharged during a 24 h run. The answer to sleeks is simple: because the polishing rate is low, the mechanical processes appear to be dominated by the chemical, so aquaplaning may be a virtue. The polyurethane rigid foam, §3.2.6, is therefore allowed to glaze (contrary to the practice detailed in §7.3.3) but not to dry out and remains ungrooved and unscraped. Flatnesses obtained are usually better than $\lambda/4_{(He)}\,cm^{-1}$, and thicknesses less than 20 μm have been achieved. Twenty-five micrometres of material is

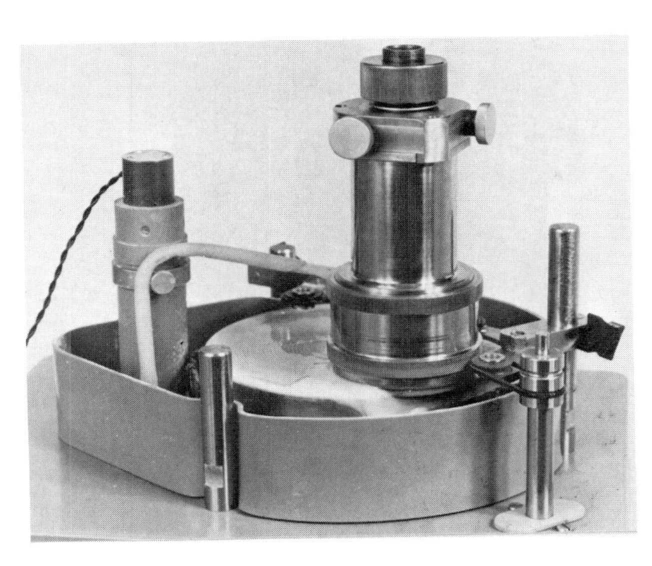

7.59 Eccentric roller for imparting a small amount of indirectly driven sweep to a block or jig on an RCMP.

usually polished away, during an overnight process. It is considered to be at least twice the amount necessary to remove damage, but it fits conveniently into the routine. Figure 7.59 shows a jig, on a RCMP, with a single specimen so that a positively driven eccentric wheel on the roller bar is needed to provide sweep. Polisher rotation is 60 rpm, the load applied less than $25\,g\,cm^{-2}$, and the slurry an alkaline silica sol + $7\frac{1}{2}\%$. H_2O_2 recirculated at about $2\,l\,min^{-1}$ over the polyurethane foam pad.

7.9.3 Cleaning

An unwanted characteristic of alkaline silica sols is that it is difficult to remove all traces of the fluid from a polished surface. The run-down method is described in §7.3, where a soft cloth is used in conjunction with the sol which is progressively reduced in concentration until water only is recirculated. However, this process cannot be applied successfully to easily damaged materials, and a method of cleaning has been devised which leaves the surface (at best) extremely clean and covered with oxides a few monolayers thick.

The first step is the removal of the work from the polishing machine without stopping the polisher-rotation—that is, work and pad must not be allowed to come into direct contact: they must remain spaced by the film of slurry which causes the aquaplaning. A firm, swift upward movement is needed to take the jig (or block) from between the rollers, than the specimens are immersed as quickly as possible in 0·01–0·001 M potassium hydroxide solution and agitated or ultrasonically vibrated. The work should not be allowed to dry, or even come into prolonged contact with air if it can be avoided and all processes should be carried out without delay. An intense, fine spray of the same solution is directed onto and around the specimens for about one minute (in a fume cupboard) followed by spraying with de-ionised water, again for at least one minute. Pressured dry nitrogen (<15 psi) is used finally to blow the water droplets away.

If, on close inspection, any traces of the sol remain on a surface, it can only be removed by repolishing, and repeating the cleaning cycle until the results are satisfactory.

7.10 Stainless Steel, Electroless Nickel Plated and Hardened Steel Surfaces

The metallographic field has developed many techniques for polishing a wide range of metals: for example, Samuels (1971) has dealt thoroughly with the subject, especially the employment of chemical–mechanical polishing processes. Though we have polished many metals at various times, of the softer ones, only pure aluminium (which polishes well on indium) has proved difficult: the harder ones are correspondingly easier to polish well because they are less susceptible to damage, but their preparation to high optical standards deserves additional description.

7.10.1 Stainless steel

Lapping many ductile materials by free-abrasive techniques is a slow operation. However, since a necessary stage is to obtain a surface with the flatness required for subsequent polishing, a pre-lapped surface as well should be shaped as accurately as possible. High-quality surface grinding or machining will help to shorten the lapping duration. Depending upon the rigidity of the workpiece the holding technique used while machining it is a critical factor. If magnetic clamping is used and the work is distorted, the surface then produced will change shape on release from the chuck. Turned faces on stainless steel flats (produced on precise lathes) can be lapped to within optical limits very quickly, thus indicating the high-quality machine and cutter. Even so, the first areas lapped are often at 120°, evidence of the strain resulting from three-jaw chucking. A reduction of the clamping pressure for the final cut reduces the distortion

Lapping of all flat work is done on a Lapmaster machine (§2.1) with 600 carborundum and Payne's No. 3 oil, producing a scratch-free finish. Finer-grades of abrasive can be used but the probability of scratching increases—or, perhaps more accurately, the machine-conditions for uniformly greying surfaces become more critical.

A variety of techniques can be used to produce specular finishes on stainless steels: polisher matrices of cloth, wax or pitch (and mixes of the last two, §3.2.3.1) with α-alumina, Linde A, slurry all produce moderate polishes. If the stainless steel can be (and has been) hardened, solder or tin polishers (§3.2.2) produce a specular surface, finely sleeked. Spiral grooving (§3.3) is recommended and 3 μm diamond abrasive is used with Hyprez OS fluid as a lubricant. The turndown produced on soft-metal polishers is negligible. Two common causes of degraded edge-definition should be noted: firstly, excessive load on specimen relative to that on the conditioning ring can produce a channel on the polisher; secondly, and less obvious, a build-up of black-coloured detritus on the surface of the polisher which erodes the edge as the work ploughs through it. The remedies are to balance the loads more carefully for the first and increase the supply of lubricant for the second until the polisher regains its metallic appearance.

The best surfaces obtained so far on stainless steel have been achieved, either by a final polish or an extended polish from the grey-state, on a three-station recirculating slurry machine (§2.2.2). An alkaline silica sol is used (unmodified) with a polyurethane rigid foam, pad, (§3.2.6) the work loaded to give 100–200 g cm^{-2}. The edge turndown is small while ×400 magnification shows only the grain of the metal.

Working non-flat surfaces makes, as usual, special demands for techniques in order to maintain figure. Electroforming provides a ready means of producing components, that require highly specular finishes—such as laser pumping cavities, usually cylindrical ellipses or ellipses of revolution. Figure 7.60 shows a stainless steel mandrel for one of the former type, machined with a single-pointed tool which rotates on

7.60 Stainless steel ellipsoidal mandrel being machined.

an inclined axis while the work is held and fed past on the machine. With care, the surface produced can have a texture as fine as 0·1 μm centre-line-average (CLA) value.

Since the form is elliptical, successively finer grades of Bramit paper are used for lapping, cemented to an angle-strip so that a two-line contact is made. In this way, at least the straight line direction of the mandrel is preserved from too much turndown. The snag with these simple arrangements is that the abrasive paper wears rapidly over the small but moving area of contact. Flexible spring-steel paper-holders have been used and these laps are more long-lived.

Polishing is carried out with Microcloth either glued to the same angle-strip or held in the same flexible metal device. If unacceptable orange-peel appears at this stage, a cast-metal matrix with 3 μm diamond, or formed pitch (cf §3.2.3.2) plus Linde A, is used immediately after lapping. The temptation to use felt pads rather than cloth should be avoided unless, for instance, the mandrel is longer than the wanted electroform—and suitably masked for electrodeposition by rubber rings—in which case the turn down produced by flexible polishers is unimportant. The

7.61 Electroform from polished mandrel.

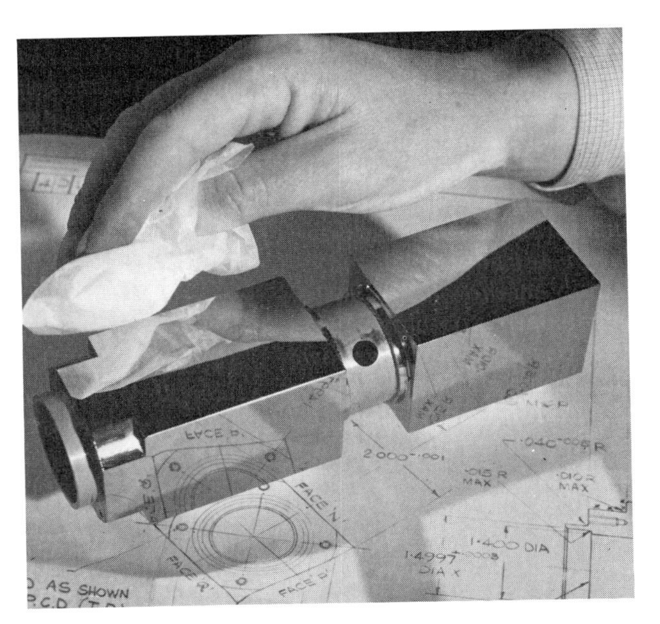

7.62 Rotating mirror made from electroless nickel coated alluminium alloy.

motion of the polisher is the same as that used in polishing any cylindrical object: stroked longitudinally by hand while a slow rotational motion is provided to the work (see figure 5.27). Figure 7.61 shows an ellipsoidal laser pumping cavity; the specular interior is a faithful replica of the mandrel's polish.

7.10.2 Electroless nickel deposits

Lightweight reflecting optical components can be made in high-stability light alloys, their surfaces covered by electroless nickel (e.g. Brenner, 1954) so that very hard surfaces are obtained on the softer metal substrates. The hardness of the surface thus obtained allows glass-polishing techniques to be used, except that α-alumina (Linde A) polishes them more quickly than cerium oxide. A component like a rotating mirror (figure 7.62) provides a typical example of the techniques employed.

A Lapmaster is used for the preparatory greying-off, and the surface produced must be proved before polishing is started. Section 2.1.2 describes the use of a grooved Tufnol polisher for producing a surface sufficiently burnished to show fringes with an optical flat under a monochromatic light source. Uniform brightening is needed and the fringe pattern must be acceptably what is required finally. It is a mistake to attempt to correct a surface by polishing when it has been insufficiently or incorrectly lapped.

Pitch polishers are prepared as described in §3.2.3.1; unmodified pitch is satisfactory, but if machine methods are to be used for surfacing and scrolling (§3.3) then a mixed preparation is preferred because of the relative ease with which it can be machined.

Most of the components produced so far have been too large for our existing general purpose machines, and therefore have been worked on a so-called hand-polishing machine which rotates at about 2 rpm. In this way, the traditional optical worker's slow walk around a polishing post is simulated—his motion helps to prevent the polisher from becoming non-uniformly worn. The direction of the hand-strokes made with the work on the rotating polisher varies from diametral to W-figures. As soon as the surface becomes specular, it should be tested so that future stroke-extent can be determined: at its most rudimentary, work going convex requires an increase in stroke-length and vice versa. The geometrical shapes of the electroless nickel-plated components worked up to date have been very unfavourable from the polishing standpoint—oblong, for example. A very good guide is obtained from the 'feel' of the optical surface sliding over the polisher. As in lapping plate testing (§2.1.2) if the work pivots on its ends these will have material abraded away from them at first;

and if it turns on its middle, its centre will be reduced in thickness. Though the errors can be in either work or polisher, if the lapping has been carefully carried out and checked, the latter is more likely to be at fault. In practice, observation of the first areas brightened is valuable in determining the polisher's shape, and re-forming of the pitch on a warm masterflat may be necessary.

Once the brightening is seen to be uniform, and the flatness checked, the process is continued until a highly specular surface is obtained. During this time the work can be expected to gradually lose flatness because of its unfavourable aspect, and the pitch polisher may need re-flattening several times. In the later stages, some small areas may lag in polishing and therefore remain slightly higher than the general levels. If they are slow to disappear, local polishing (figuring) with small pitch polishers may be needed. These tactics have been found necessary only on oblong surfaces, 2:1 length to width.

7.10.3 Square rollers

Hardened, precision steel balls, 'square' cylindrical rollers and needle rollers have uses in metrological and optical fields: for the latter applications in particular, their forms and surfaces need improving. The most useful of these bearing components are the 'square' rollers (length = diameter)—for example, as plane-parallel reflectors (§6.4.3) and, with suitable treatment of their peripheries, as alignment jigs for vee arrangements. Section 6.1.3 describes the uses of balls, rollers and cylinders in kinematic design.

Adjustable-angle jigs (§2.3) are conveniently used for polishing the ends of rollers and a DTI (figure 6.11) may be used for alignment purposes. However, when auto-collimation is available a slight pre-polish is required on one end of a roller. With care, it can be held on a slowly revolving solder polisher (§3.2.2.3) for a few seconds and supplied with 3 μm diamond abrasive. An alternative, less exacting, technique is to use a Tufnol collar, bored centrally to take the roller, running within the conditioning ring on a GPPM (§2.2.1) or a Multipol (§2.2.4). Pressure is applied via a push rod which is located in a dimpled cap on the exposed end of the roller (see figure 2.30).

Parallelism of the ends to better than a few seconds of arc can be obtained by mounting the roller on a workplate in a Mark II adjustable-angle jig (§2.3.3). For the maximum sensitivity of angular setting by auto-collimation (§6.1.4) the workplate should have the largest viewing hole consistent with leaving enough surface available for mating with the roller. Tan wax is a suitable cement for bonding firmly between small areas.

Polishing follows the routine practice for all plane-parallel work: auto-collimate the rear surface to be parallel to the polishing plane; polish (lap if required) the front face; clean and reverse the roller and re-wax it in the jig; auto-collimate; and re-polish the face temporarily polished at first.

A machined solder (or electroplated tin) polisher with 3 μm diamond is adequate for polishing all but the most critical surfaces. Improved surface finishes may be obtained after a careful clean-down by using 0·3 μm alumina on soft pitch finally. Where the bearing components are hardened stainless steel the alkaline silica sol process described in §7.10.1 is applicable.

The cylindrical surfaces of commercial roller bearing components can be improved by semi-hand polishing techniques similar to those used for laser rods (§7.1). Several rollers are glued together axially while held in alignment either by means of a surface table and an angle plate, or in a vee block; however, it is easier to achieve accuracy with two independently precise surfaces rather than the two integral faces of a vee block. The stick of cemented rollers may be stuck simultaneously to a rod of the same diameter to provide a handle for rotation in a small lathe (cf mandrel polishing, §5.2.3 and §7.9.1).

7.11 Gallium Arsenide Modulator Rods

Gallium arsenide, GaAs (Rowland, 1969) is a semiconducting compound which has found many applications in the electro-optics field. Techniques for cutting and polishing it are suitable, too, for gallium phosphide. The etch-polishing techniques (usually involving halogen/alcohol systems (Strehlow, 1969)) that are employed on large volumes of material, are not dealt with herein. Preparing substrate slices follows procedures described in §7.6 in general; and dicing may be carried out by several techniques, one of which is explicitly detailed in §1.5, another implicitly in §1.4. Making modulator rods (Kaminov and Turner,

1966) involves the description of processes in addition to optical polishing and it therefore forms the subject of this final section.

7.11.1 Sawing

An oriented boule is sawn to provide a bar of about 8 mm square—large enough to supply four rods, each 3 mm square and 45 mm long (cf §7.2). This bar is demounted and, for optical alignment purposes, temporarily polished faces are worked on its ends and one side-face, usually by rubbing them lightly on a solder polisher fed with 3 μm diamond abrasive. It is remounted on a glass platform (with Tan wax cement) with its polished side uppermost and fixed to a saw worktable (see §1.6.3). Re-establishment of the oriented face can be carried out by DTI methods (§1.3.3); but the specular faces permit alignment to be done from the side face with the complete workarm assembly removed from the saw (§1.6.1). The shaft is retained vertically on a surface table (§6.1.1) while the crystal is viewed through an auto-collimator, §6.1.4, and adjusted to be parallel to the table.

Three cuts are made with a peripheral diamond saw disc on a Microslice II saw, producing two slabs of material, each having oriented ends and one side-face plus the two newly sawn. With the possible exception of the waxed-down side, the three axes should be closely orthogonal. The assembly is removed from the saw and dewaxed. In order to preserve squareness of the rods finally cut from these slabs, each is mounted separately and aligned as before. The rods are sawn to size, demounted and cleaned.

7.11.2 Mounting and polishing

The end faces of the four rods produced can be polished separately or in a bundle; the latter process will be described, since the information for polishing a single rod is implicit in it. They are bonded together in a vee-block with a clear high-temperature cement (such as Lakeside—a shellac). Two sheets of thin polyester film are used to prevent adhesion to this alignment jig. Since each rod already possesses an adequately specular face, a brief inspection of the cementing accuracy is made through an auto-collimator. If the multiple images produced by the slightly misaligned ends show that the error is considered to be satisfactory (say ~1 min) the second stage of mounting can proceed. A jig workplate is needed with a tubular extension, its bore made to receive the bonded bundle of GaAs rods. The front face of the plate carries four slabs of packing (gallium arsenide of the same orientation, if possible), bonded with a rapid-setting epoxide adhesive and positioned so as to form a close fit on the rod-bundle, (figure 7.63).

Dental or 'Paraplas' wax is used to hold the bundle of rods in the tubular part of the work-holder: by using the paraffin wax at a temperature just above its melting point (see table 4.1) the bond between the rods themselves is not disturbed. When cool, the assembly is bolted into an adjustable-angle jig (§2.3), aligned (§6.1.4), lapped with 600 carborundum on a ring-lapping machine until an even grey surface is produced on all rods and packing.

Polishing is carried out using the ring mode of a Multipol machine (§2.2.4) or a GPPM (§2.2.1) (the ring being formed by the glass-faced conditioning ring of an adjustable-angle jig), with the stroke limited to about 5 mm. The polisher is a standard wax matrix (§3.2.4) spirally grooved (§3.3), though any other form of reticulation by netting or grooving should serve. Linde A (α-alumina 0·3 μm particle size) is used as the abrasive mixed with weak chromic acid as an etch/lubricant. Since 'chromic acid' is often loosely applied to its anhydride, chromium (V1) oxide CrO_3, our preparative method for weak chromic acid (a mixture of H_2CrO_3 and $H_2Cr_2O_7$) is given. There are, of course, many other possible routes to the solution. Potassium dichromate ($K_2Cr_2O_7$) is added in excess to 600 cm^3 distilled water in a litre beaker and the

7.63 Gallium arsenide modulator rod with surrounding packing, mounted in tabular work holder.

contents heated to about 60 °C and stirred until the crystals dissolve. The solution is allowed to cool and the fluid above the crystals which are precipitated (the supernatant fluid) is decanted into a clean beaker. (Further additions of water are made to the crystals remaining so that they can be recycled). Fifty cubic centimetres of concentrated sulphuric acid (H_2SO_4) are added carefully to the solution, preferably in two portions via a 25 cm^3 pipette. A fairly safe procedure is to insert the pipette into the beaker's contents and release the acid gradually near to the bottom of the beaker whilst maintaining a fairly rapid stirring motion. It is essential to wear rubber gloves, visor and laboratory coat/apron during the mixing operation.

Because of the increased possibility of turndown on the rod's edges due to an etch-polishing process used in conjunction with light pressures, the applied load is more critical than usual. At the upper limit, too great a pressure causes tracking in the wax which in its turn results in turned-down edges and material damage. In practice, the load control nut should be adjusted to give 60–80 g cm^{-2} of presented work and waster area. The abrasive slurry can be either squeezed on from a wash bottle constantly shaken, or recirculated by a small submersible pump (see figure 2.38). Alumina powders settle-out very quickly in aqueous slurries so that a vigorous pumping is required to keep the abrasive particles active. Some form of paddle attached to the underside of the polisher plate helps, too—either bolted on or held magnetically.

When the first ends of the rods have been finished, the rod bundle has to be reversed in the tubular workholder without disturbing the alignment of the four faces. Although the two waxes used differ by 40 °C in their melting points, the work must be manipulated without raising the higher melting point shellac to the lower end of its softening range. This is best achieved by providing a short pulse of heat to the workplate and tube by means of a pre-heated block of bored brass. The workplate is lowered in it and when the dental wax is seen to become transparent, the bundle is pushed out and returned reversed. A similar block of cold brass is then used for cooling down the assembly and, at the same time, the axial position of the bundle is adjusted so that the second set of faces is in line with the packing. Alignment, lapping and polishing is a repeat of the earlier stages.

When etch-polishes are being used, it is important not to allow any of the slurry to pass through the gaps between the rods and the tubular mount and thence to the faces first-polished. The general practice of always cleaning the polished faces with the jig and work facing downwards should be observed, at least to remove the majority of the slurry. (Cleaning the ring face-downwards is good practice since it keeps abrasive slurries from the screw-threads which adjust the conditioning-ring position). Cleaning and testing can be done finally with the ring facing upwards.

The heated brass block can be used conveniently for removing the finished work although the technique is not important at this stage since overheating, causing relative drift between the rods, no longer matters. A two-stage cleaning process is necessary: firstly, with warm trichloroethane or toluene to remove the paraffin wax; and secondly with an alcohol to remove the shellac. A camel-hair brush, dipped in the appropriate solvent, serves as a gentle pressure-limiting method for finally cleaning the polished faces. All intermediate cleaning (that is, for surface finish and flatness inspections) can be done with paper tissues since further polishing will remove the sleeks that they cause.

Where etch-polishing techniques are used, it is usually unnecessary to chamfer the edges of the work. However, in the interests of robustness in subsequent handling, chamfering will reduce chipping. The process is the same as that used for proustite, (§7.5), and other soft materials.

Table 7.1 Summary of typical polishing conditions (other than diamond abrasives and soft metal matrices)

Material	Scratch hardness (Mohs's scale)	For grey surface: abrasive (on cast-iron laps)	For polished surfaces		Lubricant	Load (g cm^{-2})	Remarks	Related section
			Polisher	Abrasive				
ADP	2	(From sawn surface) or Carborundum 600	Wax	Linde A	Ethane diol	50	Water soluble: finish on dry Selvyt cloth	7.7
Antimony sulphur iodide	2	(From sawn surface) or Carborundum 600	Wax	Linde A	Water + 5% Na_2CO_3	50	Can be polished on sides or ends of needles	
Arsenic: selenium glasses	3	Carborundum 600	Wax	Linde A	Water + 5% Na_2CO_3	80		
Barium strontium niobate	5	Carborundum 600	0·5 mm pitch	Linde A	Water + 5% Na_2CO_3	75		
Cadmium mercury telluride	2	(From sawn surface)	(1) Wax (2) Polyfoam	(1) Linde A (2) Colloidal SiO_2	(1) Water/glycol (2) Alkaline silica sol + H_2O_2	50	Cloths can improve finish with increased turndown	7.9
Calcite	3	Carborundum 600	Wax	Linde A	Water + 5% Na_2CO_3	80	Avoid thermal shock	7.1
Calcium fluoride	4	Carborundum 600	2·5 mm pitch or wax	Linde A	—	50	Avoid thermal shock	7.1
Calcium molybdate	5	Carborundum 600	2·5 mm pitch	Linde A	—	75	Wax polishers remove final sleeks. Avoid thermal shock	7.1
Calcium tungstate	4·5	Carborundum 600	2·5 mm pitch or wax	Linde A	—	60	Avoid thermal shock	7.1
Cerium fluoride	4	Carborundum 600	Wood loaded 2·5 mm pitch	Linde A	—	50		7.1
Electroless nickel	8	Carborundum 600	4 mm pure pitch	Linde A	—	150		7.10
Ferrites	7	Carborundum 600	0·5 mm pure pitch	Linde A	—	100	Alkaline silica sol improves surfaces	7.11
Gallium arsenide	3	Carborundum 600	Wax	Linde A	10% chromic acid	60–80	Benefits from alkaline silica sol polishing	(cf 7.3)
Gallium gadolinium garnet	7	Carborundum 600	2 mm pitch wood loaded	Cerium oxide	Water + 5% Na_2CO_3	100		
Gallium phosphide	4·5	Carborundum 600	Wax	Linde A	Oil or ethane diol	80	Usually polished with Br_2/Methanol (Strehlow, 1969)	
Germanium	5	Carborundum 600	Wax or cloths	Linde A	Water + 5% Na_2CO_3	75	Alkaline silica sols + H_2O_2 improves finish	(cf 7.9)
Glasses (general)	6	Carborundum 600	2·5 mm pitch wood loaded	Cerium oxide	Water + 5% Na_2CO_3	60		(cf 7.2)
Hardened steel	8	Carborundum 600	4 mm pure pitch	Linde A	Water + 5% Na_2CO_3	100	Benefits from alkaline silica sol treatment	7.10
Indium antimonide	3·5	Carborundum 600	Wax or cloths	Linde A	Water + 5% Na_2CO_3	60	10% chromic acid improves finish	(cf 7.11)
KDP	2·5	(From sawn or Carb surface) 600	Wax	Linde A	Ethane diol	50	Water soluble finish on dry selvyt cloth	(cf 7.7)
Lanthanum fluoride	4	Carborundum 600	2·5 mm pitch wood loaded	Linde A	Water + 5% Na_2CO_3	50		7.1
Laser glasses	6	Carborundum 600	2·5 mm pitch wood loaded	Cerium oxide	Water + 5% Na_2CO_3	60		7.1

Table 7.1—*continued*

Material	Scratch hardness (Mohs's scale)	For grey surface: abrasive (on cast-iron laps)	For polished surfaces: Polisher	For polished surfaces: Abrasive	Lubricant	Load (g cm^{-2})	Remarks	Related section
Lead germanate	~3	(From sawn surface) or Carborundum 600	Wax	(1) Linde A (2) Linde B	Ethane diol	100		7.6
Lead telluride	2·5	Silicon carbide paper Grade 600	Wax	Linde A	Water + 5% Na_2CO_3	50	Benefits from Br_2/methanol etch polishing (Strehlow, 1969)	(cf 7.9)
Lithium fluoride		(From cleaved surfaces)	(1) Pitch (2) Wax	Linde A	Water + 5% Na_2CO_3	100		7.4
Magnesium oxide (periclase)	6	Carborundum 600	3 mm pitch wood loaded	Linde A	Water + 5% Na_2CO_3	100		
Perspex	—	(Silicon carbide Paper Grade 600)	Wax	Linde A	Ethane diol	100		
Potassium bromide	2·5	(From sawn surface) or Carborundum 600	Wax	Linde A	Ethane diol	50	Water soluble. Finish on dry Selvyt cloth	(cf 7.7)
Potassium chloride	2	(From sawn surface) or Carborundum 600	Wax	Linde A	Ethane diol	50		(cf 7.7)
Proustite	2·5	(From sawn surfaces) or Carborundum 600	Wax	Linde A	Ethane diol	75	For better edge definition see table 6.2. Clean with chamois leather	7.5
Quartz (+ fused quartz)	7	Carborundum 600	(1) 2 mm pitch (2) Polyfoam	(1) Cerium oxide (2) Colloidal SiO_2	(1) Water (2) Alkaline sol	100		7.3
Rutile	6·5–7	Carborundum 600	0·5 mm pitch	Linde A	Water + 5% Na_2CO_3	100	Alkaline silica sol helps finally	7.1
Silicon	7	Carborundum 600	(1) Wax (2) Polyfoam	(1) Linde A (2) Colloidal SiO_2	(1) Water (2) Alkaline sol	100		(cf 7.3)
Silver chloride	<2	(From sawn surface) or Carborundum 600	Wax	Linde A	Ethane diol	50	Sparingly soluble in water	(cf 7.7)
Sodium chloride	2·5	(Silicon carbide paper Grade 600)	Wax	Linde A	Ethane diol	50	Very water-soluble. Finish on dry Selvyt	(cf 7.7)
Spinel	8	Carborundum 600	1 mm pitch wood loaded	Linde A	Water + 5% Na_2CO_3	100	Chamfer edges	7.1
Stainless steels	6–8	Carborundum 600	4 mm pure pitch	Linde A	Water + 5% Na_2CO_3		Improves with alkaline silica sol treatment	7.10
Tri-glycine sulphate	2	(Silicon carbide 600 Grade Paper) or Carborundum 600	Wax or cloths	Linde A	Ethane diol	50	Very hygroscopic. Mount with paraffin wax; remove with 'dry' toluene	7.7
Yttrium aluminium garnet	6·5–7	Carborundum 600	0·5 mm pitch	Linde A	Water + 5% Na_2CO_3	100	Alkaline silica sol improves finish	7.1
Yttrium lithium fluoride	3·5	(From sawn surfaces) or Carborundum 600	Wax	Linde A	Water + 5% Na_2CO_3	80	Improves with short duration polish with alkaline silica sol	
Zinc selenide	3	(From sawn surfaces) or Carborundum 600	Wax	Linde A	Ethane diol	80	For optimum finish and slight loss of edge definition	7.8
Zinc sulphide	3·5	(From sawn surfaces) or Carborundum 600	Wax	Linde A	Ethane diol	80		7.8

Table 7.2 Summary of typical polishing conditions for diamond abrasives and metal polishers

Material	Scratch hardness (Mohs's scale)	For grey finish: final abrasive (on cast-iron laps)	Specular finish: Polisher	Specular finish: Abrasive (with Hyprez of fluid)	Approx. load ($g\,cm^{-2}$)	Remarks	Related section
α-alumina ceramics	~9	Carborundum 600 or 15 μm diamond	(1) Copper (2) Solder or tin	3 μm diamond	100	Specularity limited by grain size of alumina	
Aluminium (single crystal)	2	Silicon carbide paper. Grade 600	Indium	3 μm diamond	50	Clean with chamois leather	7.10
Barium strontium niobate (BSN)	5	Carborundum 600	(1) Solder or tin (2) Indium	3 μm diamond	100 60		
Calcite	3·0	Carborundum 600	(1) Solder or tin (2) Indium	3 μm diamond	100 60	Avoid thermal shock	7.1
Copper (OFHC)	~3·0	Silicon carbide paper 600 grade	Solder or tin	3 μm diamond	60	Alkaline silica sol and hydrogen peroxide gives good finish	
Ferrites	7	Carborundum 600	Solder or tin	3 μm diamond	100	Alkaline silica sol improves surfaces	
Germanium	~5	Carborundum 600	Solder or tin	3 μm diamond	100	Alkaline silica sol and hydrogen peroxide for best finishes	(cf 7.9)
Hardened steel	~8	Carborundum 600	Solder or tin	3 μm diamond	100	Benefits from alkaline silica sols	7.10
KDP	~2·5	(From sawn surface) or Carborundum 600	Indium	3 μm diamond	50	Good edge definition but improved finishes on wax. Water soluble	(cf 7.7)
Lithium niobate	4·5	Carborundum 600	(1) Solder or tin (2) Indium	3 μm diamond	100 60	Excellent finishes with alkaline silica sols	(cf 7.3)
Lithium sulphate	2·5	(From sawn surface) or Carborundum 600	Indium	3 μm diamond	60	Water soluble	(cf 7.7)
Lithium tantalate		Carborundum 600	Solder or tin	3 μm diamond	100	Very short duration finish on alkaline silica sol improves it	(cf 7.3)
Magnesium oxide (periclase)	6	Carborundum 600	(1) Solder or tin (2) Indium	3 μm diamond	100		
Oilstone (porous substrate)	~9	Carborundum 600	Copper	3 μm diamond	100		4.16
PLZT		Carborundum 600	Solder or tin	3 μm diamond	100	Alkaline silica sol for improved finish	
Proustite	2–2·5	(From sawn surface) or Carborundum 600	Indium	3 μm diamond	75	Clean with chamois leather	7.5
Ruby	9	15 μm diamond	Copper	8, 3 μm diamond	100	Chamfer polish needed: alkaline silica sol improves finish	7.1
Rutile	6·5–7	Carborundum 600	Solder or tin	3 μm diamond	100	Alkaline silica sol helps finally	7.1
Sapphire	9	15 μm diamond	Copper	8, 3 μm diamond	100	Chamfer polish needed: alkaline silica sol improves finish	7.1
Silicon	~7	Carborundum 600	Solder or tin	3 μm diamond	100	Alkaline silica sol gives best finishes	(cf 7.3)
Spinel	8	Carborundum 600	Solder or tin	3 μm diamond	100	Chamfer edges	7.1
Stainless steel	5–8	Carborundum 600	Solder or tin	3 μm diamond	100	Alkaline silica sol optimum finishes	7.10
Yttrium aluminium garnet (YAG)	6·5–7	Carborundum 600	Solder or tin	3 μm diamond	100	Alkaline silica sol improves finish	7.1
Yttrium iron garnet (YIG)	6·5–7	Carborundum 600	Solder or tin	3 μm diamond	100	Alkaline silica sol improves finish	7.1
Zinc oxide	~4·5	Carborundum 600	Solder or tin	3 μm diamond	100	Prolonged alkaline silica sol improves it	
Zinc selenide	3·0	(From sawn surfaces)	Indium	3 μm diamond	80	Moderate finish and extreme edge definition (Zinc selenide and Zinc sulphide)	7.8
Zinc sulphide	2·5	or Carborundum 600	Indium	3 μm diamond	80		

References

Bardsley W *et al* 1969 *J. Opt. Electron.* **1** 29–31

Birch K G 1973 *J. Phys. E: Sci. Instrum.* **6** 1045–8

Bond W L 1962 *Rev. Sci. Instrum.* **33** 372–5

Brenner A 1954a *Metal Finishing* November, 68–76

—— 1954b *Metal Finishing* December, 61–8

Forrester P A 1974 *Agard Lecture Series* No. 71

Fynn G W and Powell W J A 1969a *RRE TN709* (2nd ed) (London: HMSO)

—— 1969b *J. Phys. E: Sci. Instrum.* **2** 756–7

—— 1970 *J. Phys. E: Sci. Instrum.* **4** 248–50

Fynn G W, Powell W J A and Jenkins G C J 1970 *RRE TN749* (London: HMSO)

Hume K *et al* 1967 *App. Phys. Lett.* **10** 133

Jones G R, Shaw N and Vere A W 1972 *Electron. Lett.* **8** 345–6

Kaminow I P and Turner E H 1966 *App. Optics* **5** 1612

Makarov V L and Novikov M 1967 *Sov. J. Opt. Technol.* **34** 463–5

Morgan D V and Bøgh E 1972 Surface Sci. **32** 278–86

Putley E H 1970 *Semiconductors and Semimetals* vol 5, (London: Academic) ch 6

Rabinowicz E 1968 *Sci. Am.* 91–8

Rich G J 1976 *Wave Electron.* **2** 219–37

Rowland M C 1969 *Gallium Arsenide Lasers* (New York: Wiley) pp 139–92

Samuels L E 1971 *Metallographic Polishing by Mechanical Methods* (London: Pitman)

Smith H I 1974 *Proc. IEEE 62* **10** 1361–87

Strehlow W H 1969 *J. App. Phys.* **7** 2928–32

Twyman F 1957 *Prisms and Lens Making* (London: Hilger and Watts) pp 174–81

Watton R, Burgess D and Harper B 1977 *J. App. Sci. Eng. A* **2** 47–63

Glossary

Special instruments and components mentioned in the text are accompanied by their manufacturers' or suppliers' names. Most of the widely available components need no labelling, but the following short list itemises some which either constantly recur or may otherwise prove difficult to locate and obtain. Their sources are not, in every case, unique and alternatives are undoubtedly equally satisfactory.

Item	Source
Abralap 2A oil	Lawrence Industries
Abraslice saw	Malvern Instruments
Acrulite	Engineering Plastics and Manufacturing Ltd
Adjustable-angle jigs	Metals Research Ltd: Logitech Ltd
Aloxite abrasives	Carborundum Co Ltd
Auto-collimators	Rank Taylor Hobson Optical Tools for Industry: Metals Research Ltd
BAO abrasives	British Americal Optical Co Ltd
Blade packs	Varian Associates Ltd
Bramit papers	Engis Ltd
Bunter centreless grinder	Bunter SA
Capco Q.35	Caplin Engineering Co Ltd
Glycol phthalate cement	Malvern Instruments
Habit diamond drills	Habit Diamond Tooling Ltd
Hyprez fluids and diamond compounds	Engis Ltd
Lakeside cement	Cutrock Engineering Co Ltd
Lapmaster machine	Payne Products (International) Ltd
Linde A, B and C	Union Carbide
Macrotome saws	Metals Research Ltd
Magnetic bases	Eclipse Tools Ltd
Microcloth	Engis Ltd
Microslice saws	Metals Research Ltd
Monochromatic light source	Payne Products (International) Ltd
Multipols	Metals Research Ltd

Okerin 100 wax	Aster Petrochemicals Ltd
Paraplast wax	BDH Ltd
Parvalux motors	Parvalux Ltd
Polyester film (Melinex)	ICI Ltd
Polyurethane foam	HV Skan Ltd
Porous optical flats	Metals Research Ltd
Precision mirror mounts	Malvern Instruments
Silicon carbide	RPI Ltd
Swedish pitch	HV Skan Ltd TA Hutchinson Ltd
Talyrond	RPI Ltd
Tan wax	Fred Lee and Co Ltd
Transcut oil	A Duckham and Co Ltd
Tufnol (SRBF)	Tufnol Ltd

Index